Prosimian Behaviour

Prosimian Behaviour

edited by
R.D. MARTIN, G.A. DOYLE and
A.C. WALKER

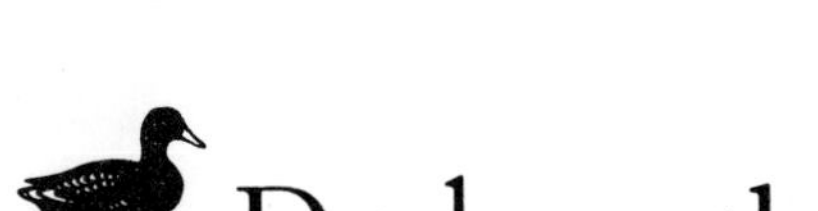

Publisher's Note

This book, *Prosimian Behaviour*, is the first half of *Prosimian Biology*, first published in 1974. The second half, *Prosimian Anatomy, Biochemistry and Evolution*, and the complete work *Prosimian Biology* are also available as separate publications.

For convenience of reference the preliminary pages (Contents, Preface, List of Contributors and Foreword) and the Index are retained from the complete work in full. All references to pages after p. 413 apply only to the second volume.

This volume first published 1976
Published by Gerald Duckworth and Co. Ltd.
The Old Piano Factory
43 Gloucester Crescent, London N.W.1

ISBN 0 7156 1178 8 (hardcover)
ISBN 0 7156 0981 3 (paperback)

Printed in Great Britain by
Redwood Burn Limited
Trowbridge & Esher

CONTENTS

Section B: *Laboratory Studies of Behaviour*

Section C: *Olfactory Communication in Prosimians*

Section D: *Physiology of Behaviour in Captivity*

PART II: PROSIMIAN ANATOMY, BIOCHEMISTRY AND EVOLUTION

Section A: *General Studies of Prosimian Evolution*

Section B: *Anatomy and Function of Skull and Teeth*

Section C: *Morphology of the Brain*

Section D: *Prosimian Locomotion*

Section E: *Chromosomes, Proteins and Evolution*

Preface

On 14, 15, 16 and 17 April 1972, an international meeting was held at the Institute of Archaeology, University of London, to discuss the general field of prosimian biology. Following a suggestion made by Dr P.J. Ucko, coupled with much valuable advice and encouragement, the meeting was held within the framework of the series of Research Seminars in Archaeology and Related Subjects, and the organisers of the present meeting were accordingly able to benefit from the accumulated expertise of what has so far proved to be a highly successful pattern of seminars. One happy result was that the Research Seminar on Prosimian Biology profited greatly from the hospitality and ready assistance of the Institute of Archaeology, for which thanks are especially due to Professor W.F. Grimes. In particular, the success of the meeting depended heavily on the invaluable services of Miss S.E. Johnson, who undertook almost all of the secretarial work for the meeting, provided seasoned advice on the running of the Research Seminar, and assisted in the preparation of this book. Thanks are also due to Miss Lorna Mullings and Mrs Penny Wyatt for their willing assistance in the Research Seminar secretariat, which dealt with all the practical problems of running a meeting with over fifty active participants.

The meeting – like its three predecessors in the Research Seminar series – was designed to enable research workers concerned with different specialised aspects of the subject to meet in an attempt to bridge gaps and achieve a synthetic approach through personal contact and lively discussion. For this reason, emphasis was placed on informality of presentation and debate. Following the practice established in previous Research Seminars, all papers were circulated prior to the meeting in order to reduce the stifling influence of formal reading of papers during periods which could be more profitably used for informal discussion. Several sessions were arranged in which particular topics were thrown open to general discussion, and participants made full use of the opportunities provided. Judging from the various comments which were made after the meeting, the air of informality was generally greatly appreciated, and many participants felt that they had gained far more than would have been possible with more formal conference organisation. It must be admitted, however, that the general friendly tone of the meeting was set from the beginning by the warm official welcome provided by Professor Grimes in the form of a reception to open the Research Seminar.

The important task of chairing discussion sessions was undertaken by Professor R. Andrew, Professor N.A. Barnicot, Professor F. Bourlière, Dr W. Bishop, Professor M. Day, Dr K. Hiiemae, Dr J.R. Napier, Professor W.C. Osman Hill, Dr L. Rosenson, Dr E.L. Simons and Dr L. Wolin. All these people deserve our deep gratitude for encouraging constructive debate and ensuring that the participants did not wander too far from the subject of each session. In addition, the editors would like to express their thanks to all of the more than fifty participants who came from various parts of the world to contribute papers and to join in the discussion of prosimian biology. They have helped to produce a text which represents one of the first really broad-based attempts to consider the general biology of the prosimians. The success of this attempt is reflected in the fact that many of the participants extensively expanded and modified their contributions as a result of discussion held during the meeting, and that a number of new lines of research seem to have developed following the meeting.

The general organisation of the Research Seminar was dependent upon the assistance of Mr J.F. Dahl, who operated the tape-recorder, and of Mr D.G. Lewis, who was in charge of the projection facilities. Additional secretarial assistance was provided by Mrs P.M. Blair. The Institute of Archaeology provided many different services, which were invaluable to the smooth running of the meeting, and thanks are due to the many un-named people who provided assistance on behalf of the Institute.

Finally, a special note of thanks must be made to the initiating publishers, Duckworth & Co. Ltd., without whose help the Research Seminar could not have been held. Among the several unique features of this meeting, it was notable that our publisher attended the greater part of the proceedings himself, and in the preparation of the book even insisted on the inclusion of *more* material in order to make the text as comprehensive as it is.

R.D. Martin
G.A. Doyle
A.C. Walker

List of Contributors (present addresses)

B.C. Albright, Department of Anatomy, Medical College of Virginia, Health Science Division of Virginia Commonwealth University, Richmond, Virginia 23298, USA

R. Andrew, School of Biological Sciences, Sussex University, Falmer, Brighton, Sussex

N.A. Barnicot, Department of Anthropology, University College London, Gower Street, London WC1

K. Bauer, 69 Heidelberg 1, Postfach 100.560, West Germany

S.K. Bearder, Department of Psychology, University of the Witwatersrand, Jan Smuts Avanue, Johannesburg, South Africa

J.H. Bergeron, Department of Anatomy, Duke University Medical Center, Durham, North Carolina 27706, USA

A.C. Berry, Royal Free Medical School, 8 Hunter Street, London WC1

F. Bourlière, Faculté de Médecine, 45 Rue des St Pères, Paris 5e, France

F.M. Bush, Department of Anatomy, Medical College of Virginia, Health Science Division of Virginia Commonwealth University, Richmond, Virginia 23298, USA

M. Cartmill, Department of Anatomy, Duke University Medical Center, Durham, North Carolina 27706, USA

P. Charles-Dominique, Muséum National d'Histoire Naturelle, Ecologie Générale, 4 Avenue du Petit Château, 91 Brunoy, France

B. Chiarelli, Centro di Primatologia, Istituto di Antropologia, Università di Torino, Via Accademia Albertina 17, 10123 Torino, Italy

C.N. Cook, Department of Anthropology, University College, Gower Street, London WC1

H.M. Cooper, Department of Biological Sciences, Biology Unit I, Florida State University, Tallahassee, Florida 32306, USA

U.M. Cowgill, Department of Biology, Faculty of Arts and Sciences, University of Pittsburgh, Pittsburgh, Pennsylvania 15260, USA

M. Delombas, Muséum National d'Histoire Naturelle, Laboratoire d'Anatomie Comparée, 55 Rue de Buffon, Paris 5e

R.F. Doolittle, Department of Chemistry, University of California, San Diego, Box 109, La Jolla, California 92037, USA

G.A. Doyle, Department of Psychology, University of the Witwatersrand, Jan Smuts Avenue, Johannesburg, South Africa

F. D'Souza, Wellcome Institute of Comparative Physiology, Regent's Park, London NW1

J. Egozcuc, Instituto de Biologia Fundamental, Universidad Autónoma de Barcelona, Avda, San Antonio M.a Claret 167, Barcelona 13, Spain

Judy Epps, Department of Zoology, University of New England, Armidale, New South Wales, Australia

R.G. Every, Centre for the Study of Conflict, 25 Clifton Terrace, Sumner, Christchurch 8; and Dept. of Zoology, University of Canterbury, Christchurch 1, NZ

M.P.L. Fogden, Rancho Experimental La Campana, Apdo. 682, Chihuahua, Mexico

J. Gasc, Museum National d'Histoire Naturelle, Laboratoire d'Anatomie Comparée, 55 Rue de Buffon, Paris 5e

P.D. Gingerich, Division of Vertebrate Palaeontology, Peabody Museum, Yale University, New Haven, Connecticut 06520, USA

G.E. Goode, Medical College of Virginia, Health Science Division of Virginia Commonwealth University, Richmond, Virginia 23298, USA

M. Goodman, Department of Anatomy, School of Medecine, Wayne State University, Detroit, Michigan 48207, USA

A. Gorton, Department of Anthropology, New York University, Washington Square, New York, NY 10003, USA

C.P. Groves, Duckworth Laboratory, Faculty of Archaeology and Anthropology, Downing Street, Cambridge

Duane E. Haines, Department of Anatomy, West Virginia University School of Medicine, Morgantown, West Virginia 26506, USA

E.C.B. Hall-Craggs, Department of Anatomy and Embryology, University College, Gower Street, London WC1

J. Harrington, The Rockefeller University, New York, NY 10021

D. Hewett-Emmett, Department of Physiology, Wayne State University Medical School, Detroit, Michigan 48201, USA

M. Hladik, Muséum National d'Histoire Naturelle, Ecologie Générale, 4 Avenue du Petit Château, 91 Brunoy, France

K.M. Hiiemae, Unit of Anatomy with Relation to Dentistry, Anatomy Department, Guy's Hospital Medical School, London SE1

K. Holmes, Department of Physiology, Southern Illinois University, Edwardsville, Illinois 62025, USA

C.J. Jolly, Department of Anthropology, New York University, Washington, Square, New York, NY 10003, USA

F.K. Jouffroy, Muséum National d'Histoire Naturelle, Laboratoire d'Anatomie Comparée, 55 Rue de Buffon, Paris 5e

W.P. Luckett, Department of Anatomy, Columbia University, College of Physicians and Surgeons, 630 West 168th Street, New York, NY 10032, USA

R.F. Kay, Museum of Comparative Zoology, The Agassiz Museum, Harvard University, Cambridge, Massachusetts 02138, USA

P.H. Klopfer, Department of Zoology, Duke University, Durham, North Carolina 27706, USA

R. Klopman, School of Biological Sciences, Sussex University, Falmer, Brighton, Sussex

G. Manley, Department of Zoology, Birkbeck College, Malet Street, London WC1

R.D. Martin, Department of Anthropology, University College, Gower Street, London WC1

H.M. Murray, Department of Anatomy, Medical College of Virginia, Health Science Division of Virginia Commonwealth University, Richmond, Virginia 23298, USA

S. Oblin, Département Audiovisuel, Université Paris 7, 2 Place Jussien, Paris 5e, France

W.C. Osman Hill, Oakhurst, Dixwell Road, Folkestone, Kent

M.G. Pariente, Muséum National d'Histoire Naturelle, Ecologie Générale, 4 Avenue du Petit Château, 91 Brunoy, France

M. Perret, Muséum National d'Histoire Naturelle, Ecologie Générale, 4 Avenue du Petit Château, 91 Brunoy, France

J.-J. Petter, Muséum National d'Histoire Naturelle, Ecologie Générale, 4 Avenue du Petit Chateau, 91 Brunoy, France

A. Petter-Rousseaux, Muséum National d'Histoire Naturelle, Ecologie Générale, 4 Avenue du Petit Château, 91 Brunoy, France.

L.B. Radinsky, Department of Anatomy, University of Chicago, 1025 East 57th Street, Chicago, Illinois 60637, USA

A. Richard, Department of Anthropology, Yale University, New Haven, Conn. 06520, USA

Y. Rumpler, Ecole Nationale de Médecine, Laboratoire d'Histologie, Embryologie et Cytogénétique, B.P.375 Tananarive, Madagascar

A. Schilling, Muséum National d'Histoire Naturelle, Ecologie Générale, 4 Avenue du Petit Château, 91 Brunoy, France

J. Schwartz, Department of Anthropology, University of Pittsburgh, Pennsylvania 15260, USA

D. Seligsohn, Department of Anthropology, Hunter College, 695 Park Avenue, New York, NY 10021, USA

E.L. Simons, Peabody Museum, Yale University, New Haven, Connecticut 06520, USA

R.W. Sussman, Department of Anthropology, Washington University, St Louis, Missouri 63130, USA

F. Szalay, Department of Anthropology, Hunter College, 695 Park Avenue, New York, NY 10021, USA

J. Tandy, c/o Mills Tandy, Department of Zoology, University of

Texas, Austin, Texas 78712, USA

I. Tattersall, Department of Anthropology, American Museum of Natural History, Central Park West at 79th Street, New York, NY 10024, USA

D. von Holst, Zoologisches Institut der Universität München, 8 München, Luisenstrasse 14, West Germany

A.C. Walker, Department of Anatomy, Harvard Medical School, 55 Shattuck Street, Boston, Mass. 02115, USA

L.R. Wolin, Laboratory of Neurophysiology, Research Division, Cleveland Psychiatric Institute, 1708 Aiken Avenue, Cleveland, Ohio 44109, USA

W.C. OSMAN HILL

Foreword

The foundation of our knowledge of the prosimians was laid during the latter half of the nineteenth century, and it is there that we first meet with monographers of the stature of Burmeister (1846) and Owen (1863) through their respective studies of the tarsier and of the aye-aye. There have also been numerous and outstanding contributions in the form of more comprehensive works, and here I should especially mention the labours of Murie and Mivart and the great explorers of Madagascar, such as Schlegel and Pollen (1867-77) and particularly Grandidier and Edwards (1876-1900), who produced a magnificent set of volumes on the natural history of the island.

While speaking of the giants of our discipline, it is opportune to include one of the more recent exponents – the late Sir Wilfred Le Gros Clark. Though he also made major contributions to many aspects of the anatomy, neuroanatomy and palaeontology of the Prosimii, he will perhaps best be remembered for his advocacy of the inclusion among them of the tree-shrews. These animals clearly still have a place in our deliberations, however controversial it may be, as is evidenced by several contributions to the present symposium.

It is now almost a quarter of a century since the first overall assessment of the status of the prosimian primates was made. At that time, I had the privilege of partaking in a discussion which was opened by my revered colleague F. Wood Jones, on the occasion of the meeting of the section on Anatomy and Anthropology at the annual meeting of the British Medical Association in Cambridge, 1948.

Since then, much has happened that is of vital importance in the field of prosimian biology, which is now not only wider, but has inevitably expanded to incorporate applications of disciplines almost unheard of at that time. It is, therefore, a matter of no little difficulty for any single author to write with authority on the whole gamut of prosimian biology. Thus the four-day research seminar on the subject held in April 1972 was particularly welcome, in that it enabled a wide range of prosimian biologists to take stock of their own advances and to profit from the researches of their colleagues from many different disciplines.

Needless to say, the problems posed by the position of *Tarsius* and its fossil allies proved to be as intriguing as ever, and they would have still delighted Wood Jones and engaged his powers of debate in no small measure, just as they did at the memorable meeting held at the London Zoo in 1918. But the problems set by the tarsiers are by no means the only ones that have engaged the interests of workers in recent decades. Our attention is automatically drawn towards 'la grande île' of Madagascar, where the activities of conservationists have emphasised time and again the pressing urgency for field-work on the behaviour of all of the lemurs. In spite of the laudable endeavours of the Malagasy Republic to protect their unique fauna, it may soon be too late to study at least some of the lemur species in the natural state. Fortunately, heed has been taken of the warnings, and there are now a number of studies that have been published or are in progress, as demonstrated by several of the contributions to the Research Seminar.

Many lemur species still demand special attention, however, and there is an even more pressing need for detailed study of their ecology so that regional schemes may be examined, which may permit the establishment of local breeding units. Madagascar is ideally situated as a natural workshop for the student of evolution, and it behoves us to adopt early measures for the protection of the existing fauna – an urgent necessity already recognised by the enlightened Malagasy government – and also to provide the means of enhancing the existing facilities for internal laboratories where local and visiting students may avail themselves of the unique opportunities to study the rich fauna.

That definite advances in prosimian biology have accrued in many different directions is amply shown by the contents of the present symposium. The wide range of problems which have now been considered makes this abundantly clear, and I have myself gained valuable knowledge both of a wealth of new facts and of reassessment of earlier problems. It would be invidious for me to single out any one paper for special praise, since all are remarkable in their various approaches and treatments; I will, therefore, deal with the different papers in an order that appears to link the subject matter in a logical fashion, as a guide to the mass of information brought together by the Research Seminar.

The Seminar was broadly divided into two sections: studies of prosimian behaviour, and studies of prosimian anatomy and evolution, including reference to cytogenetics and biochemical studies. Behaviour in the field justly received a major share of attention from the participants, and papers were given on representatives from all the major prosimian groups. Contributions by Frances D'Souza and

Michael Fogden brought in the two most controversial prosimians – namely, *Tupaia* and *Tarsius* – and thus provided welcome additions to the field of prosimian research.

In Madagascar, field studies on *Lemur catta* have been undertaken both by Alain Schilling and by Robert Sussman, the latter providing a very elegant study of the ecological distinctions between sympatric populations of *Lemur catta* and *Lemur fulvus.* Alison Richard elaborated on the patterns of mating in *Propithecus* and provided us with one of the first detailed accounts of the natural mating behaviour of any lemur species. Marcel Hladik and Pierre Charles-Dominique have significantly expanded our knowledge of lemur feeding ecology and social dynamics with their intensive study of *Lepilemur*, while Jean-Jaques Petter and A. Peyriéras have added to our understanding of *Indri*, specifically with respect to population density and home range. This latter paper is again one of particular interest, in that it deals with a little-known genus.

Turning to mainland Africa, the bushbabies and pottos were very comprehensively covered. All the prosimians outside Madagascar are nocturnal, and special techniques have been developed for studying the behaviour of these night-active forms over the last few years. These techniques are illustrated in the papers on *Tarsius* (Fogden) and *Lepilemur* (Hladik and Charles-Dominique); but in the papers on the African prosimians, the full possibilities of these techniques are displayed. Bearder and Doyle have given a comprehensive account of the South African species *Galago senegalensis* and *Galago crassicaudatus*, whilst Charles-Dominique has achieved the magnificent feat of considering the ecology and feeding behaviour of all 5 prosimian species in Gabon (*Galago demidovii, Galago alleni, Euoticus elegantulus, Arctocebus calabarensis* and *Perodicticus potto*). Finally, a link with laboratory studies is provided by Georges Pariente's able contribution dealing with two little-studied nocturnal Malagasy lemurs, *Phaner furcifer* and *Lepilemur mustelinus*, and illustrating the relationship of their activity patterns to ambient light intensities.

Laboratory studies of prosimian behaviour were equally well represented. The paper presented by Jan Bergeron on the Duke University Primate Facility, set up by John Buettner-Janusch, which I had the privilege of visiting a few years ago, summarises a very interesting large-scale project on laboratory observation of prosimians. We may expect a stream of far-reaching research projects on prosimian biology from this well-equipped facility. The lesser galago has been further studied in South Africa by Gerald Doyle in a very comprehensive programme aimed at obtaining quantifiable data over long periods, and his paper summarises our knowledge about the behavioural repertoire of this species. Judy Epps and Jocelyn Tandy, respectively, have given detailed attention to social behaviour of captive *Perodicticus potto* and captive *Galago crassicaudatus*, while

Peter Klopfer has looked specifically at aspects of lemur mother-infant relationships. Ursula Cowgill reported on some unusual observations indicating a possibility that pottos may exhibit a form of cooperative behaviour in the laboratory. On a more experimental slant, Howard Cooper investigated the possibility that *Lemur macaco* might exhibit learning sets in captivity, and reported on evidence that this capacity does in fact exist in at least some prosimians.

The functions of the glandular apparatus associated with the external genitalia of the two slow-moving African lorisoids (*Perodicticus* and *Arctocebus*) provided an interesting topic for Gilbert Manley, who exploited the subject to the full, drawing conclusions from behaviour, morphology and biochemistry. The perennial subject of urine-washing has been studied on a comparative basis by Richard Andrew and Robert Klopman, who have underlined the practical difficulties of providing a conclusive interpretation of a specific aspect of behaviour of this kind by presenting a wide-ranging examination of the various possibilities. The subject of olfactory communication, which had been considered in Alain Schilling's paper on *Lemur catta*, was also tackled in Jonathan Harrington's paper on *Lemur fulvus.*

A more physiological approach to the behaviour of captive prosimians was provided by two papers relating to a laboratory colony of Mouse Lemurs (*Microcebus murinus*). Arlette Petter-Rousseaux submitted a throughly competent study of the relationships between photoperiodicity, sexual activity and body-weight in this small nocturnal lemur, whose behaviour (like that of all other Malagasy lemurs) has a strict seasonal basis. The endocrine glands, especially the hypophysis, have been explored in detail by Martine Perret, who relates various changes to aspects of behaviour and the seasonal reproductive cycle. To round off the laboratory studies of behaviour, Dietrich von Holst provides a salutory warning for those who maintain animals of any kind in captivity. In his paper on social stress in tree-shrews (*Tupaia belangeri*), he provides ample evidence of the wide scale of pathological changes that may be brought about by interaction between conspecifics in captivity. He also raises the interesting possibility that such interaction between conspecifics – probably in a far milder form – may play a part in the natural situation.

In the second section, dealing with anatomy, biochemistry and evolutionary aspects, special praise is due to a number of papers which have a functional anatomical emphasis. This is evident in Tattersall's investigation of the mechanics of the temporomandibular joint and accessory structures of the subfossil *Archaeolemur.* Matt Cartmill casts a wide net in his behavioural and ecological consideration of anatomical features of such biological aberrants as *Daubentonia, Dactylopsila* and various woodpeckers, which are functionally linked by virtue of their adaptation to a wood-boring insectivorous regime, as a result of which they all exhibit some degree of klinorhynchy. P.D. Gingerich, in 'Dental function of the Palaeocene primate *Plesiadapis*', has produced cogent arguments for the retention of this genus within the Order Primates, basing his reasoning on imputed feeding behaviour, dental morphology and molar faceting. Dental occlusion mechanisms were likewise the subject of the joint contribution by Seligsohn and Szalay on Lemurinae, with its bearing on the systematics of *Lemur* and *Varecia*. They confirm the separation of *Varecia* from *Lemur* at the generic level and the relegation of the subfossil species *insignis* and *jullyi* to the former genus. Other significant odontological studies with a functional basis are the contribution on *Galago* presented by R.F. Kay and Karen Hiiemae, illustrating their cinefluorographic technique of analysis, and Ron Every's magnificently illustrated paper on thegosis. Jeffrey Schwartz, in a theoretical reinvestigation of premolar loss among primates, resorts to the 'field theory' to help resolve the problems posed by diminishing premolar representation in primate evolution.

Three contributions are devoted to cerebral anatomy in prosimians. One, by Leonard Radinsky, deals with functional and phylogenetic implications of prosimian brain morphology, and two others are concerned with the neuroanatomy of lorisids. These are Duane Haines' study of the cerebellum in various lorisids and the joint contribution by Haines and his colleagues from the Anatomy Department of the Medical College of Virginia, dealing with the external cerebral morphology of galagos, lorises and pottos.

There are also three papers dealing in various ways with the topic of prosimian locomotion. Clifford Jolly and Ann Gorton examined the proportions of the intrinsic foot muscles in some lorisid prosimians, while Mme F. Jouffroy used cineradiographic techniques to examine the biomechanics of jumping in *Galago alleni*. Combined in one paper, E.C.B. Hall-Craggs has collected both physiological and histochemical parameters of locomotor behaviour in *Galago* and the slow-moving lorisids, including data on the differing activities of red and white muscular tissue and the mechanics of saltation.

Passing now to evolutionary studies, there are a number of papers dealing with general palaeontological and comparative anatomical treatments of the prosimians. Elwyn Simons has provided notes on Early Tertiary prosimian fossils, thus bringing our knowledge of these forms up to date. Alan Walker tackles the other major category of fossil prosimians, presenting a review of the Miocene lorisids of East Africa, including reference to the postcranial skeleton. The new associated fossil lorisid material has confirmed the value of size analysis in allocating some of the remains. A particularly thought-provoking paper linking both behaviour and evolutionary mechanisms is that of Lee Wolin, with the intriguing title of 'What can the eye tell us about behaviour and evolution? or: The aye-ayes have it, but what is it?'. His only complaint is that *Tupaia* has, so far, not received enough attention. The aye-aye is also made the subject, among other notable matters, of Colin Groves' evocative paper on the taxonomy and phylogeny of prosimians. Finally, Patrick Luckett draws upon evidence from the morphogenesis of the foetal membranes and the placenta in order to establish the phylogenetic relationships of the prosimians, whilst Caroline Berry has utilised non-metrical features of the prosimian skull for evolutionary interpretation in the manner of Wood Jones' important review (1930-1).

A particularly valuable addition to the discussion of evolutionary relationships was provided by a series of papers on chromosomal and biochemical studies of prosimians. Chromosomal evolution in the prosimians is the subject of Josei Egozcue's paper, which provides further indication of the distinctiveness of *Tarsius*, in that it has the highest diploid number so far found in any primate. The results of Rumpler's study of lemur chromosomes with a view to revising their classification indicate that it would be justifiable to raise the cheirogaleines to family rank, with the family Cheirogaleidae containing two subfamilies – Cheirogaleinae and Phanerinae. Rumpler also confirms the distinctiveness of *Daubentonia*, suggesting that this genus should be separated from all other lemurs at least at the family level. A general survey of chromosomal features in prosimians, provided by Brunetto Chiarelli, rounds off this triplet of cytogenetic studies.

Nigel Barnicot and David Hewett-Emmet have contributed a paper on electrophoretic studies of red cell and serum proteins in prosimians, which covers a wide range of species. It was of great interest to learn that the haemoglobin, adenylate kinase and phosphatase patterns of *Tarsius* are so like those of man. The properties of the Cheirogaleinae studied are also of systematic value, emphasising the distinctiveness of *Microcebus* and *Cheirogaleus*. Electrophoretic properties of haemoglobins were also the subject of a paper by Francis Bush and colleagues. Other phylogenetically directed protein studies concern the amino acid sequences of individual proteins and the use of serological tests to determine affinity. Christopher Cook and David Hewett-Emmett provide a general paper examining the uses and limitations of tree-building techniques based on amino acid sequence data, while Russell Doolittle provided an account of recent work on the short-chain fibrinopeptides, which are of great significance because of their evident lability in evolution. Klausdieter Bauer reports on his recent work based on identification of the actual number of determinant sites on proteins which are involved in serological cross-reaction, and a paper by Morris Goodman and his colleagues rounds off the subject with a treatment of immunodiffusion systematics in *Tarsius*, Lorisidae and Tupaiidae, in which he introduces the concept of antigenic distance.

This collection of papers can thus be seen to cover an extremely broad spectrum, and the Research Seminar has accordingly provided a general basis for future discussion of the biology of prosimians and of their evolution. In conclusion, I can only commend to all the overall value of such a broad-based synthesis in dealing with a particular topic, such as that of prosimian biology.

W.C. Osman Hill
Folkestone, September 1973

PART I
PROSIMIAN BEHAVIOUR

G. A. DOYLE and R. D. MARTIN

The study of prosimian behaviour

While interest in the behaviour of the primates as a group can be traced back to Darwin's time, this interest was – until quite recently – manifested predominantly in studies of the 'higher' (simian) primates. It is now widely accepted that study of prosimian behaviour is important and of great interest in its own right, and that it is especially necessary for a reliable, overall view of the evolution of behaviour within the Order Primates. It is therefore doubly important that prosimian behaviour should now be intensively and adequately studied, both in the laboratory and in the field. One of the most significant goals of primate behavioural studies is the provision of a general framework for interpreting the evolution of primate behaviour – particularly that of man – and it is essential to recognise that extensive behavioural studies of the prosimians must play an integral part in any *comprehensive* investigation of primate evolution.

It was first pointed out by T.H. Huxley,[1] and subsequently re-emphasised by Le Gros Clark,[2] that the Order Primates is the only mammalian order in which selected extant members appear to represent a graded series approximating the sequence of evolution from early ancestral (prosimian) primates to man. The approximation to the evolutionary sequence permits us, to some extent, to use the living primates as a series of models for reconstructing our own behavioural past. This is a valuable asset, though it must be remembered that it is only an approximation. The living primates are, of course, the present end-products of an evolutionary radiation in which all species have, without exception, undergone some modification from the ancestral primate condition. This must be borne in mind when any attempt is made to reconstruct an actual evolutionary sequence in primate behaviour. The more that is known about primate behaviour in general, the more successful such attempts at reconstruction are likely to be, and due attention should therefore be directed to prosimian behaviour, as well as to that of simians.

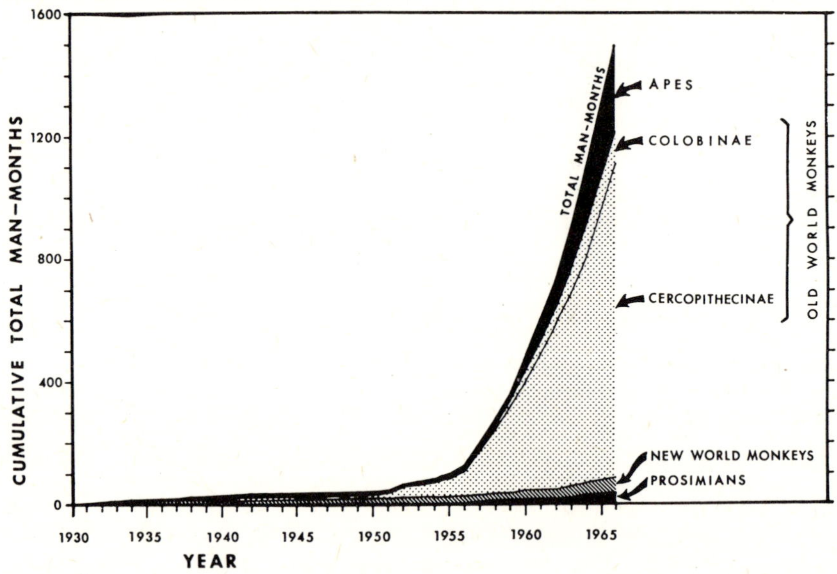

Figure 1. Cumulative man-months devoted to field studies of ecology and behaviour of non-human primates up to 1965 (after Altmann 1967).[3] The data include studies based on artificially established colonies maintained under natural or semi-natural conditions. Since 1965, the explosive growth of primate field studies has continued, and somewhat more attention has been devoted to the behaviour of promimians under field conditions. (Reprinted with the kind permission of the University of Chicago Press, © 1967 by the University of Chicago.)

Studies of primate behaviour have attracted growing interdisciplinary interest within the last few decades and, particularly from the early 1950s onwards, the growth rate in research output has increased enormously. In field studies alone the output has doubled every five years since 1955 (Fig. 1).[3] The prosimians, relatively speaking, have been generally neglected in this advancing front of research. There are many reasons for this. The simian primates are diurnal (except for *Aotus*) and live in a sensory world similar to our own, with vision predominating. Many species are terrestrial and thus more easily accessible to observation. Simians are not only easier to study; they are also closer to man in an evolutionary sense, and as a consequence more similar to man in morphology and behaviour. Altmann[3] notes that the focus of behavioural studies, within the 'higher' primates, has largely concerned the Old World monkeys and great apes, which satisfy these criteria more than do the New World monkeys. The prosimians, on the other hand, are almost exclusively arboreal, usually nocturnal, and live in a sensory world rather different from our own, with olfaction and audition still competing with vision as primary sensory modalities. They are, therefore, more difficult to study, and we cannot always be sure that our interpretations of observations are accurate because of the difference in sensory priorities. Since prosimians are farther removed from man in the evolutionary sense, studies of their behaviour are less likely to reveal interesting behavioural homologies with man than are studies of the simians. Yet an overall picture of primate evolution is just as incomplete without data on the prosimians as it would be without data on the simians.

The bias of primate behaviour studies towards the simian primates continued without modification until the early 1960s, after which time there was a notable increase of research interest in prosimians. Of the total number of papers in scientific journals that have been concerned wholly or partly with prosimian behaviour, approximately 80% have been published since 1960.

The 23 field and laboratory studies of prosimian behaviour represented in this present volume range over 19 different species (including 2 species of the Tupaiidae, regarded as 'honorary' prosimians for the purposes of the conference). Of a total of 18 generally recognised genera of Lemuriformes, Lorisiformes and Tarsiiformes, only 5 are not represented, although the lively discussions that followed all papers managed to bring in *all* genera. Many of the suspected difficulties involved in the study of small, nocturnal and arboreal primates in their natural habitat have been surprisingly easy to overcome, as is testified by a number of papers on field-studies in the present volume. Some of the difficulties have been overcome through indirect methods of studying behaviour (e.g. through various trapping techniques and analysis of stomach contents); but it has also proved possible to conduct extensive direct observation of behaviour by use of battery-operated headlamps to locate animals (through reflection of light from the tapetum of the eye) and to observe various aspects of ongoing behaviour. In captivity, because of their small size, the prosimians lend themselves more easily to accommodation under semi-natural conditions, and various devices (e.g. reversal of the light cycle and hence of the normal diurnal rhythm; use of weak red or blue observation lights) permit relatively easy observation of their behaviour in captivity.

In fact, the distinction usually made between field and laboratory studies of primates is often less valid for the prosimians than for the simians. In addition, although the distinction between studies conducted in the natural habitat and those conducted in the laboratory is nonetheless a vital one, consideration of the types of information sought indicates that a tripartite distinction is more relevant; namely, between studies in the natural habitat, investigations in semi-natural conditions, and studies in unnatural and restricted laboratory environments. This view is fully supported by Washburn and Hamburg[4] for 'higher' primates as well, the Rhesus monkeys of Cayo Santiago and the Japanese macaques in the large enclosed corral at the Oregon Primate Research Center providing good examples of semi-natural facilities for studying simian primates, intermediate between the typical field and laboratory situations.

Ideally, all three approaches to the study of prosimian behaviour should be utilised to the full in order to provide a sound basis for discussion. Each approach can be used to clarify particular aspects of behaviour for an eventual synthesis. However, it is essential to remember that all behavioural studies are subject to certain sources of error, and that great caution must be exercised in interpreting observations. Both field and laboratory studies are open to methodological errors, and in both areas the development of reliable techniques is a matter for priority. Some of these techniques are uniquely necessary for prosimians, but some have already been developed for field study of simians. For example, under field conditions it is becoming increasingly clear that long-term studies of prosimians and simians alike are necessary in order to take account of seasonal effects and in order to eliminate erroneous first impressions (e.g. underestimation of social group size; oversimplification of dominance relationships). It is also valuable to study the same species in different areas – as was the case with Alison Richard's study of *Propithecus verreauxi* – in order to determine natural variation in patterns of behaviour. Studies in captivity, on the other hand, are subject to a particular additional disadvantage. Whereas the field-worker is primarily concerned with the effect of his (or her) presence on the animals in their natural habitat, the laboratory investigator is faced with the dual danger of disturbing the captive animals directly through techniques of observation, and indirectly through provision of inadequate environmental conditions. There is also the combined drawback that animals confined in relatively small spaces may react far more violently to observer presence. A single example will place this in perspective: the lesser mouse lemur (*Microcebus murinus*) in its natural habitat in Madagascar[5] occupies a home-range which is probably comparable in volume to that of a cylinder of radius 25m. and height 15m., viz. approximately 30,000 cubic m. In the laboratory, it is customary to confine Lesser Mouse Lemurs in cages of 1-5 cubic m. capacity. Thus, the animal's effective living space is reduced by four orders of magnitude when it is confined in captivity. The laboratory observer must also choose what will be presented in the form of a social and physical environment within this restricted living-space, and the nature of this choice will necessarily influence the information which is collected.

This does not mean, however, that laboratory studies are less valuable than field studies; the two approaches provide different kinds of information which are equally valuable. Field observations are indispensable whenever *evolutionary processes* are considered, for the evolution of each prosimian species has taken place under a given set of environmental conditions, which may have changed gradually with time. Without knowledge of the present natural habitat, and some impression of the past history of that habitat, it is impossible to provide an accurate interpretation of the survival value of specific behaviour patterns or of a general evolutionary development. On the other hand, laboratory studies are necessary whenever any detailed analysis of *behavioural mechanisms* is required, and in such cases the research worker benefits greatly from the fact that the environmental conditions can be controlled. In this respect, studies conducted under semi-natural conditions perform a valuable intermediate role by indicating the behavioural effects achieved by the attempt to replicate the natural situation as far as is possible. Obviously, it is of great value to combine the results of field and laboratory studies in order to extract the special advantages of each approach and to eliminate sources of error. By bringing together both field and laboratory workers within the framework of the present conference, the authors hoped to further this kind of fruitful interchange of information about prosimian behaviour. The field worker can assist the laboratory worker by providing illustrations of the operation of behaviour patterns under natural conditions, while the laboratory worker can assist the field worker by indicating possible mechanisms and suggesting pertinent observations to be conducted under natural conditions.

Confinement *per se* away from the natural habitat may give rise to stress responses, as was suggested during the discussion (Martin). Stress can result both from the physical conditions of captivity and from the social environment of conspecifics caged in the same area. Of course, it is quite likely that stress responses – particularly when evoked by conspecifics – are present under natural conditions. However, confinement in captivity must almost always produce much higher stress levels, particularly since animals in caged groups are forced to interact far more frequently than would be the case under natural conditions. The laboratory observer should accordingly be on the lookout for symptoms of stress and consequent distortion of behaviour in caged animals. It was shown by von Holst that in *Tupaia belangeri* crowded conditions in captivity produce psychological stress which may result in 'ethological and physiological disturbances and diseases (e.g. renal disease)'. In particular, stress under caged conditions can, in this species, give rise to modification of mating and maternal behaviour. One of the best indications of captive stress in these tree-shrews is, in fact, a poor breeding record.[6,7]

It was also shown by Perret that in *Microcebus murinus* the adrenal glands are more developed under laboratory conditions, perhaps as a result of population density in captivity, and that this is associated with a very poor reproductive record. Again, as von Holst indicated for *Tupaia belangeri*, reproduction itself is probably a good index of the level of stress in the laboratory, and it can probably be assumed that prosimian species which breed successfully in captivity are not subject to high stress levels. Both of the present authors have attempted, with their laboratory colonies, to provide conditions as close as possible to those prevailing in the natural situation, despite the relatively small cages imposed by considerations of economy. With *Galago senegalensis* (Doyle), it has proved possible to attain a reproductive rate at least as high as that found in the wild. Behavioural considerations indicate that, for *G. senegalensis* at least, such semi-natural laboratory conditions are not notably stressful. With *Microcebus murinus* (Martin), maintenance of individuals in far smaller enclosures apparently gives rise to stress levels which are only partially offset by attempted replication of key features of the natural environment. Through constitution of social groups similar to those in the wild, replication of the natural diurnal and annual light-cycles, maintenance of natural temperature and humidity levels and provision of a varied and balanced diet, limited breeding successes have been achieved.[8] Recently (Martin, unpublished data) somewhat better breeding successes have been attained, though it is not certain whether this is a result of further adaptation of the animals to laboratory conditions, or to further improvements of those conditions (additional attention to diet; provision of a diurnal variation in temperature). Experience with *Microcebus murinus* generally indicates that this prosimian species is extremely sensitive to conditions in captivity, especially when housed in small cages (1 cubic m. capacity). Judging from the relative ease with which *Galago senegalensis* has been bred in captivity (Doyle), it seems likely that *M. murinus* would breed far more reliably if housed in cages of greater capacity (8 cubic m. and above).

Optimal conditions of hygiene must be at least as important as spatial, social and other factors, if animals are to be expected not only to survive but also to behave 'normally' in captivity. Doyle mentioned that the survival rate of his animals under semi-natural laboratory conditions is probably higher than in the wild, and the same probably applies to many laboratory colonies of prosimians. Bergeron and Buettner-Janusch's detailed description of the primate facility for prosimians at Duke University provides a model for the maintenance and care of prosimians under conditions which are eminently suitable for behavioural and other studies, and extensive veterinary precautions probably ensure that the animals are healthier in that laboratory than they would be in the natural habitat. Bergeron, in discussion, suggested that we might question the wisdom of this in the light of implicit and explicit attempts to duplicate natural conditions in the laboratory. Should we really be studying the behaviour of optimally healthy animals when their wild counterparts are not, on average, as healthy? However, close confinement of prosimians in captivity is once again a crucial factor, since the natural pattern of disease transference is completely altered by forced proximity. There is also the point that animals subject to social and environmental stress may be far more prone to suffer from diseases of various kinds. Thus, it would be difficult to ensure a natural balance of disease risk and hygiene under laboratory conditions. Therefore, with the usual aims of laboratory studies of behaviour in mind, there can be no doubt that extensive measures must be taken to ensure the health of laboratory animals: it will be the healthy, rather than the unhealthy, animal which will tend to display the full repertoire of both structural and functional characteristics that have been selected in the wild, and that account for successful adaptation to the ecological niche and hence for survival.

Turning to the various papers on prosimian behaviour, one can see how these various principles apply. Schilling's study of *Lemur catta* marking behaviour under natural conditions and Epps' study of *Perodicticus potto* social behaviour in captivity demonstrate the advantages and drawbacks of the two approaches in two cases where the central interest is naturalistic behaviour. The distinction between the two types of study lies mainly in the fact that in the first (field study) there is little manipulation of environmental variables,[9,10] whilst in the second (laboratory study) there is more manipulation of these variables provided primarily by the automatic constraints of a semi-natural environment, affecting (for example) living space, the arrangement of social groups, availability of food, and so on. While the intention of the field study is to leave the animal largely undisturbed so that its behaviour will be determined almost entirely by naturally occurring phenomena and can be observed over its entire range, the aim of the semi-naturalistic laboratory study is to observe the animal under conditions of minimal observer-interference consistent with the need to investigate particular aspects of natural behaviour at close range and in greater detail than is possible under natural conditions. Observations from both types of study are equally valid and useful; but care must be taken in interpretation of behaviour in terms of its survival value and adaptation to the natural environment. These two types of study, with their common aim of describing and interpreting naturalistic behaviour, are sharply distinct from laboratory investigations of specific aspects of behaviour where certain inherent mechanisms are the focus of interest. This applies, for example, to studies of intelligent behaviour and learning ability, where the nature of the investigation requires a situation almost totally unnatural to the animal, as was the case with Cooper's study of *Lemur macaco*. Petter-Rousseaux's study of the effect of photoperiod on sexual activity and body-weight of *Microcebus murinus* similarly involved the investigation of specific aspects of behaviour and physiology without particular reference to the natural situation, and this study dovetailed very neatly with Perret's investigation of endocrine gland variation in the same species. Laboratory investigations of this kind, devoted to specific aspects of behaviour and physiology, are extremely important in their own right, and they can also shed light on factors contributing to behaviour under natural conditions. However, special caution must be exercised in the general interpretation of such restricted laboratory studies in the absence of adequate information from the natural environment. The effect of confinement, directly or indirectly, on behaviour is invariably raised during discussions at conferences such as the one on which this present volume is based. Are we, in fact, studying the same animal in the laboratory – whether in small cages or in large semi-natural environments, whether indoors or outdoors – as that which exists in a natural state in the wild?

A good deal depends, of course, on the central interest of each study. For purely behavioural studies, Klopfer assures us that, with respect to his investigation of mother-infant relationships in *Lemur*, 'conditions of captivity were not unduly distorting the behaviour'. The same is probably true of many other studies where the research worker is interested in a fairly circumscribed aspect of behaviour, and where the conditions in captivity at least satisfy certain minimal conditions. For many far-reaching aspects of behaviour, however – notably such components as intra-group behaviour, dominance relationships and territoriality – confinement in the laboratory (even in a semi-natural environment) may seriously distort behaviour and create artificial pressures which might give rise to behaviour patterns not characteristic of relatively solitary animals in the wild. For this reason, consideration of social and territorial behaviour would normally be confined to field studies; yet there are some aspects of this very broad area of study which may legitimately be the subject of laboratory investigations. Tandy's study of social behaviour in an artificial group of *Galago crassicaudatus* in captivity, in which she demonstrates (among other things) a tendency towards solitariness and the importance of olfaction, provides a good example. Bearder (represented by Doyle) suggested that because even solitary animals must and do encounter one another in the wild, a good idea of the characteristic patterns of response to such encounters can be gained by arranging paired presentations under laboratory conditions, a procedure which he utilised with *Galago senegalensis*. Epps used essentially the same procedure with *Perodicticus potto* in her outdoor enclosure.

Nevertheless, it would not be true – except in a very limited sense – to think of laboratory studies as concerned exclusively with certain types of behaviour while field studies are concerned exclusively with other types. Although each type of study serves its own particular purposes, all studies are ideally directed at achieving a total descriptive picture of the species in question, and thus field and laboratory studies must be mutually complementary. This is particularly obvious from a comparison of Charles-Dominique's field studies relating to bush-baby locomotion[9] with the laboratory data reported below by Jolly and Gorton, Jouffroy et al, and Hall-Craggs. A synthesis of various kinds of data on locomotion[11] provides an extremely broad basis for considering the evolution of prosimian morphological adaptations for locomotor function. Very often, behaviour observed in one context cannot be fully appreciated without reference to observations from other contexts. Washburn and Hamburg[4] give many good examples of the inter-relationships between field and laboratory studies in simian primates, and of the extent to which each depends on the other for a full understanding of many complex aspects of behaviour. Several other examples among prosimians are provided in this present volume. For instance, the frequent carrying of very young infants in the mouth (characteristic of Galaginae) and the baby-parking behaviour found in many lorisids (characteristic of the Lorisinae, and apparently common among Galaginae) was mentioned by a number of participants in discussion. Such behaviour is, relatively speaking, much easier to observe under laboratory conditions, yet its significance is not readily apparent in the laboratory and only emerges when the natural situation is documented.

The field of olfactory communication provides a good illustration of the inter-dependence of studies conducted in the laboratory and in the wild, and this matter is specifically raised in Doyle's paper on *Galago senegalensis moholi* and in Harrington's paper on *Lemur fulvus*, which is exclusively devoted to olfactory communication. The need to appreciate the full range of marking behaviour in the wild is exemplified by Schilling's field study of *Lemur catta*, while the advantages of laboratory study for examining the fine details of specific aspects of marking behaviour are very neatly demonstrated by Manley's paper on two lorisine species.

Preliminary information on little-known species, often of great importance, is invariably provided by field-studies, which are generally devoted to aspects for which laboratory studies are not particularly suitable. There is also the point that the little-known prosimian species are almost always notoriously difficult to maintain in captivity. Such studies are typified by Petter and Peyriéras' paper on *Indri indri*, Hladik and Charles-Dominique's paper on *Lepilemur mustelinus* and Fogden's paper on *Tarsius bancanus*. All three of these species have a poor survival record in captivity, and the papers on their behaviour under natural conditions represent major contributions to our knowledge of these prosimians. Hopefully, these pioneer studies will lead to further field studies and eventually to appropriate laboratory studies. Detailed information on their behaviour in the wild may lead to solutions for the problems of maintaining these species in captivity.

Studies of factors accounting for adaptation to preferred habitats, as for instance in Bearder and Doyle's study of *Galago senegalensis* and *G. crassicaudatus* and D'Souza's field study of *Tupaia minor* and *T. glis*, and studies of ecological and ethological relationships between closely related species where they are sympatric in the wild, can (of course) only be undertaken in the natural situation, though some of the questions asked may have been generated out of laboratory studies. The broad ecological/ethological approach is exemplified by Charles-Dominique's study of live sympatric lorisid species, and by Sussman's detailed study of sympatric *Lemur catta* and *Lemur fulvus* in Madagascar. Sussman's study shows particularly clearly the kind of phenomena which may be observed where two closely-related species occur together in one given area, yet occur separately in other areas. Klopfer's study in the laboratory of *Lemur catta* and *L. fulvus* did, in fact, raise questions regarding differences in maternal-infant relationships, and Sussman's study provides partial answers to these questions from the natural situation, once again highlighting the inter-dependence of field and laboratory investigations.

Another area in which field and laboratory studies interact is that of seasonal variation in behaviour. Richard's study of mating activity in *Propithecus verreauxi* as a function of seasonal change, apart from providing surprising new information on the mating patterns of this species under natural conditions, provides a classical example of a field study of behaviour. Since reproductive behaviour in this lemur species depends upon seasonal variations in natural factors, such as availability of food, temperature, rainfall, foliage cover and daylength – which cannot be effectively varied in a large enough enclosure under laboratory conditions – its study calls for intensive field-work. The effects of certain diurnal and annual variations in environmental factors such as temperature, humidity and photoperiod, which can be individually controlled in the laboratory with respect to some elements of behaviour, do lend themselves to studies in captivity, however. Both Perret and Petter-Rousseaux reported on the effects of annual variation in daylength on endocrine activity and reproductive behaviour in *Microcebus murinus*, and provided us with a general basis for the interpretation of seasonal variation in the behaviour of the Malagasy lemurs. Along with Pariente's paper on the effect of diurnal and seasonal variation in light availability on the behaviour of *Phaner furcifer* and *Lepilemur mustelinus*, these different studies cover many aspects of seasonal variations in lemur behaviour.

As has already been stated, studies requiring closely controlled manipulation of experimental variables can, of course, only be undertaken under highly artificial laboratory conditions, as Cooper's study of learning sets in *Lemur macaco* demonstrates. Even in this case, however, the answers sought are related to questions raised by observations in the field as well as in the laboratory, such as those conducted by Jolly on object relationships.[12] Relevant questions for detailed research may also be raised through comparative considerations of brain function and behaviour and through comparative study of similar phenomena in simian primates and non-primate mammals.

The kinds of information that the student of prosimian behaviour seeks are, in general, much the same as those sought in the study of higher primates. Each of the behavioural studies presented in this volume has its counterpart in studies of simian primate behaviour. With the prosimians, perhaps, there is a greater implicit interest in more remote evolutionary developments. In the simians, particularly the Ponginae and the more terrestrial Cercopithecinae, there is an understandably greater interest in more recent evolutionary developments, particularly in such things as complex social group behaviour and communication, for example. All primate studies are, however, directed ultimately towards increasing our general knowledge of the entire mammalian Order to which man belongs. In the same way that field and laboratory studies should be regarded as mutually complementary, studies of prosimians are complementary to those of simians. One of the aims of this volume was to bring together a number of papers about prosimian behaviour as a first step in the long-outstanding drive towards a more comprehensive knowledge of prosimian biology.

NOTES

1 Huxley, T.H. (1863), *Evidence as to Man's Place in Nature*, London.

2 Le Gros Clark, W. (1962), *The Antecedents of Man*, Edinburgh.

3 Altmann, S.A. (1967), Preface, in S.A. Altmann (ed.), *Social Communication among Primates*, Chicago.

4 Washburn, S.L. and Hamburg, D.A. (1965), 'The Study of Primate Behaviour', in I. DeVore (ed.), *Primate Behaviour*, New York.

5 Martin, R.D. (1972), 'A preliminary field-study of the Lesser Mouse Lemur (*Microcebus murinus* J.F. Miller 1777)', *Z.f. Tierpsychol.*, Beiheft 9, 43-89.

6 Martin, R.D. (1968) 'Reproduction and ontogeny in tree-shrews (*Tupaia belangeri*), with reference to their general behaviour and taxonomic relationships', *Z.f. Tierpsychol.* 25, 409-532.

7 Von Holst, D. (1969), 'Sozialer Stress bei Tupajas (*Tupaia belangeri*)', *Z.vergl.Physiol.* 63, 1-58.

8 Martin, R.D. (1972), 'A laboratory breeding colony of the Lesser Mouse Lemur', in W.I.B. Beveridge (ed.), *Breeding Primates*, Basel, pp. 161-71.

9 Charles-Dominique, P. (1971), 'Eco-éthologie des prosimiens du Gabon', *Biol. Gabon.* 7, 121-228. This paper provides a very good example of a detailed prosimian field-study, though this particular investigation did involve some manipulation.

10 Even in field-studies, a degree of manipulation of the environment may occasionally be employed, as in Charles-Dominique's study[9] of various prosimian species in Gabon, where animals were forced to take certain aboreal routes or deliberately limited in their choice of routes in order to facilitate study of locomotor behaviour. Trapping techniques used under natural conditions, as in the separate investigations carried out by both Charles-Dominique and Fogden

(this volume) also represent a major intrusion into the natural situation.

11 Martin, R.D. (1972), 'Adaptive radiation and behaviour of the Malagasy lemurs', *Phil. Trans. Roy. Soc. (Lond.)*, B, 264, 295-352.

12 Jolly, A. (1964), 'Prosimians' manipulation of certain object problems', *Anim. Behav.* 12, 560-70. Jolly, A. (1966), *Lemur Behaviour: A Madagascar Field-Study*, Chicago.

PART I SECTION A
Field Studies of Behaviour and Ecology

F. BOURLIÈRE

How to remain a prosimian in a simian world

When compared with monkeys and apes, prosimians display definite behavioural handicaps. Jolly describes *Lemur* and *Propithecus* as 'hopelessly stupid toward unknown inanimate objects'[1] and she found that *Lemur catta* and *L.fulvus* were scoring 'far below both Old and New World monkeys on every type of problem, from the simplest of insight tests in object manipulation to object discrimination and delayed-response learning'.[2] Similar conclusions have been reached by Andrew,[3] Arnold and Rumbaugh,[4] Klüver[5] and Rumbaugh and Arnold.[6] Behaviourally speaking, as well as morphologically and physiologically, the Prosimii have therefore been considered as more 'primitive' than Anthropoidea and one has wondered how these two sub-orders could still coexist in nature.

The papers presented during the first two sessions of this Research Seminar, together with the ensuing discussions, provided a wealth of ecological and behavioural data which, when confronted with the findings of the morphologists, point towards a more balanced evaluation of the situation and lead to a better understanding of prosimian evolution. At last it has become possible to provide plausible answers to the following two questions: How did these primitive primates manage to survive, and in some cases to thrive, in a tropical arboreal world dominated by the 'more gifted' monkeys and apes? Why did some ancestral prosimians successfully radiate to give descendant forms parallel to their 'more advanced' relatives (i.e. in Madagascar)?

The answer to the first question now seems quite obvious. Wherever prosimians share the same habitat as monkeys and apes, they occupy entirely different ecological niches and therefore avoid any direct competition with higher primates. On the whole they have become nocturnal, arboreal fruit-and-insect-eaters, more or less narrowly specialised according to the presence or absence of sympatric competitors (e.g. arboreal marsupials, rodents or even carnivores). Furthermore, they did not content themselves with quietly living their obscure lives in obscure places – to paraphrase Harrison Matthews' evaluation of the Insectivoran way of life[7] – they have also taken advantage of some of their 'primitive' morphological and physiological characteristics to ensure a highly efficient occupation of their particular niches. For instance, the capacity of some lower forms to undergo prolonged seasonal torpor (Bourlière and Petter-Rousseaux[8]) plays the role of an 'energy-saving adaptation' enabling them to endure seasonal food shortages which would be incompatible with the mere survival of monkeys of similar size. The same comment also applies to the ability of some species to live on a low-calorie intake or a poorly-digestible diet, and also to many locomotor adaptations. Thus, the nocturnal life-styles of many nocturnal prosimians are not merely a way of 'making themselves forget'; they represent highly efficient adaptations to very particular niches – so successful indeed that they might well have prevented the emergence of competitive simian forms. This could explain why nocturnal monkeys only exist in the neotropics (*Aotus*), where prosimians disappeared long ago, and where the African galagine niche is only partly filled by some arboreal didelphids.

The major attributes of the typical African and Asiatic prosimian ecological niche emerge quite clearly from the recent field studies. Nocturnality ranks first; the circadian activity rhythms of sympatric species of prosimians and monkeys never overlap anywhere in continental Africa or tropical Asia. Although some species of the two sub-orders may frequent the same trees, they never actually meet; the temporal separation is complete. It is also worth mentioning that other mammalian competitors are few. Nocturnal arboreal didelphids are lacking in the Old World, and nocturnal arboreal rodents are rather scarce. The major potential competitors of prosimians in African and Asian forests are the nocturnal bats – mainly the frugivorous flying foxes – and some partly frugivorous scansorial carnivores. At this point, one may wonder whether temporal separation of activities is an absolute means of avoiding competition. This is certainly so in the case of insectivorous animals, as the kinds of insects (and other invertebrates) available by night to vertebrate predators are not the same as those available by day. The situation is slightly different for fruits, which remain available to frugivorous consumers after, as well as before, dusk. But in this case ecological separation between nocturnal and diurnal frugivorous species living in the same area is brought about by other factors, particularly differential location of fruits within the various forest layers, and on a variety of supports.

Preferences in the types of support used by nocturnal prosimians do indeed enable them largely to avoid direct competition with other vertebrate consumers, or even with sympatric species of their own sub-order. This is particularly obvious in the case of the five forest species studied by Charles-Dominique in north-east Gabon.[9,10] This is also apparently the case for animals living in more open environments (Sauer and Sauer, Bearder and Doyle).[11,12] The ability of *Galago senegalensis* to move with ease in extremely thorny bushes where very few other vertebrates dare to venture is also noteworthy (Doyle, verbal communications), as is the restriction of *Tarsier bancanus* to the forest understory (Fogden).[13] The use of small, and predominantly vertical, supports is made possible by specialised adaptations of the hands and, particularly, the feet. The occurrence of different species within different layers of the same forest environment is quite probably associated with their different abilities to use the various supports available and to negotiate obstacles (difference between 'leapers' and 'non-leapers').

The data on dietary specialisation of the different prosimian species presented at the Research Seminar are also particularly significant. Whereas overlap in dietary regimes can be the rule at times when food is plentiful (seasonality has been observed in most rain-forest types so far investigated), definite species-typical preferences for certain food items are observed during periods of food shortage. The exploitation of vegetable gums by both galagines and lorisines, which is apparently restricted to the African species according to available data (Rahm;[14] Charles-Dominique;[9,10] Bearder and Doyle[12]), is unusual among mammals. The preferences of African galagines and lorisines for different types of insect prey is also noteworthy – and clearly related to their different sensory abilities. The extremely active galagines hunt primarily by sight and sound and feed on a variety of active insects; passive echo-location is possibly used for detection of their animal prey by night. The sluggish African lorisines, on the other hand, have specialised upon prey generally considered as unpalatable by most predators: urticant caterpillars, centipedes, ants, etc., which are detected by olfactory means. In Ceylon, *Loris tardigradus* also feeds upon ants, spiders and bugs, in addition to more palatable insects (Petter and Hladik).[15] In so doing, the lorisines make use of food sources generally unexploited by reptiles, birds and mammals. They are, so to speak, earning their living upon a link of the food-chain more typically used by invertebrate predators and parasites.

The ecological niches of nocturnal prosimians are therefore both highly specialised and very efficiently filled. Though confined to rather marginal habitats and driven to nocturnality by the strong competition exerted by vegetarian monkeys and apes, prosimians have succeeded in maintaining their stand for millions of years by exploiting and developing a number of structural and functional adaptations.

One may be tempted at this stage to wonder which came first and what was the prime characteristic which enabled the nocturnal prosimians to survive so successfully to the present day; whether it was their special types of locomotion, their diverse sensory abilities or their nutritional plasticity which permitted them so efficiently to hold their own. Opinions apparently vary according to the particular interests of the various investigators, as exemplified during the seminar discussions. However, such a question might well be pointless. Primitive, 'generalised' prosimians were quite probably not so highly specialised as their modern relatives. Under various selective pressures they probably developed simultaneously their battery of present-day adaptations. After all, a living organism is an integrated whole, not merely a carcass, a set of teeth or a bunch of sense organs.

The answer to the second question is not so obvious as that to the first one. Recent field studies have definitely established that prosimians were able, on the island of Madagascar, to undergo an adaptative radiation which, in many ways, parallels the evolution of monkeys in Africa, Asia and America (see for instance Sussman).[16] But the environmental factors which initiated and made possible this impressive radiation have not yet been clearly established. However, one thing seems certain: the fauna of this mini-continent is largely 'unbalanced', or 'disharmonic'; a number of important taxonomic groups (e.g. Felidae; Ungulata, with the sole exception of the bushpig) being conspicuously absent. As a result of Madagascar's early separation from the African mainland and of the 'filtering effect' upon potential invaders exerted by the Mozambique Channel, the terrestrial and arboreal vertebrate groups have been able to evolve during millions of years without the competitive and predation pressures which were so strong elsewhere in the tropics. For instance, almost no other arboreal mammal was competing for food with the Madagascar prosimians, true arboreal rodents being virtually absent. Arboreal mammalian predators were also scarce. Even birds of prey were few (11 Aquilidae, 3 Falconidae, 6 Strigidae and Tytonidae) and the large lemurs of the island might have been expected to support some powerful 'monkey-eagle', such as *Stephanoaetus* of the African forests or *Pithecophaga* of the Philippines, as pointed out by Moreau.[17] Large arboreal lizards were lacking, as were large arboreal snakes. The predation pressure exerted upon prosimians in Madagascar ecosystems must therefore have been relatively slight. This situation is apparently quite general on this island: such anti-predator devices as cryptic coloration and mimicry are rare among insects here, and the local race of *Papilio dardanus* is the only one, within the entire Ethiopian region, which lacks pronounced sexual dimorphism and does not possess females which mimic aposematic butterflies (Paulian, Ford).[18,19] This apparent lack of strong predation pressure probably accounts in part for the amazing diversity of coat patterns and the conspicuous coloration of many diurnal species, as well as for the frequent occurrence of some unusual behaviour patterns, such as the sunning behaviour of *Lemur, Propithecus* and *Indri.* It seems highly likely that this sunning posture is also correlated with the imperfect homoeothermy of these animals and the necessity for them to warm up rapidly in the early morning; but such overt behaviour would have been terribly risky if large aerial predators hunting by sight had been present.

The emptiness of some ecological niches in Madagascar has, on the other hand, permitted their unexpected occupation by primates. Thus, the absence of any representative of the woodpecker family has made possible the extreme dietary specialisation of the aye-aye, which feeds at least partly upon wood-boring insect larvae (Petter and Peyriéras).[20] One may also wonder whether it is perhaps the absence of any endemic vertebrate grazer which has allowed *Hapalemur* to specialise upon certain grasses (reeds and bamboos; Petter and Peyriéras).[21]

The nocturnal lemurs of Madagascar have developed quite remarkable adaptations to the pronounced seasonality of their habitat. Some Cheirogaleinae are able to store fat in their tails before the dry season (Bourlière and Petter-Rousseaux).[22] In the *Didierea* bush of the South the folivorous *Lepilemur mustelinus* displays remarkable adaptations (economy of movement, enlarged caecum, caecotrophy) which permit an efficient utilisation of the scanty food resources (Charles-Dominique and Hladik).[23] In the Western coastal forest, at the end of the prolonged dry season, *Phaner furcifer* and *Microcebus coquereli* eat gums, resins and insect exudates (Petter, Schilling and Pariente).[24]

To sum up: present-day prosimians can no longer be considered as mere living fossils, more or less identical to their Eocene ancestors. Some of them have indeed kept up to modern times a 'primitive' morphology, but they have at the same time become so well-adapted physiologically and behaviourally to their marginal ways of life that they are very efficiently filling their role in tropical ecosystems. When given the chance to evolve in isolation from higher primates, under low predator pressure and in an unbalanced environment, as in Madagascar, other prosimians have been able to replicate extensively the ways of life of monkeys and apes. Sometimes they have even taken advantage of the emptiness of other vertebrate niches to give rise to such a bizarre and 'unlikely' primates as the aye-aye!

NOTES

1 Jolly, A. (1966), *Lemur Behaviour: A Madagascar Field Study*, Chicago.

2 Jolly, A. (1964), 'Choice of cue in prosimian learning', *Animal Behaviour* 12, 571-7.

3 Andrew, R.J. (1962), 'Evolution of intelligence and vocal mimicking', *Science* 137, 585-9.

4 Arnold, R.C. and Rumbaugh, D.M. (1971), 'Extinction: A comparative primate study of *Lemur* and *Cercopithecus*', *Folia primat.* 14, 161-70.

5 Klüver, H. (1933), *Behavioral Mechanisms in Monkeys*, Chicago.

6 Rumbaugh, D.M. and Arnold, R.C. (1971), 'Learning: A comparative study of *Lemur* and *Cercopithecus*', *Folia primat.* 14, 154-60.

7 Harrison Matthews, L. (1971), *The life of mammals*, London, vol. 2.

8 Bourlière, F. and Petter-Rousseaux, A. (1953), 'L'homéothermie imparfaite de certains prosimiens', *C.R.Soc. Biol.* 147, 1594-5.

9 Charles-Dominique, P. (1971), 'Eco-éthologie des prosimiens du Gabon', *Biol. Gabon.* 7, 121-228.

10 Charles-Dominique, P. (1972), 'Ecologie et vie sociale de *Galago demidovii* (Fischer 1808; Prosimii)', *Z.f. Tierpsychol.*, Beiheft 9, 7-41.

11 Sauer, E.G.F. and Sauer, E.M. (1963), 'The South West African bush-baby of the *Galago senegalensis* group', *J.S.W. Afri. Sci. Soc.* 16, 5-36.

12 Bearder, S.K. and Doyle, G.A. (this volume), 'Ecology of bushbabies, *Galago senegalensis* and *Galago crassicaudatus*, with some notes on their behaviour in the field'.

13 Fogden, M.P.L. (this volume), 'A preliminary field-study of the western tarsier *Tarsius bancanus* Horsefield'.

14 Rahm, U. (1960), 'Quelques notes sur le Potto de Bosman', *Bull. IFANA*, 22, 331-42.

15 Petter, J-J. and Hladik, C.M. (1970), 'Observations sur le domaine vital et la densité de la population de *Loris tardigradus* dans les forêts de Ceylan', *Mammalia* 34, 394-409.

16 Sussman, R.W. (this volume), 'Ecological distinctions in sympatric species of *Lemur*'.

17 Moreau, R.E. (1966), *The Bird Faunas of Africa and Its Islands*, New York.

18 Paulian, R. (1961), '*La zoogéographie de Madagascar et des iles voisines*', Tananarive, Faune de Madagascar, 13, 1-485.

19 Ford, E.B. (1964), *Ecological Genetics*, London.

20 Petter, J.J. and Peyriéras, A. (1970), 'Nouvelle contribution à l'étude d'un lémurien malgache, le Aye-aye (*Daubentonia madagascariensis* E. Geoffroy)', *Mammalia* 34, 167-93.

21 Petter J.J. and Peyriéras, A. (1970), 'Observations éco-éthologiques sur les lémuriens malgaches du genre *Hapalemur*', *Terre et Vie* 24, 356-82.

22 Bourlière, F. and Petter-Rousseaux, A. (1966), 'Existence probable d'un rythme métabolique saisonnier chez les Cheirogaleinae (Lemuroidea)', *Folia primat.* 4, 249-356.

23 Charles-Dominique, P. and Hladik, C.M. (1971). 'Le *Lepilemur* du Sud de Madagascar: ecologie, alimentation et vie sociale', *Terre et Vie*, 25, 3-66.

24 Petter, J.J., Schilling, A. and Pariente, G. (1971), 'Observations éthologiques sur deux lémuriens malgaches nocturnes, *Phaner furcifer* et *Microcebus coquereli*', *Terre et Vie*, 25, 287-327.

C. M. HLADIK and
P. CHARLES-DOMINIQUE

*The behaviour and ecology of the sportive lemur (*Lepilemur mustelinus*) in relation to its dietary peculiarities*

Introduction

In the course of a field-visit to the south of Madagascar (September/October 1970), *Lepilemur mustelinus leucopus* (F. Major 1894) was intensively observed in order to define the ecological characteristics of this nocturnal folivorous species (Fig. 1*a*), which is adapted to particularly arid conditions in this region. Study of one population provided us with an understanding of the social structure, permitting comparison with other prosimian species. Subsequently, a laboratory investigation was conducted on various samples from a digestive tract collected in Madagascar. This permitted more detailed examination of the phenomenon of *caecotrophy*, which had previously been observed in the field.

Field conditions

The sportive lemur,* which is still the subject of taxonomic debate, is found in almost all Malagasy forest areas. Most of the observations made during the present study were carried out in South Madagascar in a quite distinctive forest zone – Didiereaceae bush. This spiny bush forest is characterised by dense undergrowth composed of bushes, lianes and shrubs, dominated by species belonging to the Didiereaceae (a primitive endemic family intermediate between the Cactaceae and the Euphorbiaceae). *Alluaudia procera* and *A. ascendens*, which are the principal representatives of the family in

* The name 'sportive lemur' does not refer to the animal's motor activity, but to the defensive attitude which it adopts when threatened. The hands are used in the manner of a boxer to strike at the aggressor.

Figure 1 (*a*). The Sportive Lemur *Lepilemur mustelinus leucopus* (F. Major), photographed whilst active at night.

Figure 1 (*b*). The Didiereaceae bush occupied by the Sportive Lemur in the main study area (Berenty region; South Madagascar). The predominant plant representatives are the tall, taper-like *Alluaudia procera* and *A. ascendens.*

the Berenty region (S. Madagascar), constitute 65 per cent of the vegetation (Fig. 1*b*). These species resemble large, more-or-less ramifying tapers, about 12 m. in height, with a covering of fleshy leaves set at the bases of hard, sharp spines. Here and there, baobabs and some arborescent leguminous species project above the bush, which lacks a true canopy.

A different plant formation – *alluvial forest* or *gallery forest* – lines the rivers in the same area as a band 100-800 m. in width. This classical forest type is dominated by tamarind trees, which form the closed canopy at a height of 25-30 m., above a relatively scanty under-growth. Observations were conducted in this second type of forest purely for comparative reasons. The transition from gallery forest to Didiereaceae bush is very abrupt.

There is a long dry season, and rainfall (500 mm./year) occurs mainly during the austral summer (December/January/February). The monthly average of the maximum daily temperatures fluctuates between 37°C (austral summer) and 28°C (austral winter), and the monthly average minimum temperature fluctuates between 26°C (austral summer) and 16°C (austral winter).

In this xerophile forest formation, most of the plant species store reserves of water for the austral winter, which is particularly dry. In fact, our observations were conducted during part of the austral winter, and in that particular year (1970) there was an exceptional drought more pronounced than any experienced over the last ten years, at least.

Diet

In the Didiereaceae bush, the sportive lemur feeds primarily on the tough foliage of the two *Alluaudia* species, and on their inflorescences when the latter are available. *Alluaudia* flowers appear at the end of the dryest period, at a time when almost all of the leaves have fallen. Thus, the sportive lemur survives for some time on the basis of these flowers, which disappear when the first leaves begin to appear. The leaves of certain shrubs and lianes (*Salvadora augustifolia, Xerosicyos perrieri, Marsdenia cordifolia, Boscia longifolia*) are eaten in far smaller quantities. In addition, there may be a large number of other species whose leaves are utilised occasionally, along with certain fruits, since the present list is entirely derived from the end of the austral winter.

The sportive lemur is distinguished from other prosimians by its restricted motor activity during the night, which may obviously be correlated with the poor calorific value of the diet. As soon as the sun sets, the animal – which spends the entire day hidden in a tree-hollow, a fork of *Alluaudia ascendens*, or (more rarely) in a bundle of lianes or *Euphorbia* foliage – makes a number of rapid leaps to arrive at a leafy ramus, where it will 'browse' for short periods of 1-10 minutes (about 10 sessions per night). Apart from these feeding periods, the sportive lemur remains immobile almost continuously, rarely changing its place. When the animal does move, it makes a few rapid leaps and then remains immobile 10-20 m. away.

In the section on 'social life', it will be seen how we were able to study an entire population by capture, marking and release. The individual animals, which were exceptionally lacking in timidity, could be followed continuously with the light of a head-lamp. After a few days, we were able to observe the animals continuously from 3-6 m. without disturbing their activities. Such favourable conditions permitted us to observe the animals clearly and continuously. We were able to identify all of the food items taken, and vegetation samples taken back to the laboratory in France allowed us to conduct detailed analyses. In addition, continuous observation of individual animals and examination of digestive tracts collected at different times permitted us to determine the average dietary intake.*

* We are indebted to the Service des Eaux et Forêts de Madagascar for a special authorisation to capture a small number of specimens for subsequent laboratory analysis.

In collaboration with the Institut National de la Recherche Agronomique (Laboratorie d'Analyse et d'Essai des Aliments), we have analysed and described the average composition of the diet,[1] which is essentially composed of crassulescent leaves. This represents the 'poorest' diet hitherto observed in a primate: 13.6% vegetal proteins; 1.8% lipids; 4.9% reducing sugars; 15.1% cellulose and a 'complementary fraction' representing 64.6% (Fig. 2). The latter fraction is essentially composed of long-chain sugars (ligno-celluloses and hemicclluloscs), which cannot be digested by most mammals, together with a large mineral component.

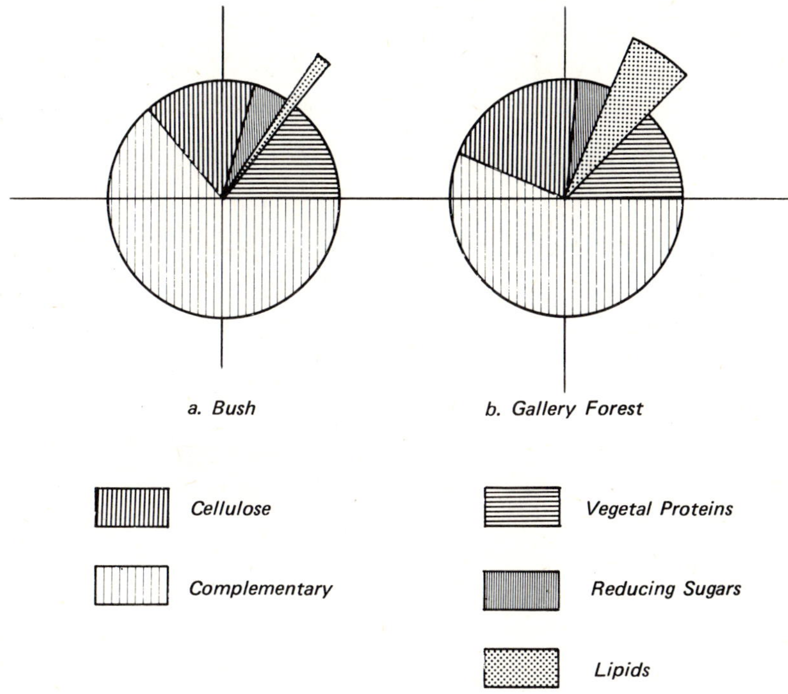

Figure 2. Diagrammatic representation of the dietary components of the Sportive Lemur in Didiereaceae bush and in gallery forest of South Madagascar.

In the gallery forest lining the river Mandrare, other individuals belonging to the same sportive lemur subspecies were, during the same period, feeding primarily on the leaves of *Tamarindus indica*, which contain 12.5% proteins, 6.0% lipids, 4.8% reducing sugars, 20.5% cellulose and a 'complementary fraction' of 56.2%. This composition is little different from that available to the animals feeding in the dryest areas of the bush (Fig. 2).

If this diet is compared with that of other known folivorous primate species, there is a considerable difference. In particular, the howler monkey (*Alouatta palliata*), which is the most folivorous of the South American monkeys, has a diet under natural conditions which includes a large proportion of fruits (60%), and thus has a relatively higher proportion of reducing sugars (21.7%), which are easily assimilable.[2] Similarly, among the Colobinae of the Old World the proportion of fruits included in the diet is quite high: 28% in *Presbytis senex* and 45% in *P. entellus.*[3] It is therefore certain that conditioning to soluble components of the diet, which plays an important part in the feeding behaviour of the 'higher' primates, is of negligible importance for *Lepilemur.* An immediate response to certain soluble components (principally short-chain sugars) can lead to rapid conditioning, which increases the efficiency of the animal in its natural environment, where it must detect its food with maximum yield. This constitutes motivation of the hedonic type. By contrast, conditioning through factors independent of the 'olfactory-gustatory' sense (physiological factors corresponding to an increased feeling of 'well-being' arising after the animal has taken its food) may play an important part in determining the sportive lemur's activity rhythm.

Dietary utilisation (caecotrophy) and the energy budget

Calculation of the energetic value of the dietary intake (about 60 gm. of fresh food per night), which was determined as part of the study already mentioned, shows that those nutritive elements which are readily assimilable (proteins, lipids and reducing sugars) represent only a weak energy source (13.5 Kcal. per day) for an animal weighing 600 gm. on average.

In fact, the cellulose fraction of the diet is degraded in the course of its sojourn in the caecum and colon of *Lepilemur.* The products of the components thus degraded are at least partially resorbed by the animal during a second passage of the food through the intestinal tract, which is achieved through *caecotrophy* (ingestion of certain faeces).

The behaviour pattern of re-ingesting certain faeces is reminiscent in several respects of similar behaviour in the rabbit, which is now well documented.[4] At about the mid-point of its diurnal resting period, the sportive lemur exhibits a phase of hyper-excitability. The animal begins by licking its fur and then concentrates on licking the anogenital region. When doing this, the thighs are spread apart and the tail is curled upwards, such that the pelvic area forms a kind of basin. The animal licks its anus and raises its head from time to time in order to swallow. This behaviour was observed directly on several occasions, always at the hottest time of the day (14.00-15.00 hrs.); but it would seem that it can also occur in the morning, as soon as the animal has returned to its daytime retreat. In fact, in the stomach contents of an animal captured at 7.00 hrs. we found fatty acids of bacterial origin, which could have originated from contamination by material re-ingested the day before, but which were more probably produced in the caecum and re-ingested as soon as the animal returned to its diurnal resting-place.

Analysis of the digestive tract contents of animals captured at different times of the day and night permitted us to follow the actual transformations which take place. In order to conduct this analysis, we profited from the period when the animals were feeding essentially on the leaves and flowers of the two *Alluaudia* species (which are very similar in composition), in order to have a 'basic diet' formed of a homogeneous mixture of these dietary samples. The differences in composition along the digestive tract are very marked, and there is no possibility of confusion. Although the basic diet only contained long-chain fatty acids, gas chromatography demonstrated that ramifying short-chain fatty acids ($C \leqslant 16$) with uneven numbers of carbon atoms were present in the digestive tract (analysis carried out at INRA).[5] The bacterial origin of these latter acids, and their abundance in the caecum, is not surprising. Electron micrographs of the caecal mucous membrane showed a dozen species of small-sized bacteria in contact with the microvilli. The structure of the caecum itself, which is lined with long villi supported by prolongations of the muscularis mucosae, shows that there must be mixing movements allowing rapid fermentation. The caecal contents become more and more fluid as the transformations progress.

The ingested food initially undergoes rapid absorption of the soluble fractions, which results in an apparent augmentation of the proportions of ligno-cellulose (rising from 21.4% to 43.0%). In the caecum, there is first of all hydrolysis of the cellulose fraction, which brings about a reduction in the ligno-cellulose level (17.5% in the caecal contents by the beginning of the night). The hemicelluloses are subjected to much slower degradation; their concentration amounts to 51.8% of the dry weight of the caecal contents, drops to 34.5% in the colon by the beginning of the night and reaches 20.6% in the faeces. Thus there is a progressive inversion of the hemicellulose/ligno-cellulose ratio, which changes from 3.9 in the caecum to 0.45 in the faeces.

The caecal contents are enriched with proteins (up to 36.0%, whereas the basic diet contents incorporate only 15.1%). The surplus protein is derived from desquamation of the mucous membrane and proliferation of bacteria. The importance of caecotrophy thus lies in limitation of nitrogen-loss. In the rabbit, the caecal matter which is re-ingested similarly contains a much higher concentration of proteins than the ingested food, and there is a marked difference in appearance between the caecotrophe (re-ingested caecal material) and the true faeces. In *Lepilemur*, however, we did not find any clear-cut morphological difference between the 'true' faeces and the caecotrophe.

In order to calculate an approximate energetic balance for *Lepilemur*, we selected an average example provided by an adult male which was continuously followed from the onset of activity until its return to rest. This individual ingested 61.5 gm. of leaves and flowers, corresponding to 8.0 Kcal. of proteins, 2.7 Kcal. of lipids and 2.8 Kcal. of carbohydrates, and thus to a total of 13.5 Kcal. of directly assimilable elements. The energetic contribution based on bacterial degradation of celluloses can be estimated at 10-15 Kcal. Thus, the overall energetic product is less than *30 Kcal/24 hr.*

In the example studied (Fig. 3), the animal which was followed covered 270 m. in the course of the night, making 180 leaps of about 1.5 m. Standard calculation of the energy expenditure indicates a *minimum muscular expenditure of 2.16 Kcal.*[6]

According to standard calculations of the basal metabolism, an animal of 600 gm. would require 40-60 Kcal. per day. In fact, it is known that many tropical and equatorial animals have imperfect thermoregulation and a basal metabolism inferior to the norm calculated for palaearctic animals adapted to cold climates. As has been demonstrated by Kayser (see also Bourlière and Petter-Rousseau)[7,8] the metabolic level of homoeothermic animals increases, with increasing perfection of thermoregulation, whilst some primitive homoeotherms in regions with a uniform climate have a basal metabolism at half the standard level. Research in progress in Gabon has confirmed this marked difference at the physiological level between primitive species derived from tropical and equatorial areas and species derived from temperate regions, on which the standard values have been based. For example, Hildwein[9] has shown that the basal metabolism of certain lorisids is inferior to that of 'true' homoeotherms, amounting to a deficit of 20-30% in *Arctocebus calabarensis*, and of 50% in *Perodicticus potto*.

Thus, one can expect that *Lepilemur* has a basal metabolism requiring in the order of *20-30 Kcal.*

This crude calculation is not intended as an exact indication of the energetic balance. It simply demonstrates, for an animal living under natural conditions, a system permitting it to survive with a minimal energy-input.

During the night, movement is fragmented into brief, regularly spaced periods (approximately twenty per night, see Fig. 3). Thus, chemical thermoregulation is augmented by the production of muscular heat energy at the time when it is most necessary. By contrast, during the diurnal resting phase, when the external temperature remains in the region of 30°C, energy expenditure must be minimal (bearing in mind the insulation provided by the dense fur).

The sportive lemur does not have to move far to find its food: a

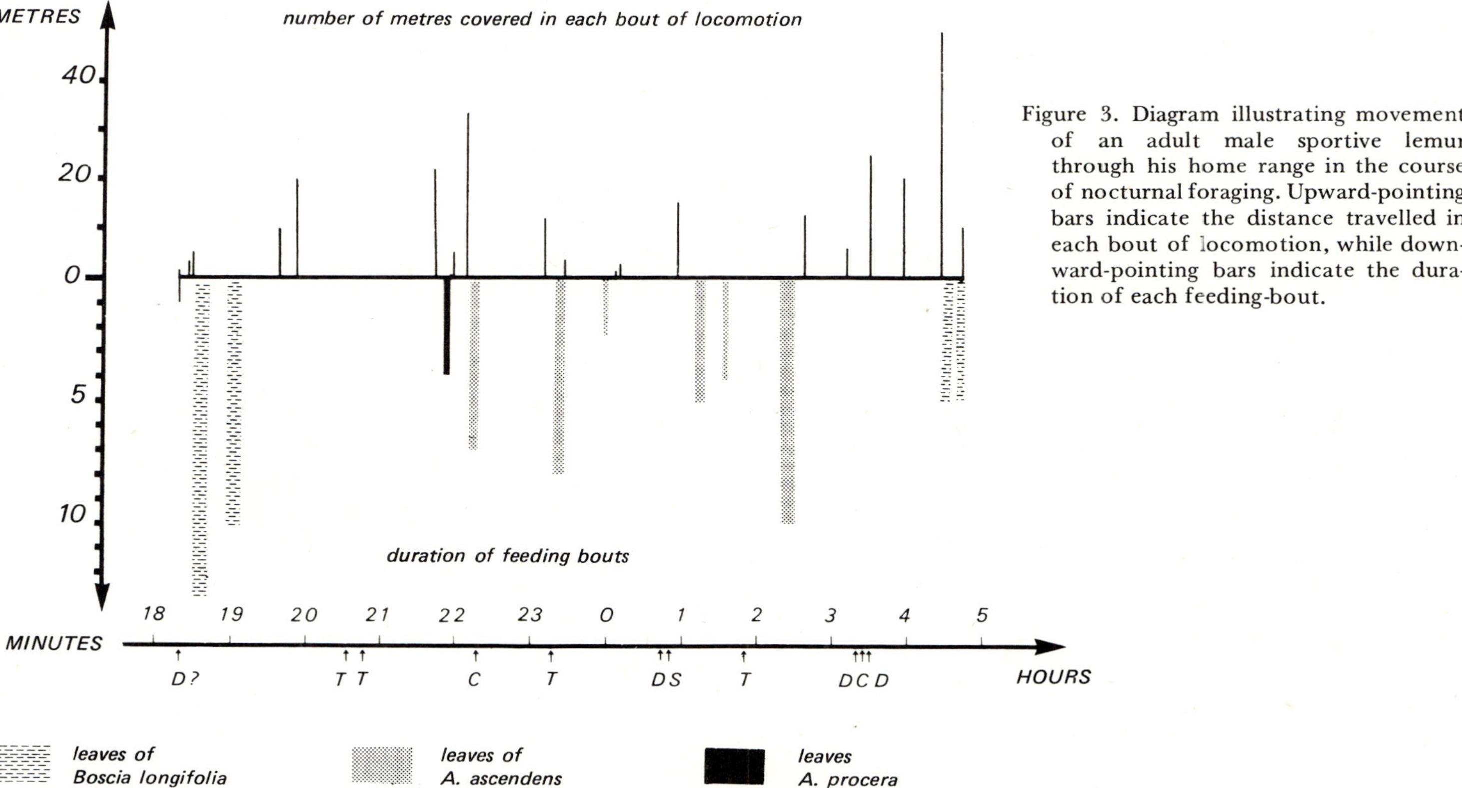

Figure 3. Diagram illustrating movement of an adult male sportive lemur through his home range in the course of nocturnal foraging. Upward-pointing bars indicate the distance travelled in each bout of locomotion, while downward-pointing bars indicate the duration of each feeding-bout.

single leaf-bearing ramus will suffice for one night. Thus, the animal performs a minimal muscular effort in moving around (scarcely more than 10% of the total energy expenditure). This economic way of life necessitates maximal utilisation of a naturally poor food-source, from which celluloses can be assimilated only after bacterial degradation. Accordingly, caecotrophy is a specialisation permitting utilisation of a food-source which normally yields little energy. This process, which is common among rodents and lagomorphs, seems to be exceptional among primates; nevertheless, it is less efficient than rumination, which permits even better utilisation of ligno-celluloses.[10]

Adaptation to the environment: population densities

The economic life-style of the sportive lemur is further expressed in the utilisation of extremely small home-ranges within which it can find its food. In the Didiereaceae bush, the average area of each home-range is 2,300 square m., and we were able to calculate under field conditions that the food available in the month of September could (theoretically) have provided for a maximum of 50 days. Bearing in mind the irregularity in distribution of plants and territorial demarcation (see next section), this figure can be regarded as a *minimum* which is explained by the drought present throughout the observation period. This exceptional drought, which was far more marked than any in the ten years preceding the study, represented a critical period, in the course of which there is doubtless a process of equilibriation between such populations and the environment.

The distribution of sportive lemurs in the Didiereaceae bush is not homogeneous, and the population counts that we carried out along the transects show that there are small population nuclei which are isolated to varying degrees. Various areas of the bush remain unoccupied, partly because of a lower density of edible plant species; but more frequently we observed that there was a scarcity of adequate diurnal shelters for the sportive lemurs to rest in.

In the gallery forest, the population density is greater than in the bush; it reaches 450 *Lepilemur* per square km., corresponding to a biomass of 2.7 kg. per hectare, as compared to 200-350 *Lepilemur* per square km. in the bush, which amounts to a biomass of 1.2-2.1 kg. per hectare. These population densities can be compared with those observed in a habitat entirely comparable to the gallery forest of southern Madagascar – that is, the alluvial forest of Ceylon. The biomass of a nocturnal prosimian (*Loris tardigradus*) found in the latter areas is extremely low, less than 0.25 kg. per hectare.[11] However, since *Loris tardigradus* is an insectivore, the 1:10 ratio of the biomasses is quite in agreement with standard ecological relationships found when passing from one trophic level to one immediately above. (*Microcebus murinus*, which is also partially insectivorous, has a biomass of 0.23 kg./hectare in the Didiereaceae bush). If we consider the folivores living in the alluvial forest of Ceylon (that is, the two colobines *Presbytis senex* and *P. entellus*), there are biomasses of 10 and 15 kg. per hectare.[3] These figures represent the maximum densities observed for animals whose 'dietary assimilation' is certainly better than that of *Lepilemur*, since their digestive system exhibits remarkable convergence with that seen in ruminants. In the gallery forest of South Madagascar, two large-bodied lemurs (the Sifaka, *Propithecus verreauxi*, and the Ringtail, *Lemur catta*) occupy ecological niches analogous to those of the colobines in Ceylon.

Therefore, the size of the *Lepilemur* home-range reflects, in the most arid area, more effective utilisation of the terrain. The biomass is very large in view of the low productivity of the Didiereaceae bush during adverse periods.

We carried out a complementary study of the distribution of the sportive lemur (*Lepilemur mustelinus mustelinus*) in the east coast rain-forest of Madagascar, which is dense and evergreen, and which doubtless has much greater primary production (study area: forestry station of Perinet). This other subspecies of *Lepilemur mustelinus*, which is 50-100% larger than the subspecies found in southern Madagascar, is sympatric with a nocturnal folivorous Indriid of comparable size – *Avahi laniger*. In the course of nocturnal counts, it was not always possible to distinguish between the two lemurs; but the overall density (of *Lepilemur* and *Avahi* together) is of the same order as that of *Lepilemur* in the dry forest of south Madagascar. Thus, one might ask whether the limiting factor governing such folivore populations is, in certain cases, something other than the level of food availability. However, the synecology of dense forests is much more complex – and hence much less well known – than that of dry forests. There are far more species in dense forests, the ecological niches are more specialised, and the conditions of observation are much more difficult.

Folivorous primate species share a large number of characters. They are the least active and the least mobile of all, though one must not confuse such slowness with that of the insectivorous lorisids (*Loris, Arctocebus* and *Perodicticus*), where it represents a cryptic mechanism.[12] It is obvious that the ease with which such folivorous species can find their leaf diet removes any necessity to move over large areas, and the home ranges are small. Among the gregarious 'higher' primate species the folivorous forms (*Alouatta, Presbytis*, etc.)[3,13] only defend small territories, compared with those of sympatric species which are frugivorous or insectivorous. Among the solitary nocturnal lemurs, individual territories are similarly smaller in folivorous forms than in frugivorous or insectivorous species. *Lepilemur* would appear to represent the lower limit of such home-range restriction.

Social life

The Didiereaceae bush constitutes a habitat which is particularly favourable for observation. During our study period, at the end of the dry season, the bush had the appearance of a forest of dry trunks and branches without leaves, in which visibility was very good. In addition to this, it was relatively easy to drive sportive lemurs from their retreats during the daytime, by beating the vegetation, and subsequently to capture them. When disturbed, sportive lemurs take refuge at a height of 5-8 m. on a ramus of *Alluaudia* and observe the intruder without moving. With a little patience it is possible to approach the animal very cautiously with a noose suspended at the end of a long pole and to pass the noose around the animal's neck. Some captures took only 10 minutes, whilst others required more than two hours; but we were nevertheless able to capture 12 of a population of 13 sportive lemurs. After a standard examination (genital organs, mammary glands, teeth, body-weight, etc.), the animals were marked by clipping the ears and shaving the tail according to various patterns, and then they were released and observed. In this study area, strings attached at a height of 1 m. in rows 10 m. apart, each marked with numbers, permitted us to localise any observation to within 2-3 m. on a map, and thus to delimit individual territories.

1. Territories

As has already been mentioned, the territories are small.[1] Adult females range over 0.18 hectares (0.15-0.32), adult males over 0.30 hectares (0.20-0.46), and juvenile females over 0.19 hectares (0.18-0.20). For comparison, one can consider the following values taken for 2 lorisid species (insectivorous and frugivorous prosimians): *Galago demidovii* – 0.8-2.7 hectares;[14] *Perodicticus potto* – 7.5-15 hectares.[15]

The female territories are distinct from one another. However, we noted a certain range overlap and mutual tolerance between an old female and two young females of one and two years of age. (All the females were gestating in October, and reproduction takes place only once a year). The territories of the males are similarly separate from one another, but we did observe one case of overlap between the two smallest males in the population. However, these two individuals were never observed together in this common area. Although there is quite clear separation of territories between males and between females, overlap of female territories with those of the males is the rule. The same applies, in fact, to the Lorisidae and to *Microcebus murinus.*[14,15,16]

The largest of the males was associated in this way with 5 females, whilst the other males were associated with two, one and one female, respectively.

2. *Territorial defence*

All adult males and females exhibit scars on their ears, snouts and tails as witness to intra-specific fights. We have never observed such traces of former wounds with young females at one year of age. Although olfactory marking constitutes the principal means of territorial defence with most nocturnal species, *Lepilemur* exhibits a process which is quite remarkable. In the sportive lemur, which has such small territories of approximately 50 m. diameter, the territory-owner can easily survey the limits, which are absolutely rigid. The occupant of a territory, usually whilst squatting on a high ramus of *Alluaudia*, surveys his immediate neighbours, which are also located on elevated branches. Such mutual surveillance is particularly frequent with the males, which spend hours observing one another, often with only a few metres between them. When one of them moves, it is common to see the neighbour move in his turn along the territorial limit and come to rest facing his opposite number once again.

From time to time, two – and sometimes three – males exchange vocalisations as a duet or a trio. A duet consists of a rapid series of 'Heh, heh . . . ' calls (aggression), followed by an abrupt, high-pitched 'Hiii' call (signalling). These two vocalisations are subsequently exchanged for a period of 10-20 seconds in a less precise order, and calm then returns. In fact, we counted many more duets of this kind when the moon was visible, and this indicates quite clearly that contact between neighbours is largely based on vision. In the course of reciprocal surveillance, we also saw a display consisting of powerful leaps against a support, or a simulation of jumping reminiscent (in a slow-motion form) of the rapid branch-shaking exhibited by monkeys. Side-to-side head shaking also occurs in territorial encounters.

The sportive lemurs deposited urine on the branches of the 'surveillance posts', but we never noticed any particular behaviour associated with such urination.

3. *Social relationships*

Even if there is nothing more than superposition of male and female territories, there must be relatively frequent encounters between individuals of opposite sex. When they do meet, one individual utters a 'Hiii' vocalisation, and then both continue on their way.

During our period of observation, it was seen that males and females sleep in separate retreats each night; but that the same retreat may be used by a male and a female on successive nights.

A few days before our observations were completed, we removed – by way of experiment – the largest male, whose territory overlapped with that of five females. The very next night, all of the other males penetrated into this zone, and each one began to associate with one of the females. It is important to note that the males were all in a sexually inactive state (regression of the testes) and that the females were either gestating or immature. Two days after the experiment, one female was already accepting the presence of one of the males, which was seen sniffing at her ano-genital region.

Courtship behaviour, as with surveillance of territories where females occur, is thus independent of direct sexual behaviour, since it can occur outside the reproductive period. Obviously, this represents some kind of social behaviour, as is found in other nocturnal prosimians.

Conclusions

Lepilemur, whose phylogenetic relationship to other lemurs is difficult to establish, is distinguished first and foremost by its extremely specialised diet. Alongside primitive anatomical and behavioural characters resembling those of the Cheirogaleinae and the Lorisidae (closure of the vulva during sexual inactivity; transport of the infant in the mouth; social organisation; overlapping of male and female territories), one can see a series of adaptations directly or indirectly associated with the dietary regime: specialised dentition, modified digestive tract, caecotrophy, small territories. Social organisation is of a primitive type; but the system of visual territorial surveillance, which is unusual for a nocturnal mammal, is doubtless related to the very small area of the defended territories.

Acknowledgments

This study was carried out in the private reserve of Mr H. de Heaulme in Berenty, where we benefited greatly from his generous hospitality. We would also like to thank our colleague, Dr R.D. Martin, who managed to find time to translate this article.

NOTES

1 Charles-Dominique, P., and Hladik, C.M. (1971), 'Le *Lepilemur* du sud de Madagascar: ecologie, alimentation et vie sociale', *Terre et Vie*, 25, 3-66.

2 Hladik, C.M., Hladik, A., Bousset, J., Valdebouze, P., Viroben, G. and Delort-Laval, J. (1971), 'Le régime alimentaire des Primates de l'île de Barro Colorado (Panama); résultats des analyses quantitatives', *Folia primat.* 16, 85-122.

3 Hladik, C.M. and Hladik, A. (1972), 'Disponibilités alimentaires et domaines vitaux des Primates à Ceylan', *Terre et Vie* 26, 149-215.

4 Morot, C. (1882), *Rec. Med. Vet.* 59, 635-46. [This was the first reference to caecotrophy in the rabbit; see reference 10 for more recent work.]

5 Hladik, C.M., Charles-Dominique, P., Valdebouze, P., Delort-Laval, J. and Flanzy, J. (1971), 'La caecotrophie chez un Primate phyllophage du genre *Lepilemur* et les correlations avec les particularités de son appareil digestif', *C.R. Acad. Sci. Paris* 272, 3191-4.

6 Kayser, C. (1963), 'Bioénergétique', in *Physiologie*, Paris, vol. 1.

7 Kayser, C. (1967), 'Evolution de l'homéothermie incomplète', in *Colloque pour le centenaire de Claude Bernard*, Fondation Singer-Polygnac, pp. 285-323.

8 Bourlière, F. and Petter-Rousseaux, A. (1953), 'L'homéothermie imparfaite de certains prosimiens', *C.R.S. Soc. Biol.* 147, 1594.

9 Hildwein, G. (1972), personal communication.

10 Thompson, H.V. and Worden, A.N. (1956), *The Rabbit*, London.

11 Petter, J.J. and Hladik, C.M. (1970), 'Observations sur le domaine vital et la densité de populations de *Loris tardigradus* dans les forêts de Ceylan', *Mammalia* 3, 394-409.

12 Charles-Dominique, P. (1971), 'Eco-éthologie des prosimiens du Gabon', *Biologia Gabonica* 7, 121-228.

13 Hladik, A. and Hladik, C.M. (1969), 'Rapports trophiques entre végétation et Primates, dans la forêt de Barro-Colorado (Panama)', *Terre et Vie* 23, 25-117.

14 Charles-Dominique, P. (1971), 'Eco-éthologie et vie sociale des prosimiens du Gabon', Doctoral thesis, no. A.O. 5816, Paris.

15 Charles-Dominique, P. (1972). 'Ecologie et vie sociale de *Galago demidovii* (Fischer 1808, Prosimii)', *Z.f. Tierpsychol. Suppl.* 9, 7-41.

16 Martin, R.D. (1972), 'A preliminary field-study of the lesser mouse lemur (*Microcebus murinus* J.F. Miller 1777)', *Z.f. Tierpsychol. Suppl.* 9, 43-89.

J.-J. PETTER and A. PEYRIÉRAS

A study of population density and home ranges of Indri indri *in Madagascar*

Introduction

The Indri (Fig. 1) is the largest of the extant Malagasy lemurs and the most specialised living representative of the family Indriidae, which also contains the Sifakas (*Propithecus verreauxi* and *P. diadema*) and the Avahis (*Avahi laniger*). All members of the family typically move around by leaping from trunk to trunk, with the body held vertical.

Although it is always difficult to observe Indri because of their unobtrusiveness, they are well known from their far-carrying, melodious vocalisations. They only live in mountainous areas in the northern half of the East Coast forest of Madagascar, which is characterised by an extremely humid climate. Because of the difficulties involved in observing these animals, largely as a result of their timid nature and the rugged terrain of their mountainous habitat, many features of their biology are still unknown. It is important that we should learn more about the Indri under natural conditions, while there is still time, particularly since such knowledge could assist us in protecting them more effectively. Data are particularly lacking for characteristics of diet, reproduction and home range.

In August 1968 and November 1969 we conducted expeditions into primary forest zones far from human influence, with the main aim of obtaining information about home range size. These two areas visited were: (1) a zone to the north-east of Maroantsetra and (2) the forest of Fierenana, to the north of Lake Alaotra. In 1971 and 1972, visits were made to zones close to forestry concessions in the forests of Perinet and Lakato.

Figure 1. The Indri (*Indri indri*), photographed in east coast rain-forest.

Working conditions

Indri occur in the forest of Analamazotra, at Perinet, which is a well-known forestry station on the road between Tananarive and Tamatave. In this area, they are more readily observed than anywhere else. In fact, the Service des Eaux et Forêts of Madagascar has recently created a special reserve in this zone for protection of the Indri. Numerous pathways, which are regularly serviced, permit easy penetration into the forest, and with a little luck it is possible to observe or film Indri under satisfactory conditions. Most of the photos, films and tape-recordings which have been obtained to date stem from the forest at Perinet.

Unfortunately, the forest at Perinet is already heavily exploited, and animals have been hunted in this area for some time. The fact that it is still possible to find Indri (whose Malagasy name 'Babakoto' can be colloquially translated as 'little father') is due to partial protection by local beliefs. However, collections of various kinds, exploitation and reforestation have modified their habitat, and the other diurnal lemurs have almost all been wiped out.

Thus, this area – despite the advantages which it offered for research on certain features of Indri biology – was not satisfactory for the basic requirements of this study. The same applied to various other forest zones in this region (around Mount Anketrambe on the road to Lakato, South of Perinet).

A number of surveys eventually indicated a favourable area, sufficiently remote from inhabited regions, roughly 20 km. to the east of Maroantsetra, in the forest of Fampanambo. The fringe of this forest could be approached by boat in August, following the River Antanambalana, and during the dry season (in November) it was accessible along a forest track from Maroantsetra. A three-hour march into the forest zone was necessary to reach the selected study area, where a camp-site was set up.

In July 1969, we worked there for eleven days in almost continuous rain. Subsequently, in November 1969, we profited from a period of more clement weather to work for seven days in alternating rainfall and sunshine.

On a map (Fig. 2) showing the topography of the region studied, one can distinguish a series of ridges which are mainly oriented in a north-south direction and have an average altitude of 400 m. They are separated by valleys not exceeding 200 m. in depth. Dense forest covers all the slopes, and vision is limited by foliage everywhere.

Another suitable area, though more difficult to investigate, was the forest of Fierenana, which is accessible along a pathway branching off the road from Moramanga to Lake Alaotra. Atmospheric conditions in this area are similarly extremely unsuitable for prolonged visits.

Methods of study

In the course of the first two expeditions, we were able to explore in some detail a vast forest area by spending the days, and parts of the nights, walking along pathways on the ridges in a rugged mountainous region. The animals were always very difficult to observe. In fact, over and above the inaccessibility of their habitat, Indri are almost invisible when they are resting in the fork of a tree, and they are almost impossible to detect if they remain motionless, as was recently observed by P. Charles-Dominique and M. Hladik (personal communication). The retracted dark limbs, surmounted by the white 'V' of the dorsal pelage, can easily be confused with the fork of a trunk seen against a cloudy sky, or with a patch of dark foliage beneath a fork covered with lichens. The white patch on the forehead, which gives the impression of a gap in the vegetation, adds to this effect.

We were able to locate the different Indri groups primarily from their vocalisations, by determining the direction with simultaneous estimation from a number of observation points. In this way, we were able to locate vocalising animals on a map by plotting and estimation of position (Fig. 2). In the course of this study, we were also able to collect information on the population density of *Varecia variegata*, whose powerful vocalisations are also easily recognised.

As a rule, after one Indri group had stopped calling, another neighbouring group took its turn, and successive groups took up the sequence into the distance. Thus, it is possible to make exact localisations, and there is no danger of counting a moving group twice.

Types of vocalisations, frequency and carriage

The vocalisations of the Indri are extremely characteristic; they are very resonant and are amplified by a dorsal laryngeal sac.[1] A number of observations on their structure and frequency have already been published.[2] Two types can be distinguished:

1. Repeated calls like blasts of a horn, uttered occasionally by disturbed or isolated individuals. These calls appear to be relatively powerful when one is close to these animals, but their carriage seems to be relatively weak.
2. Barks, followed by long, modulated howls (frequency range: 0.5-4 kilocycles), uttered by all individuals in a family. It is this latter type of vocalisation, which is extremely powerful, that primarily permitted localisation of Indri groups.

The number of calls is usually extremely variable on different days and at different times of the day. In general, more vocalisations are heard during good weather than during rainfall. The calls are most numerous in the morning, just after sunrise or towards midday, and they seem to be markedly more frequent in the period December/ January.

The carriage of the calls is very difficult to define precisely, since the sound emitted is extremely variable in intensity and tonality. In addition, the animals can change position whilst vocalising, and the direction of the calling seems to vary.

We attempted to evaluate the carriage of vocalisations using two

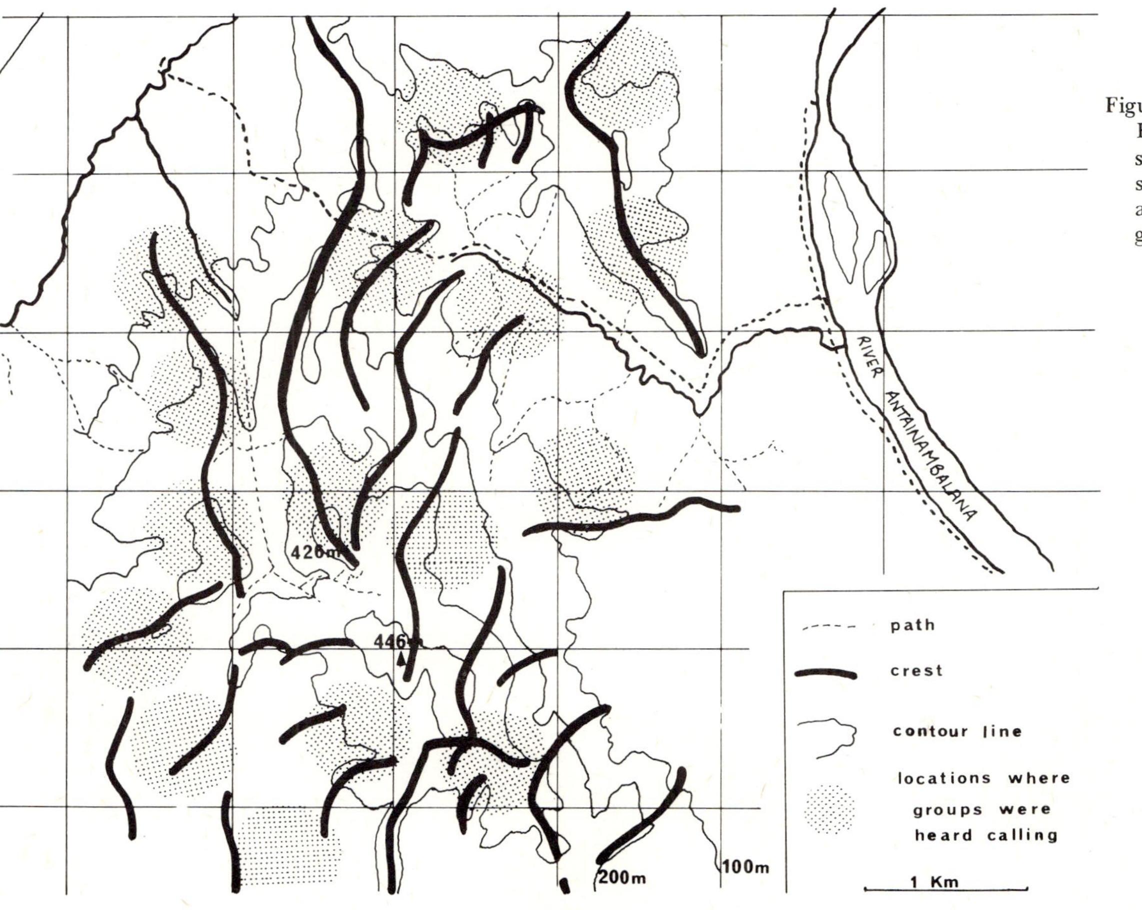

Figure 2. Map of the forest area of Fampanambo (near Maroantsetra, north-east Madagascar), showing the prominent ridges and the locations of calling Indri groups.

methods: the first method involved direct comparison of the impressions of several observers, who happened to be at differing distances from a well-localised group at the time of vocalisation. The second, indirect, method was based on recording of calls and subsequent comparison of the calls with others recorded under identical conditions, but with directly measured localisation.

Under normal conditions of population concentration, most of the Indri which we saw uttering the second type of call were near to the crests of the trees at the top of a ridge. It is possible that this position is optimal for vocalisation and reception of calls.

Under conditions where concentration is abnormally high, it seems that the Indri may emit their calls equally frequently from the depths of the valleys. Observation from one ridge under such conditions permitted localisation of three groups in three valleys surrounding the ridge.

When one is close to the ground in the middle of the forest, the vocalisations are generally audible over more than one kilometre in good weather, despite the density of the trees. When calls cross a valley, they may carry even farther, whilst calls emitted from a valley probably do not carry further than 500 metres.

The Indri possesses large ears with much better developed auricles than those found in other Indriids.

We have no idea of the auditory acuity of the Indri, but it is probably greatly superior to our own and sufficient for perception of the calls of other groups over great distances, despite the wind, the rain, and the screening effect of the rugged terrain.

The structure of the call, one part of which consists of a howl whose frequency varies in the course of emission, is certainly adapted for transmission over great distance in the forest environment. It is interesting to note that other primates living in dense forest, such as the gibbons (for example), have similar vocalisations, and that even certain birds in Madagascar (e.g. *Leptosomus discolor*) utter their calls from the tops of the trees and have a modulated, plaintive vocalisation. Incidentally, Betsimisaraka woodcutters communicate with one another in the forest – often over great distances – with modulated calls of this type.

Group density and composition

By plotting on a map (Fig. 2) the localisations obtained by listening to the calls, it is possible to calculate the approximate concentration of the Indri groups. Taking into account the imprecise nature of the localisation methods, one can trace circles covering a surface of about 100 hectares around each of the groups.

Given the present state of our knowledge, it is difficult to determine whether these 100 hectare areas correspond with a 'home-range' for each group, or whether they simply indicate reserved zones corresponding to the rather unusual relief of the eastern rain-forest. In this humid, rainy climate, it is possible that each group occupies an entire hill for thermo-regulation, seeking out optimal sun-exposure during the mornings and afternoons.

Estimations carried out in the forest zone to the east of Maroantsetra and in the forest of Fierenana are entirely comparable. They would therefore seem to be representative of the natural condition in primary forest which is still untouched by human interference.

Observations carried out at the forestry station of Perinet (P. Charles-Dominique and M. Hladik, personal communication) have shown that in this area there is a far greater density of Indri groups. The same would seem to apply to other zones, such as the region studied close to the Lakato road, where there has been – or still is – human interference (Fig. 3). This high density of groups would appear to be abnormal and associated with forest degradation. The Indri, fleeing from the presence of woodcutters, concentrate in quieter areas, and there may be a resulting temporary crowding of groups. In such areas, it is noticeable that the animals vocalise more often and reply to one another more frequently, as if continually excited by the presence of neighbours too close at hand.

In areas of high densities of Indri, the forest always exhibits a certain degree of degradation, at least in neighbouring zones. This is the case in Perinet, where – apart from the core area of the forestry reserve – there is not much left of the magnificent forest which once existed there. The same applies to the forest on the Lakato road. There is a vast zone streaked with pathways used for past or present exploitation, and virtually devoid of Indri. But on the fringe, where there is some forest which has scarcely been touched – if at all – one finds exceptionally high densities of these animals.

Near to the road to Anosibe, which once resounded with Indri calls, there are no Indri left at present. In the course of a 4-day stay in this forest, not a single vocalisation was heard. Probably, all of the Indri have been frightened away because of a heavily mechanised forestry company which is very active in this region.

Indri are indeed extremely sensitive to disturbance, and they flee rapidly when one tries to approach them; so this is probably the cause of the concentrations observed. Disturbances which occur too frequently (even in the absence of complete destruction of the biotope) induce the animals to flee. An emigrating group of this kind was observed and followed for 2 days near to the Lakato road. The Indri were fleeing from a disturbed zone where they had previously been resident. They moved along in gradual stages quite silently, and they covered about a kilometre in this way, eventually settling in an area with less disturbance, where they were subsequently observed on several occasions. In this context, it is interesting to note that

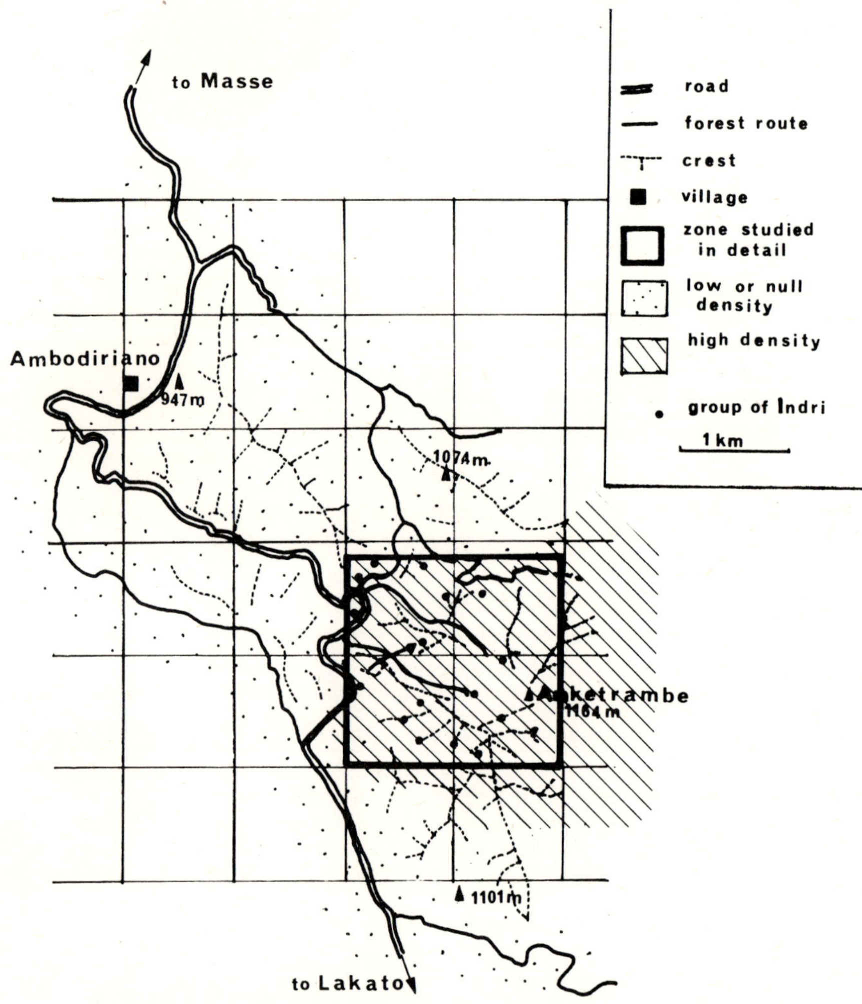

Figure 3. Map showing a forest area close to the Lakato road (near Perinet, east Madagascar), indicating the study area containing an unusually high concentration of Indri groups.

when one group of Indri crosses the 'zone of influence' of another, the invaders move along in silence, and it is the invaded group which can be heard calling frequently. Two groups observed in this region happened to come into confrontation. On ten successive occasions, they exchanged series of isolated barks, although this vocalisation is – under normal conditions – typically followed by long, modulated howls.

Another observation clearly demonstrates the sensitivity of the Indri. In the forest to the west of Mananara, Indri were once very common. The initiation of forest exploitation with heavy machinery belonging to the 'Moulins de Dakar' company drove the Indri completely from vast areas. Abandonment of exploitation two years ago has been followed by progressive re-establishment of these animals, and it seems likely that they were not wiped out by hunting forays made by forestry prospectors, but simply forced to concentrate further to the west in more elevated, calmer forest areas, after they had fled from noise and other disturbance.

Indri generally live in groups of two, three or (exceptionally) four in areas of natural forest, as we have observed. However, we have received reports of larger groups containing five to six animals. Such groups have apparently always been seen in areas disturbed by forestry exploitation, or in adjacent regions, and it would seem that these groups are abnormal products of hyperconcentration.

Conclusions

Our investigation of forest areas far from and near to zones of exploitation indicates that the Indri are normally extremely sensitive animals with respect to human presence. The 'home range' would appear to be extremely large for each group, covering about one square kilometre, and we do not yet know the survival value of such large ranges in view of the fact that the forest would seem to be rich enough to feed larger numbers. The limits of each range seem to be advertised by a powerful call which may carry further than one kilometre.

A greater concentration of such groups seems to be always associated with excessive frequency of human presence, and the animals flee from areas where they are disturbed. The number of animals in each group, normally strictly limited to three or four, seems to increase on occasions under abnormal conditions. It is interesting to compare these observations with similar reports on Sifakas in the west. With the Sifakas, as with the Avahi (i.e. in all Indriidae), the groups would normally have a strictly family basis, as we have observed in numerous areas in the west which are free of human presence. But in zones which are more-or-less degraded or transformed – notably in certain areas of the south – it is possible that the family groups are modified to form larger groups. This would explain the numbers of five to nine animals, instead of three to four, reported for groups by various observers. As with the Indri, it is degradation of the environment which favours the formation of larger groups. This modification probably results in reduced fecundity, since the family is probably the most efficient unit for favouring population growth, as J.H. Crook has shown.[3]

Acknowledgments

This study was carried out with the assistance of the Service des Eaux et Forêts of Madagascar and with the support of the World Wildlife Fund. Georges Pariente, Alain Schilling, Roland Albignac, N.S. Malcolm and Regis Prevôt all participated in this work at various times, and their assistance was extremely valuable. Translation of the article was carried out by Robert Martin.

NOTES

1 Milne-Edwards, A. and Grandidier, A. (1875), *Histoire Physique, Naturelle et Politique de Madagascar*, 6: *Mammifères*, Paris.

2 Petter, J.-J. (1962), 'Recherches sur l'écologie et l'éthologie des Lémuriens malgaches', *Mém. Mus. nat. Hist. nat.* (sér. A) 27, 1-146.

3 Crook, J.H. (1967), 'Evolutionary change in Primate societies', *Sci. J.* 2, 7.

A. RICHARD

Patterns of mating in Propithecus verreauxi verreauxi

Introduction

The earliest written account of *Propithecus verreauxi* was given by Sieur Etienne de Flacourt,[1] a French colonist who disembarked at the site of present-day Fort Dauphin in the seventeenth century. The species was not described in any detail until the 'Histoire naturelle des mammifères: Histoire physique, naturelle, et politique de Madagascar' was produced by Milne-Edwards and Grandidier in 1876, 1890 and 1896.[2] In the twentieth century, brief accounts of field observations have been published by various scientists, such as Kaudern and Rand.[3,4] However, the first systematic field study conducted was by Petter.[5] During that study, a general survey of lemur behaviour in the wild was undertaken, providing a valuable basis for further research on many species, including *P. verreauxi.* Petter-Rousseaux's laboratory work on prosimian reproductive behaviour[6] has also been of great assistance to the present study. Finally, the most recent major field study has concentrated on *P. verreauxi* and *Lemur catta* living in gallery forest alongside the River Mandrary at Berenty, in the south of Madagascar; this work, providing considerable insight into the social organisation and ecology of these two species, has been a most useful source of comparative material for the present study.[7] Nowhere, however, in the literature, is mating in *P. verreauxi* described, and indeed the only Malagasy prosimian species in which mating has been seen under natural conditions is *L. catta.*[7]

In this paper, a description of activities during the mating season of *Propithecus verreauxi verreauxi* is presented, indicating the extent to which they differ from activity patterns during the rest of the year. An attempt is made to interpret these results as part of a total mating pattern; but as the results from the two groups studied are very different, the final evaluation is highly tentative. I was not present during the mating season in the north of the island: in all

Malagasy prosimians, mating appears to be seasonal under natural conditions, and in *P. verreauxi* the timing of the birth period shows that mating must occur during January, February or March all over the island.

A field study was made in Madagascar between April 1970 and September 1971. The main aim of the study was to investigate the ecology and social organisation of *P. verreauxi* using quantitative techniques, with particular emphasis upon intraspecific variation and its ecological correlates. Two study areas were established, representing two extremes of environment in both of which *P. verreauxi* is abundant: thus divergent adaptations might be expected in each area comparable to the intraspecific variation found in many Old World primates living under different ecological conditions. One study area was located in the north west of the island, approximately one kilometre from the forestry station at Ampijoroa, in the ecologically rich region of the Ankarafantsika (National Reserve no. 7). *P. v. coquereli* is found in this region. The second study area, 1,500 km. from the first, lay about 80 km. from the south coast, in Reserve no. 11, 1½ km. south of the village of Hazafotsy. This extremely arid region is occupied by the sub-species *P. v. verreauxi* (Fig. 1).

In the north, with a mean annual rainfall of 1,600 mm., and maximum/minimum temperatures of 39.5°C and 14°C (recorded during the study), the forest is deciduous, growing on sandy soil in the hill-top study area. During the dry season, between April and September, which is also the cold season, most trees shed their leaves. There is no well-defined canopy, most trees reaching a height of 12-16 m., with emergents of 30-35 m. Although not as rich as some other parts of the Ankarafantsika, this particular area was selected because of the absence of hunting in the locality and consequent abundance of easily habituated animals, and because of its accessibility: the latter was vital for a study attempting to sample two areas in all seasons. The forest supports six other prosimian species: *Lemur fulvus, L. mongoz, Microcebus murinus, Cheirogaleus medius, Lepilemur mustelinus*, and *Avahi laniger.*

In the south, with a mean annual rainfall of 600 mm., and maximum/minimum temperatures of 44°C and 8°C (recorded during the study), the area is covered by xerophytic vegetation, rarely exceeding 13 m. in height. The forest is dominated by species of the Didiereacae family, particularly by *Alluaudia ascendens* and *A. procera*; in both, 1 cm. long spines stud the trunk, which itself divides to produce a candelabrum-like silhouette. Over 80% of the plant species in this forest are endemic, but species diversity overall is less in this region than in the northern study area, and only two other prosimian species were seen: *M. murinus* and *L. mustelinus*.

Two neighbouring groups of *P. verreauxi* were habituated in each area. Many of the activities discussed in this paper concern social communication between two of these groups. Ellefson[8] emphasised the importance of inter-group social communication in *Hylobates lar*, pointing out that this usage does not conform with Altmann's[9] definition of social communication as being intra- and not inter-group. It is thus stressed here that although extensive social communication takes place between *P. verreauxi* groups during the mating season, they are still considered to be discrete groups as defined by Struhsaker.[10]

Table 1 Composition of the groups studied

Northern Study Area		*Southern Study Area*	
Group I	*Group II*	*Group III*	*Group IV*
A♂ N (June-Aug. '70)	A♂ H	A♂ F	A♂ R (Jan-March '71)
A♂ GOP (June-Aug. '70)	A♀ NF	A♂ P	A♂ INT (April-June '71)
A♂ STR (Oct.-Dec. '70)	SA ♂D	SA ♂Y	SA♂
A♂ BE (Nov. '70-July '71)	Juvenile (♂)	A♀ FD	A♀ FNI
A♀ BT		A♀ NFD	A♀ FI
A♀ RT		Juvenile (♂)	
A♀ FT			
A♀ SL			

A=Adult, SA=Sub-adult. Initials after age/sex classes are for individual identification. Dates are given in parentheses for periods when animals did not move with groups throughout the study.

The composition of the four groups is given in Table 1. The method used to habituate the animals was similar to that used by Stoltz and Saayman with *Papio ursinus.*[11] When first contacted, an unhabituated group characteristically split up, its members fleeing in different directions. However, they rarely fled far, but rather hid themselves most effectively high up in tree forks. After constantly searching for and pursuing a group for a week, it generally stopped fleeing, and after three weeks animals would approach to feed within 1 m. of me. This was taken as the criterion of habituation, and the quantitative record was only begun after this point was reached. In Group IV the process took longer, two weeks passing before their immediate flight response disappeared. Animals were never provisioned, and although occasional glances at me demonstrated their awareness of my presence, they appeared largely to ignore and be unaffected by it.

In order to facilitate precise location of groups, the home-range of each (varying between groups from approximately 6.75 to 8.5 hectares) was divided up by a grid system of marked paths running north/south and east/west at 50 m. intervals. Quantitative observations were collected, over twelve months, covering three months in the dry season, and three months in the wet, in each area. Seventy-two hours of quantitative data were collected per group, per month, and observations were spread evenly between 06.00 hrs. and

Figure 1 (*a*). Verreaux's Sifaka, *Propithecus verreauxi verreauxi*. Female FD (Group III – Table 1) photographed together with her infant in September 1971 (Hazafotsy; South Madagascar).

18.00 hrs., and between the age/sex classes; in Groups II, III and IV, all animals were recognised individually. One animal was taken as the subject during each 12-hour observation period, in order to establish a continuous record of one individual's activities; a 72-hour sample taken in June 1970 and September 1971, made at half-hourly intervals on all group members, suggests that the timing of the various activities is closely synchronised between animals, so that the daily individual record is considered to be a fairly accurate indicator of group activities. At timed minute intervals during the observation period the subject was described in terms of a number of categories concerning, for example, its height above ground, activity, and the proximity of its nearest neighbour. A full description of recording techniques is given elsewhere.[12] As far as possible, this record was maintained during the mating season, and complemented by extensive descriptive notes.

Figure 1(*b*). Sub-adult male Y (Group III) standing bipedally on the ground whilst feeding. Note the very long hind limbs, associated with the typical vertical-clinging-and-leaping mode of locomotion.

Results

Nature of behavioural changes in the pre-copulatory period

There were two adult females in Group IV. Slight flushing of ♀ FNI(IV)'s* vulva coincided with a sudden, significant increase in certain activities in Group IV in late January 1971. Similarly increased frequencies were noted throughout February and during the first ten days in March, although there was no visible unusual coloration of either female's vulva in Group IV at this time. Copulation took place in March, and the six weeks preceding it are

* The age-class of the animal referred to is henceforth omitted, and the group to which it belongs has been added in brackets after the identifying initials. See Table 1.

henceforth referred to as the pre-copulatory period. The five activities described below occurred occasionally throughout the year, but only in the mating season were they common. The emphasis is thus upon a quantitative rather than a qualitative change in behaviour during these months. The five types of activity and their frequency changes are listed below. Where no figures are quoted, it is to be assumed that no change was found. It should be noted at this point that, although there are striking discrepancies between the data collected on each group throughout this section, infants were born to known females in both Groups III and IV in August 1971. (The implications of this are discussed below.)

1. *'Endorsing' by adult males*

Although adult animals of both sexes scent-mark in a number of contexts, during this period there were increases only in scent-marking by males. While females mark either by rubbing the ano-genital region or by urinating on a trunk or branch, marking by adult males may include rubbing a branch or trunk with the scent gland on the ventral surface of the throat (Fig.2), then with the tip of the penis, usually urinating slightly as this is done, and finally with the perineal area. Although an adult male may perform any one part of this sequence when marking, the sequence was commonly performed in its entirety. Both sexes have a highly developed, almost tubular perineal area. The term 'endorsing' was applied when a male marked a spot within five minutes of a female having been there. Frequency changes in endorsing are shown in Table 2. The figures for Group IV are highly significant,* but those for Group III are not significant.

2 *'Sniff-approach and mark' by adult males*

This occurs when an adult male approaches a female and marks the tree trunk just below her tail: the male climbs the trunk under the female and touches her anus with his nose. He then throat-marks, and finally marks with his ano-genital areas. This sequence is frequently incomplete, for the female may lunge at the male as he thrusts his nose under her tail, forcing him to retreat without marking: in such cases endorsing usually follows when the female moves off. Summing the data, sniff-approach and mark sequences were recorded 6 times in 216 hours of quantitative observation of Group IV outside the pre-copulatory period. The sequence was observed (in complete or incomplete form) 65 times in 216 hours between 24 January and 15 March 1971. The difference between

* Unless otherwise stated, 'highly significant' means significant at the .001 level, and 'significant' means significant at the .05 level, using a Chi Square Test.

Figure 2 (*a*). Adult male R *Propithecus verreauxi verreauxi* (Group IV), showing the dark throat gland area (arrow).

Table 2 Changes in frequency of endorsing in Groups III and IV

Group	*Jan. 1971* *Vulval Change* *Pre-*	*Post-*	*Feb. 1971*	*March 1971*	*April 1971*	*Sept. 1971*
III	1.95	–	2.35	1.08	0.67	1.41
IV	1.90	6.75	6.26	7.00	1.04	–

Figures are expressed as a frequency per adult male hour. Only data recorded when an adult male was the day's subject were used in calculations.

Figure 2(*b*). Adult male R (Group IV – Table 1) marking a vertical trunk with his throat gland.

these figures is highly significant.

3. *'Roaming' behaviour*

Evidence from the study group males, and from the arrival of unknown males in the study area, indicates an increase in male 'roaming' behaviour at this time: males, singly or in pairs, detach themselves from their own groups and make long forays into the home-ranges of other groups. During the mating season, these excursions sometimes culminated in fierce fights between adult males and in copulation with adult females in the groups encountered. During the rest of the study, no solitary or paired males were encountered in the home-ranges of the study groups; on only four occasions was an adult male from Group III or IV recorded as probably being out of immediate audio-visual contact with all other members of his group for over two hours. However, since it was usually the absence of these animals that was noticed, it was rarely possible to confirm that they had detached themselves from the group rather than having moved to rest in a position out of my sight.

4. *Intra-group agonistic encounters*

Three features of intra-group aggression changed during the mating season: there was a highly significant increase in the frequency of aggressive encounters in both Groups III and IV during the mating season. Secondly, although most animals contributed to this increase (See Table 3), the Index of Increased Aggression shows that the proportionate increase was higher in some animals than in others. This index is an expression of the ratio between the number of aggressive incidents an animal initiates in the mating and non-mating

Table 3 Changes in frequency of aggression in the mating season

Group	*Initiator*	*Non-mating season*	*Mating season*	*Index of increased aggression*
III	A♂ F	7	38	5.4
	A♂ P	2	2	1.0
	A♀ FD	13	13	1.0
	A♀ NFD	5	24	4.8
	SA♂ Y	–	3	–
	Juvenile	–	2	–
IV	A♂ R)* A♂ INT)	6	49	8.1
	SA♂ Q	–	6	–
	A♀ FI	24	85	3.5
	A♀ FNI	9	12	1.3

* In order to provide data over the whole six months, results from these two adult males were combined.

Figure 3. Subordinate male P (Group III) being displaced by adult male F (Group III), from above. Male P (below) is displaying characteristic submissive gestures: a special facial expression ('grin'), rolled tail and hunched back.

seasons. Where the first value is zero, the value of the index is inevitably infinity and is consequently not shown.

In Group III the highest proportionate increase was seen in ♂ F. Comparison of the combined frequencies for ♂ R(IV) and ♂ INT(IV) between the mating and the non-mating season shows much higher frequencies during the mating season. It is of interest that there was no frequency increase in ♂ P(III), but that there was a substantial increase in ♂ Q, although an index cannot be calculated for this latter male.

Outside the mating season, a linear dominance hierarchy can be determined, incorporating all adults. The hierarchy was defined on the basis of agonistic encounters over access to feeding and resting stations. Initiation of aggression and displacements (Fig. 3) were found to occur consistently in one direction.[13] For convenience, this hierarchy is henceforth referred to as the non-mating season (NMS) dominance hierarchy. During the copulatory period, a breakdown in this structuring occurred in Group IV: although ♂ Q, an immature subordinate male, took little part in events during the period of copulation, it was the only time when he was seen initiating aggression against other members of the group. This period was also associated with persistent invasion of Group IV by 'roaming' males (see below). In Group III, ♂ F retained his position as NMS dominant male, unchallenged by intruders, and no reversal in the linearity of the NMS dominance hierarchy was seen, nor was there any increase in aggression by subordinate ♂ P within the group.

5. *Inter-group encounters*

During the pre-copulatory period and the mating season, Groups III and IV were both involved in many more encounters with neighbouring groups than during subsequent months. The difference was highly significant for both groups. Fig. 4 shows this change in the frequency of inter-group encounters for each group.

During the pre-copulatory period and mating season, 79% (11/14) of Group III's encounters were with Group IV, and 65% (11/17) of Group IV's encounters were with Group III. Since the home-ranges of at least two groups in addition to Group IV's were known to lie alongside that of Group IV, the frequency of Group III/IV interactions was significantly high, assuming that all neighbouring groups would otherwise interact equally often. There was no significant difference between the frequencies with which Groups III and IV interacted with their various neighbouring groups after the mating season; it is thus unlikely that my presence accounts for the discrepancy during the mating season, in that I was equally likely to inhibit the approach of unhabituated neighbouring groups throughout the study.

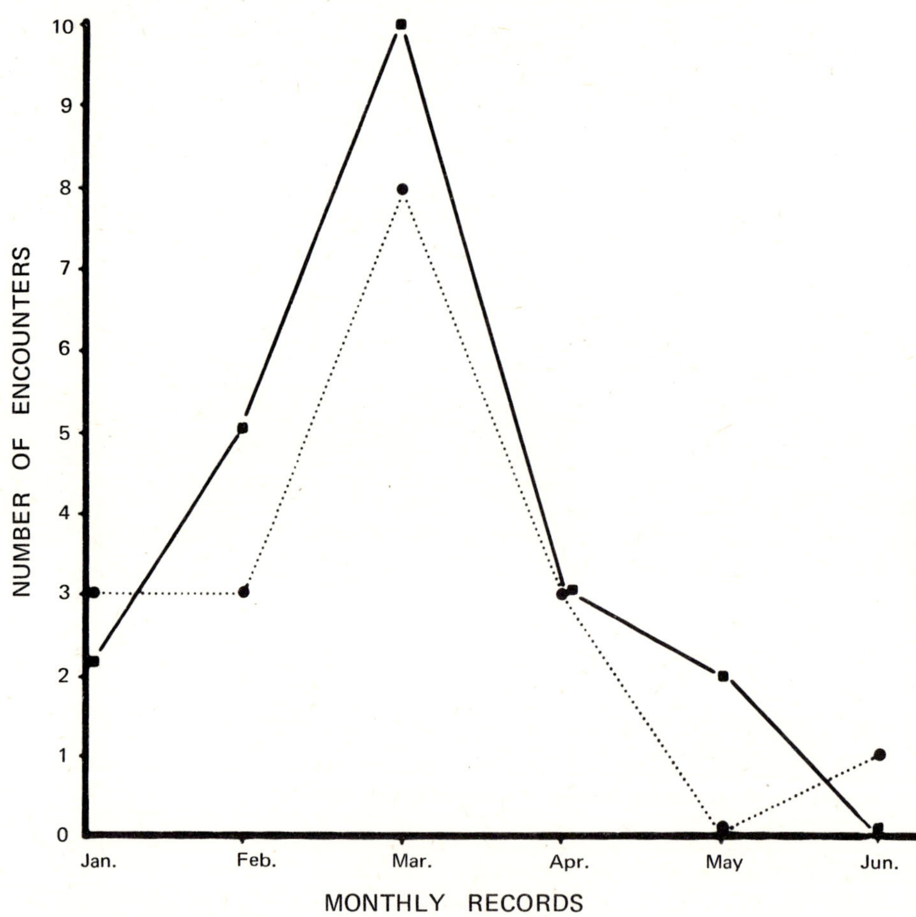

Figure 4. Changes in the frequency with which Groups III and IV encountered neighbouring groups.
Dotted line = Group III
Solid line = Group IV

This specific increase in the frequency of encounters between Groups III and IV is unlikely simply to have been a function of increased food availability in the overlap area of their home-ranges, resulting in both groups spending longer in this area. The Group III/IV overlap area was a strip about 100 m. wide, running about 150 m. along the edge of the home-ranges of the two groups; it constituted approximately 5% of the home-range of each group. Group III spent more total time in the overlap area during the pre-copulatory period and the mating season than in subsequent months, but their allocation of time to different activities did not increase uniformly. This is shown in Fig. 5. During the pre-copulatory period and mating season, 56% of the time Group III spent in the overlap area was devoted to activities other than feeding. This dropped to 38% after the mating season. Thus the results

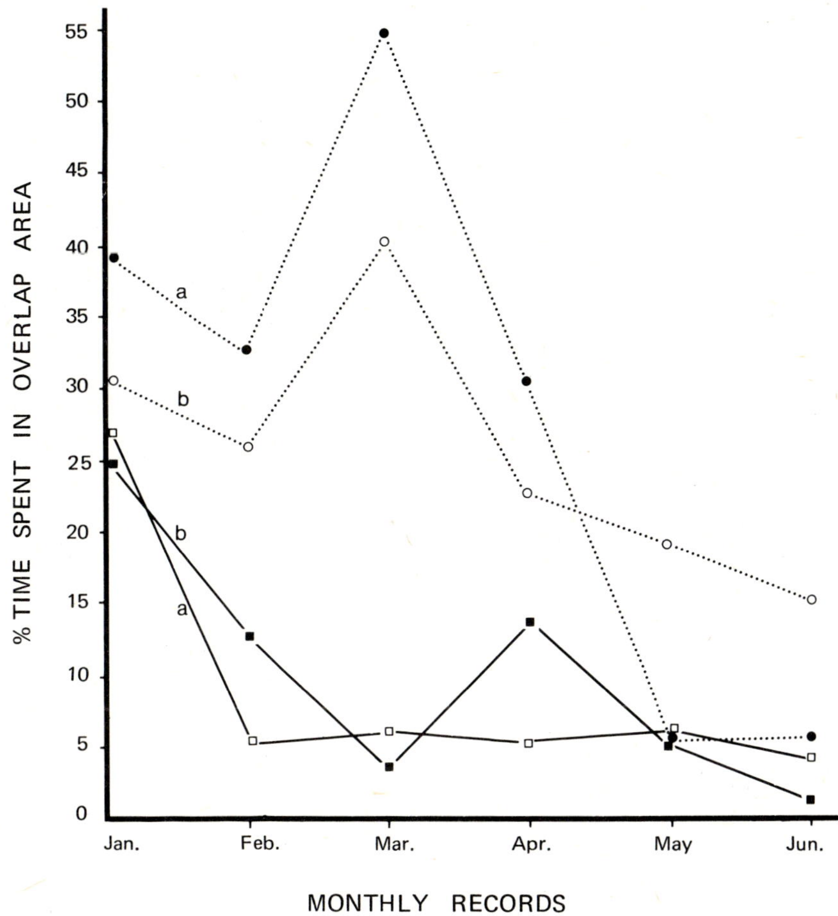

Figure 5. Number of minute records (expressed as a percentage of total number of such records) in which Group III and IV occupied the area of mutual overlap.
a = time spent in activities other than feeding
b = time spent feeding
Dotted line = Group III
Solid line = Group IV

suggest that Group III was spending longer periods in the overlap area not simply because there was abundant food there, but rather because they were involved in other activities there. Further, neither the vegetational analysis,[12] nor subjective impressions indicated that there was in fact an increase in food availability in that area at that time. Evidently, the results cannot be conclusive: it is possible that Group III entered the overlap area because of an abundance of food there, and that ensuing encounters with Group IV were incidental, although they did reduce the amount of time available for feeding.

Considering Group IV, only in January was significantly more time spent in the overlap area: when the January data are excluded, no significant difference remains between the other months (Kruskal Wallis One-Way Analysis of Variance, N=30). There was no significant difference in the proportion of time this group spent in feeding and activities other than feeding between the pre-copulatory period and mating season, and subsequent months. (Mann-Whitney U Test, N=36).

Throughout the study there was a highly significant difference between the amounts of time each group spent in the overlap area (Mann-Whitney U Test, $N_1=N_2=6$, U=0), Group IV spending much less time there than Group III.

No pattern of inter-group dominance emerged from inter-group encounters.[12] Group III did consistently spend longer than Group IV in the overlap area, particularly during the pre-copulatory period and mating season. The latter increase was associated with a disproportionate increase in activities other than feeding by Group III, and with an increased frequency of encounters between Groups III and IV. Thus, although a close association is apparent between the two groups during the mating season, the underlying mechanisms remain obscure.

Description of behavioural changes in the pre-copulatory period

The following account is of daily behavioural changes seen during the pre-copulatory period. Where no comment is made it can be assumed that activities were similar to those seen outside the mating season. Changes seen in Group IV are considered first.

Observation of Group IV was begun on *18 January*; in the following six days, ♂ Q(IV) was twice absent from the group for over two hours.

24 January: ♂ R(IV) endorsed after ♀ FNI(IV) nine times. On the four occasions he tried to sniff-approach and mark under ♀ FI(IV), she lunged at him and he retreated.

25 January: ♂ R(IV) endorsed 7 times after ♀ FI(IV) and 11 times after ♀ FNI(IV). There were 5 incomplete sniff-approach and mark sequences between ♂ R(IV) and ♀ FI(IV).

26 January: ♀ FNI(IV)'s vulva was slightly flushed. ♂ Q(IV) was found sleeping beside a strange male, ♂ INT, near the two females. This male, who had occasionally been seen with the group in September 1970, disappeared as soon as the group moved off. ♂ Q(IV) spent most of the day away from the group and ♂ R(IV) left it twice; the second time, he encountered Group III in the III/IV overlap area and headed straight for ♀ FD(III) until intercepted and chased away by SA ♂ Y(III). ♂ R(IV) left after three such attempts to reach ♀ FD(III). ♂ R(IV) endorsed 63 times after ♀ FI(IV) and

♀ FNI(IV). (Marking continued throughout the day, where in other months no marking was seen after 14.30 hrs.) Two incomplete sniff-approach and mark sequences were seen between ♂ R(IV) and ♀ FI(IV). Five complete sequences were seen between ♂ R(IV) and ♀ RNI(IV).

27 January: ♀ FNI(IV)'s vulva was still flushed. ♂ R(IV) left the group and returned to the locality of his encounter with Group III on the 26th, but did not make contact with Group III. ♂ INT approached the group and ♀ FNI(IV) marked beside him; he sniff-approached and marked under her once. ♂ R(IV), having lunged at ♂ INT when he first appeared, then moved away and only rejoined the group when ♂ INT moved off.

28 January: ♀ FNI(IV)'s vulva was no longer flushed. Endorsing and sniff-approach and mark frequencies were at a post-mating season level.

6 February: ♂ R(IV) had lost large chunks of fur. ♂ Q(IV) moved peripherally to the group and then approached with ♂ INT. The latter moved away again at once. Endorsing frequency by ♂ R(IV) was high; but no sniff-approach and mark sequences were seen.

7 February: ♂ R(IV) chased ♂ INT away from the group 3 times, each chase ending in a fight.

8 February: ♂ Q(IV) left the group at midday and was found in the evening grooming an unknown male, ♂ LCE, 10 m. from Group IV. ♂ LCE moved off as I approached and ♂ Q(IV) stayed with the group. Incomplete sniff-approach and mark sequences were seen three times between ♂ R(IV) and ♀ FI(IV).

13 February: Group IV were found sunning themselves with ♂ LCE and two unknown females. The three unhabituated animals fled when they saw me.

Cyclonic rain cut off access to the study area for the following 7 days.

20 February: ♂ R(IV) and ♂ Q(IV) were found in the middle of Group III's home-range, but no interaction with Group III was seen. ♂ R(IV) endorsed after ♀ FI(IV) and ♀ FNI(IV) 61 times. 11 complete sniff-approach and mark sequences between ♂ R(IV) and ♀ FNI(IV) were seen, and 4 incomplete ones between ♂ R(IV) and ♀ FI(IV): the latter cuffed and bit ♂ R(IV), as she did earlier in the month and in January; but for the first time, ♂ R(IV) did not retreat in the face of this aggression.

22 February: ♂ R(IV) endorsed 35 times in 3 hours' observations.

Thus, in summary, this period was notable in Group IV for the brief flushing of ♀ FNI's vulva, the heightened frequencies of endorsing and sniff-approach and mark sequences by ♂ R, and the frequency of roaming, both in non-group, and group males. Changes of this nature were absent in Group III, with two exceptions: twice during February, ♂ P(III) spent whole days away from Group III.

Description of behavioural changes and copulation in March 1971

Copulation was said to have occurred when ejaculation took place. (Mounting and intromission with no ejaculation are described separately.) Copulation by 3 males with the 2 females in Group IV was observed between 3 and 6 March. This was the only period during the field study when male/male and male/female mounting was observed, and on only two other occasions were males observed with erections: once during an inter-group interaction, and once during a play bout between a sub-adult and juvenile.

During the first observed mounting, ♂ F grasped ♀ FI with his hands around her legs, which were doubled up in a vertical squatting position (all copulation took place on vertical trunks), and he held the trunk below the female with his feet. She curled her tail up and held it slightly to one side during copulation. In subsequent mountings, there was some variation in posture, with the male grasping the female by her upper arms with his hands, and by her doubled-up legs with his feet.

The frequencies of the five activity patterns described above, namely, endorsing, sniff-approach and mark sequences, roaming behaviour, inter-male aggression, and inter-group encounters, reached a peak during the four days when copulation occurred with Group IV females FI and FNI.

The following account details events during this period.

3 March: after a brief early morning interaction with a group to the south, Group III fed, steadily moving west until 09.45 hrs. when they encountered Group IV. An extract from field notes is given below to describe the subsequent sequence of events:

09.54: A characteristic inter-group confrontation begins between Groups III and IV. ♀ FNI(IV) is not present.

09.55: ♂ F(III) mounts ♀ FI(IV); he dismounts almost at once and joins in the general chasing (characteristic of most group confrontations).

09.56: ♂ F(III) mounts ♀ FI(IV) again, but dismounts at once to chase ♂ F(IV).

10.01: ♀ FI(IV) is grasping a vertical trunk, with ♂ P(III), ♂ F(III), and R(IV), and SA ♂ Y(III) sitting on the ground in a circle round her. Each ♂ tries to approach her but she repels them by slapping, lunging and biting them. The rest of Group III are feeding 10 m. away.

10.05: ♂ R(IV) lunges at ♂ F(III) who moves off, followed at once by ♀ FI(IV). The other 3 males follow her.

10.09: ♂ F (III) mounts ♀ FI(IV) on a vertical trunk but SA ♂ Y(III) approaches and bites ♂ F(III)'s tail. He dismounts and chases

SA ♂ Y(III) down the trunk. This happens twice more.

10.14: ♂ F(III) mounts ♀ FI(IV) and ejaculates after 24 thrusts. No more than six consecutive thrusts had occurred in earlier mounting, and it was uncertain whether there was intromission. The other 3 males sit watching. ♂ F(III) then dismounts and sits beside ♀ FI(IV) licking his genitalia.

10.17: ♂ R(IV) sniff-approaches and marks under ♀ FI(IV), and then chases ♂ P(III) and SA ♂ Y(III).

10.29: ♀ FI(IV) leaps off followed by ♂ F(III). She cuffs him and he leaps off and is chased by ♂ R(IV). ♂ R(IV) then chases ♂ P(III) away. ♂ R sniff-approaches and marks under ♀ FI(IV) and she leaps off, closely followed by ♂ R(IV) and at a distance by ♂ F(III), ♂ P(III) and SA♂ Y(III).

10.35: ♀ FI(IV) sits self-grooming beside ♂ R(IV). ♂ P(III) and SA ♂ Y(III) are 5 m. away and ♂ F(III) much further still. ♂ R(IV) tries to mount ♀ FI and she cuffs him and moves off north, followed by him. ♂ P and SA ♂ Y(III) immediately endorse the sitting-place she has just left.

10.45: ♂ F(III) has moved off and is feeding with ♀ FD(III) and ♀ NFD(III).

This sequence was typical of the copulatory period, with one or other female acting as a focus for male attention, yet rejecting the large majority of male advances, and with persistent inter-male aggression. However, it was the only time a male was seen to mount and copulate with a female without first winning a prolonged and fierce battle with another male.

During the rest of the day, on 3 March, ♂ P(III) followed Group IV; he was chased away 47 times by ♂ R(IV) and several of these chases terminated in fights. SA ♂ Y(III) stayed with ♂ P(III) until midday, and then returned to Group III. ♂ P(III) mounted SA ♀ Y(III) 3 times, while they were trailing Group IV together, and each time SA ♂ Y(III) snapped and wriggled to free himself. This was the only occasion when male/male mounting was seen. Twice, ♂ R(IV) tried to mount ♀ FI(IV), and each time she snapped and wriggled and escaped.

Checks on Group III suggested that group activities were normal.

4 March: SA ♂ Y(III) and ♂ P(III), both in good condition, were found near Group IV. In contrast, ♂ R(IV) had lost large chunks of fur and had two long gashes on his thighs. Between 0600h. and 1030h., 9 incomplete sniff-approach and mark sequences were seen between ♂ R(IV) and ♀ FI(IV); ♂ R(IV) also chased and fought ♂ P(III) 45 times, and ♂ Q(IV) did so 8 times. However, ♂ R(IV) slowed up and tired visibly, and at 10.30 hrs. ♂ P(III) chased ♂ R(IV) away from the group: the latter retreated very slowly and quadrupedally across the ground; this was the only time an animal was ever seen moving quadrupedally on the ground. After a brief period of reciprocal chasing between ♂ Q(IV) and ♂ P(III), ♂ Q(IV) on his own did not chase ♂ P(III) again. At 11.36 hrs. ♀ FI(IV) leapt over to ♂ P(III) and they touched noses.

That afternoon, ♀ FI(IV) moved on to a vertical just above ♂ P(III)'s head and rolled her tail up. He mounted her and thrust slowly for about one minute, and then dismounted to chase ♂ Q(IV) away. Two further mountings were similarly interrupted, but finally he remained on the female for approximately 4 minutes, thrusting about 40 times. After this, when ♂ P(III) tried to approach, both ♀ FI(IV) and ♀ FNI(IV) started cuffing and biting. ♂ Q(IV) remained on the periphery of the group, alternately approaching and being chased away by ♂ P(III). At 17.42 hrs. ♀ P(III) chased ♂ Q(IV) west, where ♂ R(IV) was found feeding: ♂ R(IV) and ♂ Q(IV) fled together. ♂ P(III) mounted ♂ FI(IV) once more, but there was no intromission and he leapt off almost at once to chase ♂ R(IV) and ♂ Q(IV).

5 March: ♀ FI(IV), ♀ FNI(IV) and ♂ Q(IV) were found together, the latter with a swollen lower lip and a small cut on his arm. Neither ♂ P(III) nor ♂ R(IV) were present, and ♂ P(III) was not seen again until March 10th. In the late afternoon ♂ R(IV) appeared and ♀ FNI(IV) leapt over and touched noses with him briefly; ♂ R(IV) then approached ♂ Q(IV) and presented for grooming.* ♂ Q(IV) spat-called† and groomed ♂ R(IV) briefly, but suddenly appeared to start grappling with him; one of the two screamed and ♂ R(IV) leapt off at once. ♂ Q(IV) followed, but as he approached ♂ R(IV) tail-rolled, spat-called and leapt away. ♂ R(IV) did not approach the group again. That evening ♂ INT approached the group and ♀ FNI(IV) sexually presented to him (i.e. stationed herself on a vertical trunk just above his head and rolled her tail up). He mounted her and 53 thrusts were counted in 4 minutes before he dismounted. As after previous incidences of prolonged copulation, the two animals sat near each other licking their own genitalia.

6 March: ♂ INT mounted ♀ FNI(IV) 3 times in the early morning, twice dismounting to chase ♂ Q(IV). This was the last time mounting was seen. ♂ LCE appeared but was chased away by both ♂ INT and ♂ R(IV). ♀ FI(IV), ♀ FNI(IV), ♂ INT and ♂ Q(IV) moved together all day; ♂ R(IV) moved peripherally, retreating when any other animal came near him. Once ♂ Q(IV) groomed ♂ INT briefly: ♂ INT spat-called and tail-rolled as ♂ Q(IV) approached.

7 March: An inter-group encounter, typical of those seen outside

* Presentation for grooming is made by a non-sexually dominant animal to a subordinate, and involves holding out one arm to be groomed.

† The spat vocalization, described by Jolly,[7] is one of a repertoire of signals given by the submissive animal in an agonistic encounter. Other signals are the tense-lipped grin, hunching the back and tail-rolling. The latter is indistinguishable from female presentation tail-rolling prior to copulation, and involves rolling the tail up between the legs.

Figure 6. Adult male P of Group III photographed one month after the mating season. Note the healed – but still badly scarred – nose.

the mating season, took place between Groups III and IV (the latter group comprising ♂ INT, ♂ Q, ♀ FI and ♀ FNI).

10 March: ♂ P(III) joined Group III, with an injured nose: the nasal bones had been exposed, but the flesh still hung from his nose on a piece of skin; he persistently, but unsuccessfully, tried to lick his nose, and once licked his hand and then put it to his nose. After suppurating for two or three days, the wound began to heal and within a month it had healed totally, leaving him with flattened facial features and a wheeze as he breathed (Fig. 6). Throughout the 10th, and most of the 11th, ♂ P(III) moved away if approached by any member of Group III. On the evening of the 11th, however, ♂ P entered a tree where the juvenile was feeding; they touched noses, and ♂ P(III) began feeding in the same tree.

12 March: ♂ R(IV) approached J(III), who rapidly retreated towards other members of Group III, and SA ♂ Y(III) chased ♂ R(IV) away.

13, 14, 15 March: ♀ FI, ♀ FNI, ♂ INT and ♂ Q moved as a well-integrated group, and this composition was retained until observations ended in September 1971. ♂ R(IV) intermittently moved on the periphery of the group; he retreated from ♂ INT, but was not chased by him. ♂ LCE also moved peripherally to the group, attempting to approach the adult females; he was chased away by all group members and was not seen when observations were resumed in April.

The study area was abandoned on 16 March because of local political unrest. When observations were resumed in mid-April,

♂ R(IV) had disappeared and ♂ INT had largely taken over his role in Group IV. ♂ P(III) had resumed his former position as subordinate male in Group III.

A *summary* of events in this, and the pre-copulatory period, is given in Table 4.

Discussion and conclusions

Petter-Rousseau[5] has shown that, in the laboratory, the Cheirogaleinae are seasonally polyoestrous. *Lemur catta* is also seasonally polyoestrous in captivity, non-pregnant females cycling three times.[14] Jolly[7] produced evidence for *L.catta* in the wild having a 'pseudo-oestrous period' approximately one month earlier than the true breeding season: the vulval area of four or five out of nine females went through a pink phase 3-4 weeks before the week of mating, which faded and then flushed again just prior to mating. Two of these females were seen mating in the second period of flushing.

Data from Group IV in this field study indicate that *P. verreauxi* do not have more than one full oestrous period in the breeding season, but that there was a period analogous to Jolly's 'pseudo-oestrous period' in *L.catta.* It is suggested that the flushing of ♀ FNI(IV)'s vulva and associated activities, in late January, represent a partially suppressed oestrus 37 days before full oestrus. This did not occur in Group III, despite the presence of two adult females, and it seems likely that the suppression was total in this group: between 14 January and 13 February no more than three consecutive days passed without observations being made on Group III, so it is unlikely that (if it occurred) a 'pseudo-oestrous period' would not have been seen. Both females in Group III probably came into full oestrus once, at the same time, either between the 13 and 20 February, or after 15 March.

The timing of births provides further evidence for the extreme seasonality of breeding in *P. verreauxi*, in that it reflects the incidence of female receptivity in the mating season, and the degree of synchronisation between females. *P. verreauxi* has a gestation period of about 130 days.[6] The length of the oestrous cycle is not known, but it is reasonable to assume that it is of approximately the same duration as that of *L.catta*: in captivity, this lasts 39.3 days, with a range of 33-45 days. If females in the study groups had been polyoestrous, with a fully synchronised cycle similar in length to that of *L.catta*, two birth peaks might have been expected, the second about five weeks after the first, when females fertilised during the second oestrous period gave birth. However, extrapolating from the birth-period evidence in the wild, there was only one full oestrous period with considerable synchrony between females, during which

Characteristic events	*Pre-copulatory period*			*Copulatory period*	*Post-copulation*
	24-31 January	*1-13 February*	*20-28 February*	*3-6 March*	*7-15 March*
Vulval flush	Present in ♀ FNI(IV) on 26th and 27th	–	–	–	–
Endorsing	High frequency in ♂ R(IV)	⟶	⟶	High frequency in all adult males studied	–
Sniff-approach and mark sequences	High frequency in ♂ R(IV)	⟶	⟶	High frequency in all adult males studied	–
Roaming (movements by adult males)	1. ♂ INT approaches Group IV twice 2. ♂ Q(IV) leaves Group IV 3 times 3. ♂ R(IV) leaves Group IV 3 times (once to approach Group III)	1. ♂ INT approaches Group IV twice 2. ♂ LCE approaches Group IV twice 3. ♂ Q(IV) leaves Group IV once	1. ♂♂ R(IV) and Q(IV) leave Group IV and make a foray into home range of Group III 2. ♂ P(III) leaves Group III twice	1. ♂Y(III) approaches and follows Group IV 2. ♂ P(III) approaches and follows Group IV – mates 3. ♂ INT approaches & follows Group IV – mates 4. ♀ LCE approaches & follows Group IV – is chased away 5. ♂ R(IV) ousted from Group IV by ♂♂ P (III) and INT in turn.	1. ♂ P(III) rejoins Group III 2. ♂ R(IV) approaches Group III – and is chased away 3. ♂ LCE approaches Group IV and is chased away
Intra-group encounters	High frequency in both groups	–	–		
Inter-group encounters	High frequency in both groups	–	–	[♂ F(III) mates with ♀ FI(IV) during inter-group encounter on March 3rd.]	
Copulation	–	–	–	(In order of occurrence) ♂ F(III) with ♀ FI(IV) ♂ P(III) with ♀ FI(IV) ♂ INT with ♀ FNI(IV)	

all mating and fertilisation took place: at Ampijoroa in the north of Madagascar, births of *P.v.coquereli* were scattered over a maximum of 21 days in 1970 and 1971. Jolly[7] reports a ten-day birth season for *P.v.verreauxi* at Berenty, in the south of the island. In neither case was the distribution of births through time sufficient to indicate a polyoestrous breeding system.

The effect of the photo-period on *Microcebus* has been demonstrated by Petter-Rousseaux;[15] but similar work has yet to be done on *L.catta* and *P.verreauxi* to establish the influence of day-length and of temperature on sexual behaviour in these species. Since *L.catta* is polyoestrous in captivity and not in the wild, it is likely that, in this species at least, there are proximate ecological factors which repress and then induce oestrus in the wild. The absence of these factors in captivity would permit the appearance of a full polyoestrous system.

The effect of general physical condition on females' ability to come into oestrus is not well understood. Seasonal fluctuations in food availability in the two study areas are discussed in detail elsewhere;[12] in summary, food seems to be scarce at the end of the dry season, particularly in the south where the rain starts after a six-month drought. Food availability increases to a peak towards the end of the wet season in March, and then drops off again in the dry season. It is to be supposed that the physical condition of the animals mirrors this pattern to some extent and that full oestrus, if affected by the female's physical condition, would be most likely to occur towards the end of a period of optimal ecological conditions, in this case March. It is immediately apparent, however, that other phases, such as pregnancy, and suckling or weaning infants, may be more critical and that the system is timed to provide optimum conditions during one of these periods. In ultimate terms, the periodicity and control of the cycle are likely to be closely adapted to the environment in such a way as to ensure the maximum survival of populations. The timing observed in this study was as follows: mating occurred at the end of the wet season, gestation from the end of the wet season through the first half of the dry season, births in the middle of the dry season, suckling until the end of the dry season, and then the infants were almost completely weaned at the beginning of the following wet season.

The similarity between *L.catta* and *P.verreauxi*, postulated above from limited evidence, is substantiated in other aspects of their mating systems: the receptivity of Group IV females FNI and FI lasted a maximum of 12 and 36 hours respectively, as compared with a maximum of 36 hours for *L.catta* in the wild and 10-24 hours in captivity.[7,14] The incidence of marking behaviour by females in this study was not affected by the breeding season, and a similar stability is reported in *L.catta* females in captivity.[14] Finally, Evans and Goy[14] report that 'both long and short term fluctuations in gonadal activity were associated with changes in the frequency of expression of several non-sexual patterns', although these authors do not differentiate clearly between 'sexual' and 'non-sexual' patterns. In *P.verreauxi* it has been shown that changes in the frequency of five activity patterns occurred in the pre-copulatory period and that, as in *L.catta*, only two new patterns appeared: the act of copulation itself, and fierce fighting between adult males. It is probable that these frequency changes were associated with changes in gonadal activity in *P.verreauxi* in this study.

The close relationship between Groups III and IV has already been referred to. The basis of this relationship is not understood. Interactions were frequent, but never friendly, and it is unlikely that Groups III and IV had recently split up from one larger group since, in September 1970, Group IV seemed to be in the process of splitting away from another group to the south of its range.[12] The presence of three unknown animals sunning themselves in the same tree as Group IV, in the south of their range in mid-February, suggests that the split was not complete even by then. It is postulated that the frequent encounters between Group III and IV led to an across-group recognition of individuals, although not to a stable dominance hierarchy between the two groups considered as whole units. It is further suggested that ♂ R(IV) recognised ♂ F(III) as a dominant male. This contrasts with his reaction to ♂ P, a subordinate male in Group III: ♂ R(IV) fought ♂ P(III) for 24 hours to prevent him approaching either adult female in Group IV.

In the light of the circumstantial evidence, it seems improbable that events described in Group IV were atypical. There are many anecdotal accounts of fierce fighting between *P.verreauxi* in February and March, and a reliable first-hand account of one such fight was given to Jolly by S. de Guiteaud.[7] Further, the adult males in all four study groups had torn ears and old facial scars; since only one female had a slightly torn ear it is more likely that the males acquire these scars through fighting than through falling.

Given the various interpretations possible, the following tentative hypothesis is put forward as one explanation of the differential reaction observed of females towards males, and males towards each other: during the breeding season 'roaming' behaviour by adult males occurs at the same time as some degree of breakdown in group structuring; this breakdown is marked by the appearance of aggression directed at dominant males by subordinate males. This aggression may occur with respect to access to food, resting places, or females. As in *L.catta*,[7] frequency of mating – although not demonstrably of fertilisation – is not the prerogative of NMS dominant males. This does not mean that dominance becomes a meaningless concept during the mating season, but that the actual structure of the NMS dominance hierarchy changes at the onset of that period: *P. verreauxi* females allow only males dominant in the immediate situation, that is males dominant in the mating season (MS), to mount them. This MS dominance may be established through previous mutual knowledge and be a correlate of NMS dominance; however, MS dominance can also be achieved by previously NMS subordinate males, through fighting and ousting NMS dominant males. Unlike *L.catta*, where there is a total reversion to the previous NMS hierarchy at the end of the mating season,[7] the assertion of MS dominance in *P.verreauxi* appears to have enduring effects on the structure of NMS dominance within a group. A male dominant both in and out of the mating season mates with females in one or more other groups, but remains a member of the group in which he was dominant before the mating season. The NMS subordinate male who has fought his way to MS dominance in a group by ousting the resident NMS dominant male, may stay in that group, be it his own original group or one encountered while roaming, to become NMS dominant after the mating season. Fig. 7 further illustrates this idea.

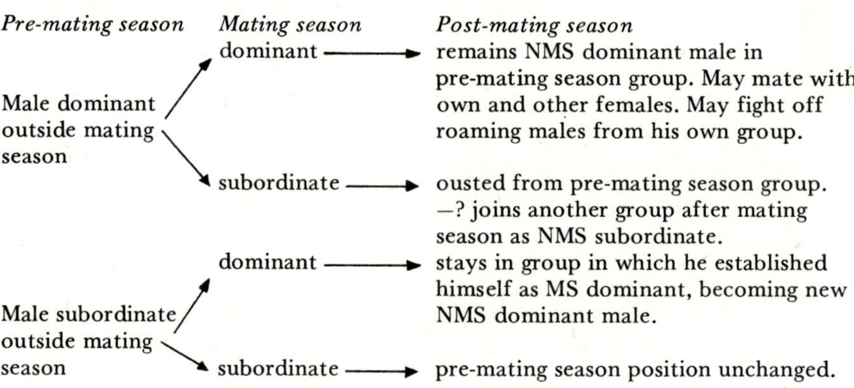

Figure 7. Possible changes in adult male status as a result of the mating season.

With reference to ♂ P(III), it is suggested that he returned to Group III only because he in turn was ousted from Group IV by ♂ INT. Under the interpretation above, ♂ R(IV) recognised ♂ F(III) as dominant both in and out of the mating season and hence as no threat to his own position of non-sexual dominance in Group IV; in contrast, ♂ P(III), a non-sexual subordinate, did constitute a challenge to ♂ R(IV)'s position, and fighting resulted.

It is not known whether ♂ F(III) mated with adult females FD(III) and NFD(III) or whether a roaming, MS and NMS dominant male did so. It is certain that ♀ FI(IV) and ♀ FNI(IV) refused to mate with ♂ R(IV), even when they were sexually receptive and prepared to mate with roaming males. According to the interpretation above, this was because the non-sexually dominant male, R, was being successfully challenged by intruders – ♂ F(III), ♂ P(III) and ♂ INT.

Finally, some consideration should be given to the selective advantage of the system. The evidence indicates that the system ensures some degree of outbreeding. Although probably this outbreeding is largely limited to neighbourhoods, it would be likely to produce some gene-flow through the population as a whole. This might be important in a species with a group size of less than 10, where chronic inbreeding might otherwise occur. New males did join Group I in the north outside the mating season, however, with no apparent fighting involved. It seems unlikely that the mating system, with its fierce inter-male fights, evolved simply as a mechanism to counter inbreeding, when more peaceful means of changing group were possible outside the mating season.

The system may also operate to produce intra-sexual selection between adult males. Access to females and, by inference, biological paternity, appears to be dependent upon the fighting ability, strength and endurance of the adult male. This may have been proved in the previous mating season or even earlier, or may only be manifested in the current mating season. Fighting ability, strength and endurance cannot be equated with overall fitness; however, it does seem unlikely that there should be intensive selection for fighting prowess specifically, during the mating season, when no fights were witnessed during the rest of the year. An alternative explanation is that the social upheaval and fights of the mating season do to some extent test the fitness of the males in terms of their ability to survive relatively prolonged periods of high energy output without a concomitantly increased energy input. This might be critical in a species subject to fluctuations in food availability.

Acknowledgments

I would like to thank J.R. Napier for his constant support throughout this research, and G. Manley, R.D. Martin, A. Jolly, and T. Clutton Brock for their comments on this paper.

The research was supported by a Royal Society Leverhulme Award, a NATO Overseas Studentship, and by grants from The Explorers Club of America, The Boise Fund, The John Spedan Lewis Trust Fund, the Society of the Sigma Xi, and Central Research Fund of London University.

NOTES

1 Flacourt, E. de (1661), *Histoire de la Grande Isle de Madagascar*, Paris.

2 Milne-Edwards, A. and Grandidier, A. (1875, 1890-1895), *Histoire naturelle des mammifères: histoire physique, naturelle, et politique de Madagascar*, Paris, vols. 6, 9, 10.

3 Kaudern, W. (1914), 'Einige Beobachtungen über die Zeit der Fortpflanzung der madagassischen Säugetiere'. *Ark. f. Zool.* 9, 1-22.

4 Rand, A.L. (1935), 'On the habits of some Madagascar mammals', *J. Mammal.* 16, 89-499.

5 Petter, J.-J. (1962), 'Recherches sur l'écologie et l'éthologie des lémuriens malgaches', *Mém. Mus. nat. Hist. nat.*, sér. A, 27, 1-146.

6 Petter-Rousseaux, A. (1962), 'Recherches sur la biologie de la reproduction des primates inférieurs', *Mammalia* 26, suppl. 1, 1-88.

7 Jolly, A. (1966), *Lemur Behaviour*, Chicago.

8 Ellefson, J.O. (1968), 'Territorial behaviour in the common white-handed gibbon, *Hylobates lar'*, in Jay, P.C., *Primates: Studies in Adaption and Variability*, New York.

9 Altmann, S.A. (1962). 'A field study of the sociobiology of rhesus monkeys, *Macaca mulatta'*, *Ann. N.Y. Acad. Sci.* 102, 460-84.

10 Struhsaker, T.T. (1969), 'Correlates of ecology and social organisation among African Cercopithecines', *Folia primat.* 11, 80-118.

11 Stoltz, L.P. and Saayman, G.S. (1970), 'Ecology and behaviour of baboons in the Northern Transvaal', *Ann. Transv. Mus.* 26, 99-143.

12 Richard, A.F. (1973), 'The social organization and ecology of *Propithecus verreauxi*', Ph.D. Thesis, London University.

13 Richard, A.F. and Heimbuch, R. (in press), 'An analysis of the social structure of three groups of *Propithecus verreauxi*', in Sussman, R.W. and Tattersall, I. (eds.), *Lemur Biology*.

14 Evans, C.S. and Goy, R.W. (1968), 'Social behaviour and reproductive cycles in captive ring-tailed lemurs (*Lemur catta*)', *J. Zool. Lond.* 156, 181-97.

15 Petter-Rousseaux, A. (1970), 'Observations sur l'influence de la photopériode sur l'activité sexuelle chez *Microcebus murinus* (Miller 1777) en captivité', *Ann. Biol. Anim.* 10, 203-8.

R. W. SUSSMAN

Ecological distinctions in sympatric species of Lemur

Introduction

In forests throughout Madagascar, there are many sympatric species of Lemuriformes. The dynamics of the interaction between these species are, however, as yet poorly understood. Generally, coexistence of related species is made possible by differential exploitation of the environment, which minimises competition. Differential exploitation of the environment, in most cases, is the result of habitat selection in which the species have particular preferences for different portions of the shared environment. Where species choose different habitats, the particular preferences may be the result of adaptations to different environmental conditions existing in places where the species are allopatric, or the result of divergence caused directly by interaction between the sympatric populations. With the latter phenomenon, referred to as *character displacement*,[1] sympatric populations of two species will tend to differ in one or more characteristics in which allopatric populations of the same two species may be similar.

A study of the ecological relations between two sympatric forms, therefore, may give indications as to the nature of the forces causing differentiation. Furthermore, once differences in ecological preferences are recognised, it is possible to formulate and test hypotheses concerning the relationship between ecology, morphophysiology, and social behaviour of the species in question.

In this paper, I will describe some of the results of a study on two species of *Lemur*: *Lemur fulvus rufus* and *Lemur catta* (Figs. 1-3). The study is discussed in more detail elsewhere.[2] Populations of *L. f. rufus* range from the north-west to the south-west of Madagascar. Populations of *L. catta* are found from the south-west to the more arid south. In many forests of the south-west, the two species are sympatric (Fig. 5).

Intensive studies were carried out in three forests: Antseran-

Figure 1. *Lemur fulvus rufus* male.

Figure 2. *Lemur fulvus rufus* female.

Figure 3. *Lemur catta.*

Figure 4. *Propithecus verreauxi verreauxi.*

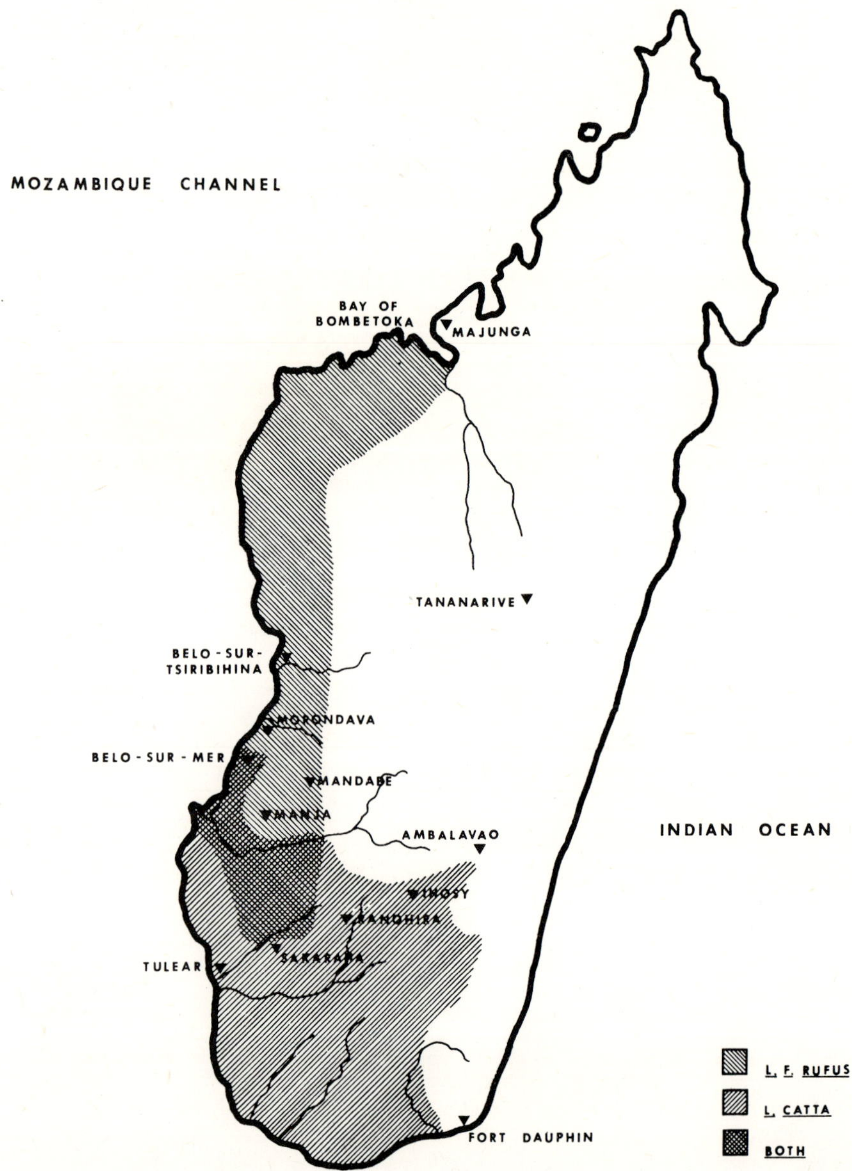

Figure 5. Geographical distribution of *Lemur fulvus rufus* and *Lemur catta.* Populations are not continuous within these areas, but are only found where suitable primary vegetation exists.

anomby, Tongobato, and Berenty (Fig. 6). At Antserananomby, *L. f. rufus* and *L. catta* are sympatric, and in the other two forests they are allopatric. Data on the utilisation of space and time, diet and social structure were collected on both the allopatric and sympatric populations of *L. f. rufus* and *L. catta.* The study was carried out

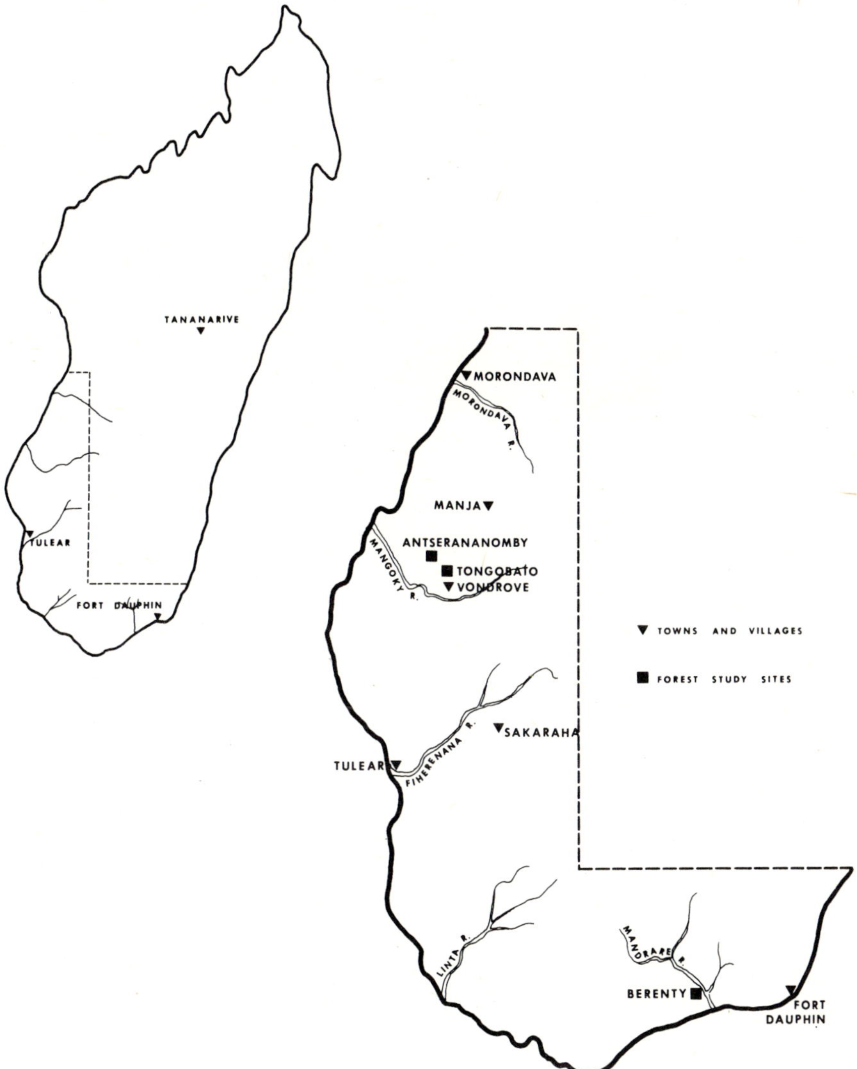

Figure 6. Study sites for intensive observations of behaviour.

between September 1969 and November 1970, during which period the animals were observed for a total of 830 hours.

General ecological distribution of the two species in the south-west of Madagascar

Populations of *Lemur fulvus rufus* and *Lemur catta* coexist in the region between 20° 44′ and 23° 12′ south latitude. These two species, along with *Propithecus verreauxi verreauxi* (Fig. 4), are the

Figure 7. The closed canopy portion of the forest at Antserananomby. Photograph taken from dry river bed.

only diurnal lemur species found in the south-west of Madagascar. The south-west is a region of transition between the moist, deciduous forests of the north-west and the desert-like vegetation of the south. In the south-west, the diurnal lemur species are found mainly in two types of primary forest: (1) closed canopy, deciduous forests and (2) brush and scrub forests. The deciduous forests are dominated by *Tamarindus indica* trees, which make up a continuous, closed canopy about 7 to 15 m. in height (Fig. 7). These forests have been classified by Perrier de la Bathie as 'bois des terrains silicieux'.[3] Forests of this type present essentially the same characters from the north-west to the south-west, and similar forests are also found along the large rivers in the south of Madagascar. The brush and scrub forests reach the height of only about 3 to 7 m. and the under-brush is very dense (Fig. 8). Perrier de la Bathie classified brush and scrub forests as 'bois des terrains calcaires'. The vegetation of Madagascar is described in detail by Perrier de la Bathie, Humbert, and Humbert and Darne.[3]

An initial survey was conducted in the south-west within the area in which populations of *L. f. rufus* and *L. catta* are sympatric. Data from the survey indicate that the ecological distribution of the two species differs within this area. *L. f. rufus* is found without *L. catta* only in small, circumscribed, continuous canopy forests. *L. catta* exists alone only in brush and scrub forests. *L. f. rufus* and *L. catta* coexist only in areas in which a continuous canopy forest merges directly into a primary brush and scrub forest. I have called these 'mixed' forests. Within the mixed forests, *L. f. rufus* was seen only in those portions with a continuous canopy, whereas *L. catta* could be found in all parts of these forests. *P. v. verreauxi* inhabits both

Figure 8. A brush and scrub forest near the Mangoky River.

continuous canopy and mixed forests, but was never seen in brush and scrub forests. This is probably due to the fact that the density and stratigraphy of these latter forests necessitate terrestrial locomotion.

The survey data, therefore, suggest that *L. f. rufus* and *L. catta* have preferences for different habitats. These preferences seem to be related to different locomotor adaptations in the two species. While *L. f. rufus* is limited to areas with a continuous, closed canopy, *L. catta*, because much of its travel is done on the ground, can exploit a number of regions which differ in ecological structure.

Intensive study

Three forests were chosen for intensive study: Antserananomby and Tongobato in the south-west, and Berenty in the south (Fig. 6). The basic physiognomy of the three forests is similar. Kily trees (*Tamarindus indica*) form a closed, continuous canopy in all cases. However, Antserananomby is a mixed forest in which only seven of the total ten hectares studied contain a closed canopy.

Lemur fulvus rufus and *Lemur catta* coexist at Antserananomby, whilst only *L. f. rufus* is found at Tongobato, and only *L. catta* is found at Berenty. *Propithecus verreauxi verreauxi* inhabits all three forests. The nocturnal lemur species found at Antserananomby are: *Lepilemur mustelinus ruficaudatus*, *Phaner furcifer*, *Cheirogaleus medius*, and *Microcebus murinus*. *L. m. ruficaudatus*, *P. furcifer*, and *M. murinus* were observed at Tongobato. *L. m. leucopus*, *C. medius*, and *M. murinus* are found at Berenty. Many of the same genera of mammals, birds, and reptiles are found in all three forests.

The study was conducted in December 1969 and in March and April 1970 at Tongobato, from July through September 1970 at Antserananomby, and in November 1970 at Berenty. The temperatures during the day were similar in all three forests. The evening temperatures, however, were higher at Tongobato and Berenty than at Antserananomby (Table 1). The days were also longer at Tongobato and Berenty (5.00 hrs. to 18.30 hrs.) than they were at Antserananomby (6.00 hrs. to 18.00 hrs.).

Table 1 Average maximum and minimum temperatures for the months of study at each forest*

Month	T_x	T_n
Tongobato		
December 1969	36.1	19.2
March 1970	36.2	19.9
April 1970	35.3	16.3
Antserananomby		
July 1970	32.1	11.8
August 1970	35.5	13.9
September 1970	35.5	15.5
Berenty		
November 1970	34.7	19.3

T_x = average maximum daily temperature in °C.
T_n = average minimum daily temperature in °C.

* The temperatures are taken from those recorded at the meteorological station closest to each forest.

1. *Utilisation of the forest strata*

Data were collected at five-minute intervals on the number of individuals engaged in each of six activities – feeding, grooming, resting, moving, travel, and other – and the levels of the forest at which the activities were performed.

Five forest levels were recognised: Level 1 is the ground layer of the forest, which includes herb and grass vegetation. Level 2 is the shrub layer, 1-3 m. above the ground. This layer is usually found in patches throughout closed canopy forests, but is much more dense and is the dominant layer in brush and scrub regions. Level 3 consists of small trees, the lower branches of larger trees, and saplings of the larger species of trees. This layer ranges from 3-7 m. above the ground. Level 4 is the continuous or closed canopy layer. It usually ranges from 5 to 15 m. in height. The dominant tree of the closed canopy, at all three forests, is the kily. Level 5 is the emergent layer and consists of the crowns of those trees which rise above the closed canopy and are higher than 15 m. All three forests in which I made intensive studies are primary forests; the tree layers are quite distinct. With observations in which the forest level could not be clearly distinguished, the level was not recorded.

Each observation of an animal constituted an individual activity record (IAR) collected in a given five-minute time sample.[4] The number of individual activity records for *L. f. rufus* was 7,084 at Antserananomby and 2,896 at Tongobato. The number of IARs for *L. catta* was 5,383 at Antserananomby and 6,861 at Berenty. The IARs were summed for each half-hour from 6.00 hrs. to 18.25 hrs. The day was thus divided into 25 half-hour periods. Comparisons were made between the quantitative data collected on the two species and between those collected on allopatric populations of the same species.

Fig. 9 shows the level at which the highest percentage of animals was observed during each of the 25 half-hour periods. Data on both species (from all three forests) are represented in this figure. Both at Tongobato and at Antserananomby, *L. f. rufus* was found in the continuous canopy for almost all of the half-hour periods. The vertical displacement of *L. catta* followed a pattern throughout the day which was related to the foraging and movement patterns of this species. The pattern for *L. catta* was similar in both forests.

Table 2 includes the mean percentage of animals observed at each level for each activity and at each level overall, regardless of activity, for the 25 half-hour periods. Again, the intraspecific comparisons show the populations of each species to be similar in their use of vertical habitat, regardless of the forest or the presence or absence of the other species. *L. f. rufus* was very specific in its choice of vertical habitat. A high percentage (over 90%) of the activities of *L. f. rufus* took place in the top layers of the forest (levels 3, 4, and 5). In all six activities, *L. f. rufus* spent the highest percentage of time in level 4, the continuous canopy of the forest. *L. catta*, on the other hand, was found in all forest levels. For the most part, *L. catta* moved and travelled on the ground, rested during the day in the low trees (levels 2 and 3), and rested at night in level 4. *L. catta* fed in all of the levels of the forest.

At both forests in which *L. catta* was studied, it spent more time on the ground than in any one of the other four levels (36% at Berenty and 30% at Antserananomby). Perhaps the most significant finding is that *L. catta* travelled on the ground 65% (Antserananomby) and 71% (Berenty) of the time. A high percentage of the individual movement of *L. catta* was also carried out on the ground. *L. f. rufus* travelled in level 4 88% of the time at Antserananomby and 83% of the time at Tongobato. *L. f. rufus* was seen on the ground in less than 2% of the observations overall.

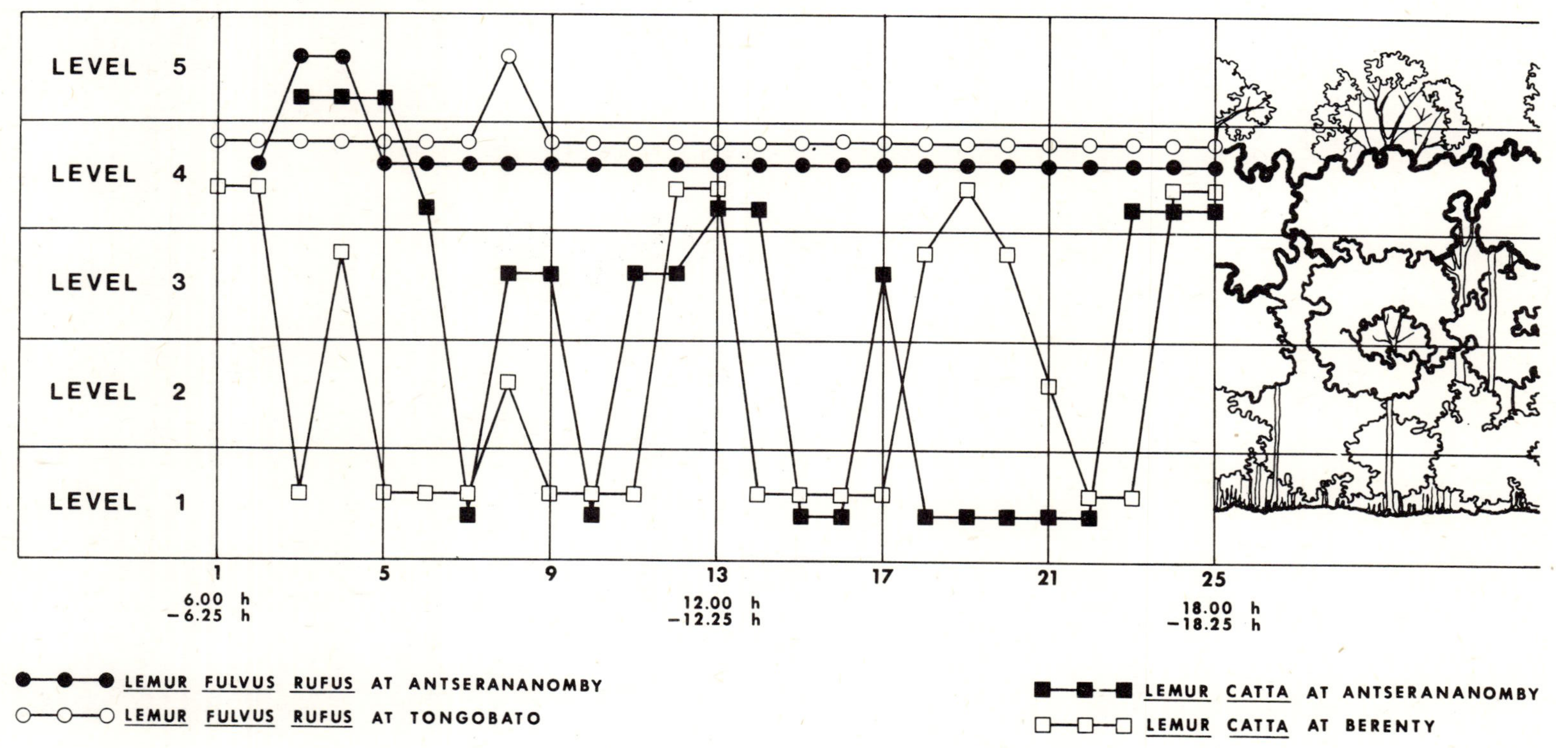

Figure 9. The forest level at which the highest percentage of animals was observed during each of the 25 half-hour periods for the species studied at each of the forests.

Table 2 Mean percentage of animals observed at each level for each activity, and regardless of activity

a. *Lemur fulvus rufus* at Antseranananomby

Activity	Level 1	2	3	4	5
Feeding	0.00	2.58	15.60	59.66	22.16
Grooming	0.00	0.54	9.34	78.24	11.88
Resting	0.00	1.20	14.05	79.60	5.15
Moving	2.61	0.60	18.14	66.16	12.48
Travel	0.00	0.00	2.60	87.84	9.56
Other	0.00	8.33	5.95	54.70	31.02
All Activities	0.44	2.21	10.95	71.03	15.38

b. *Lemur fulvus rufus* at Tongobato

Activity	Level 1	2	3	4	5
Feeding	2.17	11.40	5.01	44.66	36.76
Grooming	0.00	0.60	9.93	79.99	9.49
Resting	0.34	0.83	8.03	79.91	10.89
Moving	6.02	1.27	4.14	78.64	9.93
Travel	0.95	3.69	5.92	82.94	6.50
Other	0.00	5.00	15.00	56.18	23.82
All Activities	1.58	3.80	8.00	70.39	16.23

c. *Lemur catta* at Antseranananomby

Activity	Level 1	2	3	4	5
Feeding	27.58	12.78	25.23	19.01	15.40
Grooming	4.37	15.90	20.42	46.00	13.31
Resting	9.54	16.74	25.62	35.02	13.08
Moving	44.73	5.88	21.62	17.58	10.20
Travel	64.70	1.81	13.47	17.46	2.57
Other	28.70	5.02	27.31	22.50	16.47
All Activities	29.93	9.69	22.28	26.26	11.84

d. *Lemur catta* at Berenty

Activity	Level 1	2	3	4	5
Feeding	30.84	15.10	41.56	12.32	0.18
Grooming	10.16	23.35	27.97	38.38	0.14
Resting	14.01	26.14	28.18	31.68	0.00
Moving	55.32	13.51	15.82	15.35	0.00
Travel	70.93	6.44	5.48	17.15	0.00
Other	35.71	17.27	20.61	26.41	0.00
All Activities	36.16	16.97	23.27	23.55	0.05

When *L. catta* travelled in the trees, its locomotor behaviour differed from that of *L. f. rufus.* When groups of *L. f. rufus* moved horizontally through the trees, the animals ran along the fine terminal branches of the large trees. These branches were generally horizontal in relation to the ground and formed a continuous series of pathways throughout the closed canopy. *L. catta*, rather than moving along the fine, horizontal branches, would usually climb the large, oblique branches and leap from one of these branches to another. Thus, they moved through the trees by performing a series of runs and leaps, in each case running up an oblique branch and leaping down to another.

Morphological differences have been found between the foot of *L. f. rufus* and that of *L. catta.* In *L. f. rufus*, the heel is covered with hair, whereas in *L. catta* the heel is naked (Fig. 10). In fact, *L. catta* is the only species of prosimian in which the plantar pad extends proximally to the heel. In this feature, the foot of *L. catta* closely resembles that of monkeys. An examination of the skeletons of the feet of *L. f. rufus* and *L. catta* has shown that the mid-tarsal joints (calcaneo-cuboid and talo-navicular) are less mobile in *L. catta* and that the foot of *L. catta* may be adapted for what Morton has termed metatarsi-fulcrumation.[5] Further studies on the morphology and analyses of films of the locomotion of *L. f. rufus* and *L. catta* may allow more precise relationships to be found between the anatomy, locomotor behaviour, and vertical pattern of habitat preferences of the two species.

2. *Utilisation of the horizontal habitat*

Groups of *L. f. rufus* were identified and their locations were recorded throughout the study. Three groups of *L. f. rufus* were studied intensively at Tongobato and six groups at Antserananomby. At Antserananomby, there were twelve groups within the seven hectares of closed canopy forest. These groups were identified and their locations were noted on prepared maps whenever they were sighted. Groups of *L. f. rufus* usually could not be followed throughout the whole day because of the difficulty in remaining in continuous contact with this arboreal species. One group of *L. catta* at Antserananomby and two groups at Berenty were studied intensively. The groups of *L. catta* were usually followed throughout the day, their movements being recorded on prepared maps.

Figs. 11 and 12 illustrate the extreme differences in the size of the home-ranges and day-ranges between *L. f. rufus* and *L. catta* at Antserananomby. In this forest, the home-ranges of twelve groups of *L. f. rufus* (112 animals in all) were included within the home-range of one group of 19 *L. catta* (Fig. 11).

The day-ranges and home-ranges of *L. f. rufus* were very small, both at Antserananomby and at Tongobato. Although it was

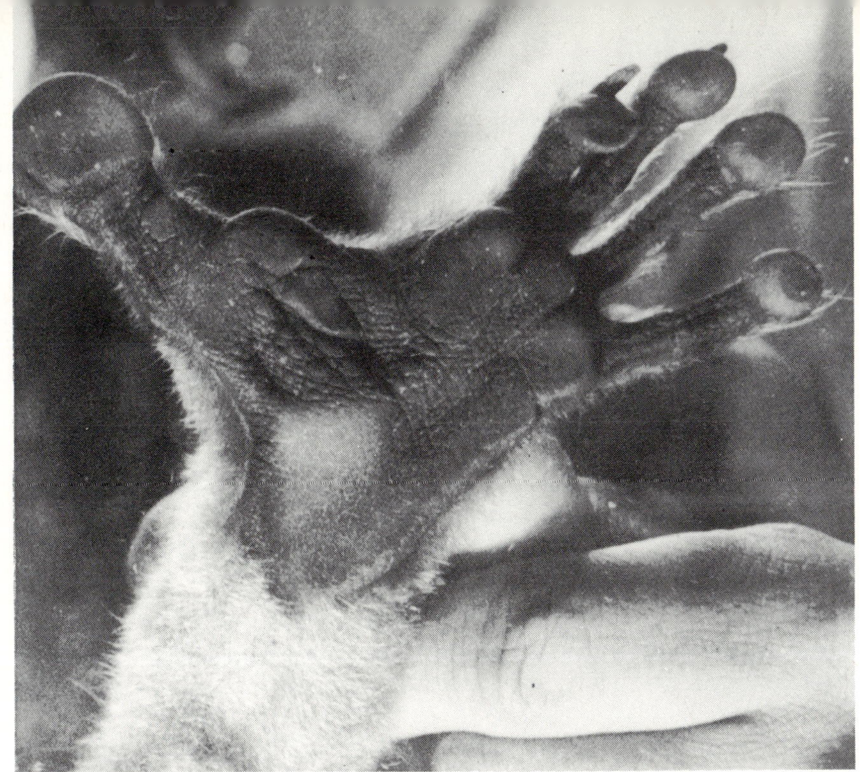

Figure 10. Above: foot of *Lemur fulrus rufus*. Below: foot of *Lemur catta*.

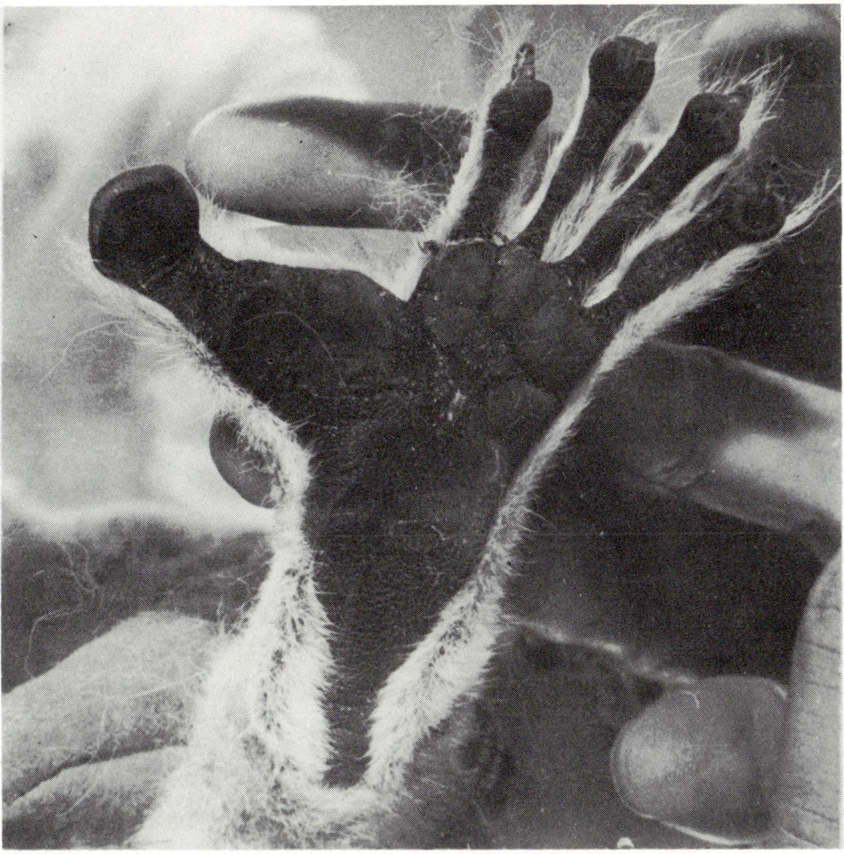

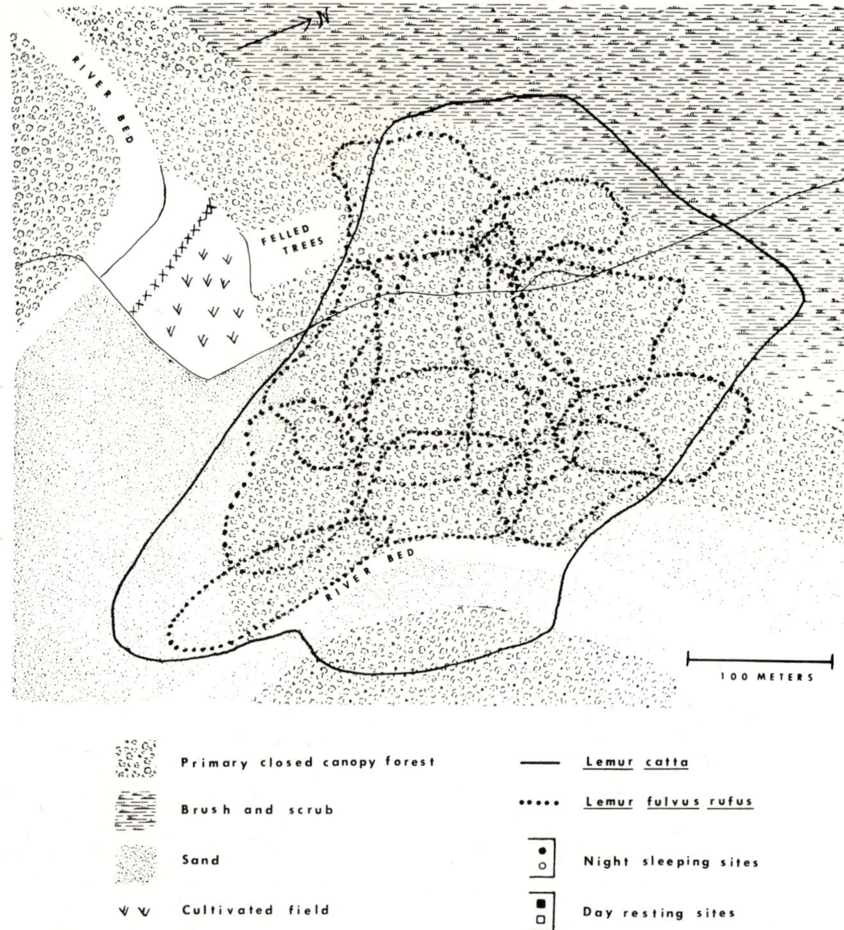

Figure 11. Antserananomby. Home-ranges of *Lemur catta* and *Lemur fulvus rufus.* This map includes the home-range of one group of *L. catta* (19 animals) and twelve groups of *L. f. rufus* (112 animals in all).

generally difficult to remain in contact with groups of *L. f. rufus*, at Antserananomby I was able to follow three groups for two days each (Fig. 12). The average day-range for the six days was between 125 and 150 m. The home-ranges of the groups studied at Antserananomby averaged 0.75 hectares (Fig. 11), while those of three groups at Tongobato averaged 1.0 hectares. The home-ranges of *L. f. rufus* were not rigidly defended, and the boundaries overlapped extensively.

L. catta had much larger day- and home-ranges than *L. f. rufus.* The average day-range for *L. catta* at Antserananomby was approximately 920 m. (Fig. 13). At Berenty, in a sample of four days, the average day-range was 965 m. The home-range of the group of *L. catta* at Antserananomby (group AC-1) was 8.8 hectares (Fig. 11). The home-range of group BC-1 at Berenty was 6.0 hectares. Jolly

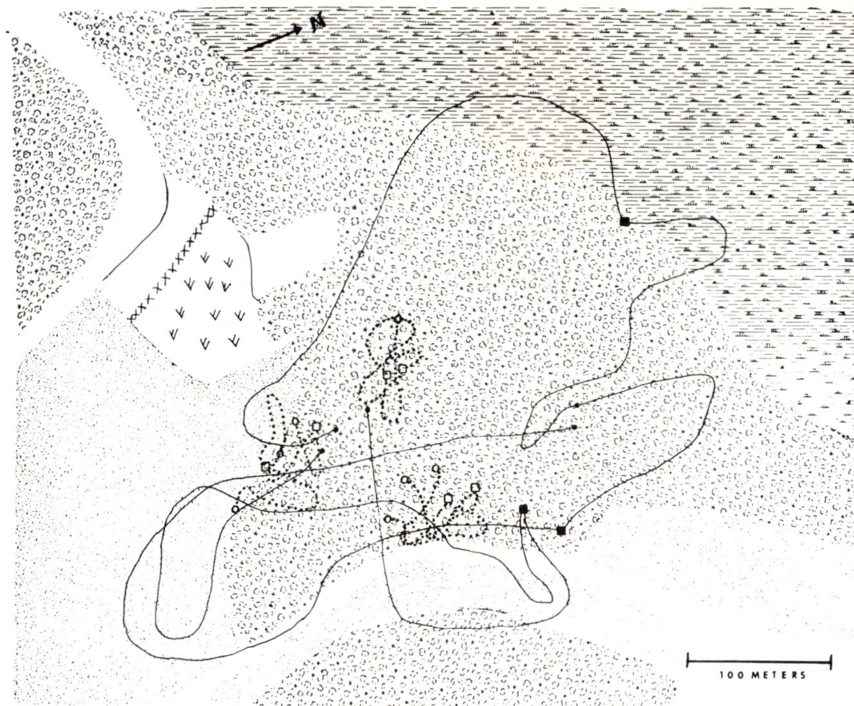

Figure 12. Antserananomby. Comparative day-ranges of *Lemur catta* and *Lemur fulvus rufus*. This map includes a sample of three day-ranges of one group of *L. catta* (group AC-1) and two day-ranges of three groups of *L. f. rufus*. (Key as for Fig. 11)

reported a home-range of 5.7 hectares for the group she studied at Berenty.[6] At Antserananomby, *L. catta* spent approximately 58% of its time in zones outside of the closed canopy area, even though these zones represented only 30% of its total home-range. By contrast, all of the home-ranges of the groups of *L. f. rufus* at this forest were located within the 7 hectares in which there is a closed canopy.

The population density of *L. f. rufus* was much higher than that of *L. catta*. The population density of each species was calculated from the average group size (see Tables 10 and 11) and the average area of the home-range. The population density of *L. f. rufus* at Tongobato was calculated only from data on groups TF2, TF3, and TF4.

Lemur catta	
Antserananomby	215 per square km.
Berenty*	250 per square km.
Average	233 per square km.

* Jolly estimated the density of *L. catta* at Berenty to be 320 per square km.[6] This would give an average density of 261 per square km. for populations of *L. catta* studied to date.

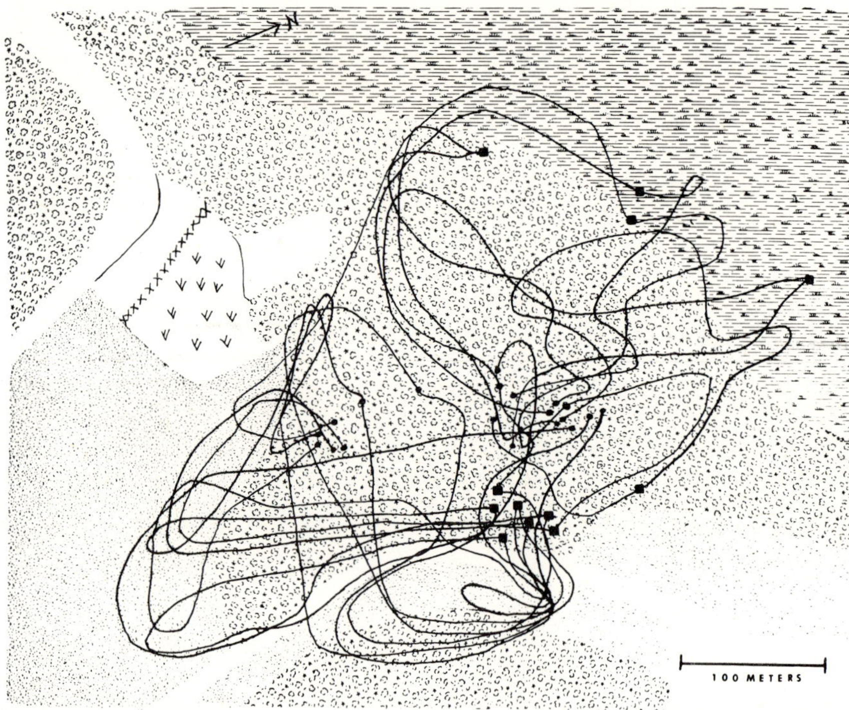

Figure 13. Antserananomby. Day-ranges of *Lemur catta* (group AC-1), including all of the ranges for which the group was followed throughout the day. (Key as for Fig. 11)

Lemur fulvus rufus

Antserananomby	1222 per square km.
Tongobato	900 per square km.
Average	1061 per square km.

3. *Diet*

The number of animals feeding and the level of the forest in which they were feeding was recorded during the collection of five-minute activity records. The plant and the part being eaten were recorded directly on the data sheet. Plant specimens were collected whenever possible and were later identified by Armand Rakotozafy of the Laboratoire Botanique of ORSTOM, Tananarive. No stomach content samples were obtained, because it was not feasible to shoot any of the animals during the field study. The dry weights of the plants and the nutritional values of the specimens have not yet been analysed. Therefore, the dietary schedules of the two species have been determined from direct observation only. Samples of faeces were collected, but have not yet been analysed.

Table 3 Plant species eaten by *Lemur fulvus rufus* at Tongobato

Acacia rovumae Olia*
Acacia sp.*
Alchornea sp.
*Flacourtia ramontchi**
Lawsonia alba Lamk
*Tamarindus indica**
*Terminalia mantaly**
*Vitex beravensis**

* Plant species eaten by *Lemur fulvus rufus* at both Tongobato and Antserananomby.

Table 4 Plant species eaten by *Lemur fulvus rufus* at Antserananomby

Acacia rovumae Olia*
Acacia sp.*
Ficus soroceoides Bak
*Flacourtia ramontchi**
Papilionaceae
Quisivianthe papinae
Rinorea greveana
*Tamarindus indica**
*Terminalia mantaly**
Tisomia sp.
*Vitex beravensis**

* Plant species eaten by *Lemur fulvus rufus* at both Tongobato and Antserananomby.

The differences between the diet of *L. catta* and *L. f. rufus* are related to the differences in habitat selection in the two species. *L. f. rufus* stayed in the continuous canopy of the forest and moved horizontally from tree to tree. It was rarely seen in areas that required movement on the ground. It fed over 90% of the time in the upper three forest levels (see Table 2). *L. catta*, on the other hand, was frequently observed feeding in all of the available forest layers. It spent about 58% of its time outside of the portion of the forest with a continuous canopy. *L. catta* fed over 65% of the time in levels 1, 2, and 3.

The restricted use of space by *L. f. rufus* is related to a less varied diet in this species than that found in *L. catta.* The plants which *L. f. rufus* was observed eating during the study are listed in Tables 3 and 4. I recorded *L. f. rufus* eating only 8 plant species at Tongobato and 11 at Antserananomby. There was a total of 13 different plant species eaten by *L. f. rufus* in the two forests. *L. catta*, on the other hand, had a much more varied diet in both of the forests in which it was studied. Both at Antserananomby and Berenty, *L. catta* was observed to feed on 24 different plant species (Tables 5 and 6). At Antserananomby, *L. catta* ate 12 plant species which were not eaten by *L. f. rufus.* Ten of these were ground plants or bushes, or plants found only outside the closed canopy portion of the forest.

L. f. rufus appears to be specialised in its choice of diet. A few species of plants made up a large proportion of its diet, and kily leaves were the main staple (Table 7). During the period of

Table 5 Plant species eaten by *Lemur catta* at Antserananomby

Acacia rovumae Olia	*Mimilopsis* sp.
Acacia sp.	*Paederia* sp.
Acalypha sp.*	*Paederia* sp.
Achyranthes aspera L.*	Papilionaceae
Adenia sp.	*Poupartia caffra* H. Perr
Alchornea sp.	*Quisivianthe papinae*
Commicarpus commersonii	*Tamarindus indica**
Ficus cocculifolia	*Terminalia mantaly*
Ficus soroceoides	*Vitex beravensis*
Ficus sp.	
Flacourtia ramontchi	
Grevia sp.	
Three species unidentified (small trees)	

* Plant species eaten by *Lemur catta* at both Antserananomby and Berenty.

observation at Tongobato, three species of plant constituted more than 80% of the diet of *L. f. rufus: Flacourtia ramontchi, Tamarindus indica*, and *Terminalia mantaly.* At Antserananomby, *Acacia rovumae, Ficus soroceoides*, and *Tamarindus indica* accounted for over 85% of the diet of *L. f. rufus.* Kily (*Tamarindus indica*) leaves made up 42% of the diet of *L. f. rufus* at Tongobato and about 75% of its diet at Antserananomby. This degree of specialisation is rare among those primate species that have been studied. It has been reported to occur, however, in *Lepilemur mustelinus leucopus.*[7]

The diet of *L. catta* was much less restricted than that of *L. f. rufus* (Table 7). At Antserananomby, the following species made up over 70% of the diet:

Achyranthes aspera	*Grevia* sp.
Alchornea sp.	*Mimilopsis* sp.
Ficus soroceoides	*Poupartia caffra*
Flacourtia ramontchi	*Tamarindus indica*

The kily tree provided 24% of the diet of *L. catta* in this forest. However, less than half of this (11%) consisted of leaves. Kily pods made up 12% of the diet, and ground plants (*Achyranthes aspera* and *Mimilopsis* sp.) 15%.

The diets of *L. catta* at Berenty and Antserananomby consisted of different plant species, but were quite similar in design. At Berenty, the following species accounted for over 80% of the observed plants eaten by *L. catta:*

Achyranthes aspera	*Phyllanthus* sp.
Boerhaavia diffusa	*Pithecelobium dulce*
Cassia sp.	*Rinorea greveana*
Melia azedarach	*Tamarindus indica*

The kily tree made up 23% of the diet. Kily leaves were eaten in 12% of the observations and kily pods in 10%. The fruit of two other

Table 6 Plant species eaten by *Lemur catta* at berenty

Acalypha sp.*	*Ehretia* sp.
Aizoaceae	*Hzima tetracantha* Lamk
Albizzia polyphylla Forven	*Mangifera indica* L.
Annona sp.	*Melia azedarach* L.
Achyranthes aspera L.*	*Opuntia vulgaris* Mill
Boerhaavia diffusa L.	*Phyllanthus* sp.
Cardiaspermum halicacabum	*Pithecelobium dulce* Benth
Cassia sp.	*Rinorea greveana* H. Bn.
Celtis philippensis Blanco	*Tamarindus indica**
Cissompelos sp.	*Zehneria* sp.
Combretum sp.	*Zizyphus jujuba* Lamk
Crateva excelsa Boj	One species unidentified (vine)

* Plant species eaten by *Lemur catta* at both Berenty and Antserananomby.

trees (*Rinorea greveana* and *Pithecelobium dulce*) together accounted for 40% of the observed feeding activity.

Both *L. catta* and *L. f. rufus* ate the fruit, leaves, flowers, bark, and sap of various species of plants (Table 8). The amount of time that they spent feeding on various parts of the plants depended upon the fruiting or blossoming seasons of the plants. As with differences in the number of plant species eaten by *L. catta* and *L. f. rufus*, the proportion of fruit or flowers fed upon by the two species is related to differences in the use of vertical and horizontal space. For example, a small tree (*Ficus cocculifolia*) produced fruits the size of large apples. I saw *L. f. rufus* feed on the fruit of this tree in a forest on the bank of the Mangoky River. There was only one of these trees in the forest at Antserananomby, and it was located on the side of a dry river bed opposite to the continuous canopy portion of the forest. *L. f. rufus* rarely crossed the river, and was never seen in this tree. *L. catta*, on the other hand, regularly crossed the river and foraged in this tree for a number of days while the fruit was ripe.

4. *Utilisation of time*

Once a group of animals was located, counts were made at five-minute intervals of the number of individuals engaged in each of six activities: feeding, grooming, resting, movement (i.e. movement of an individual), travel (i.e. movement of the group), and other (i.e. sunning, play, fighting, etc.). Each animal observed during each five-minute interval constituted an individual activity record (IAR). The IARs for each five minutes were combined for each half-hour period from 6.00 hrs to 18.25 hrs., and the percentage of IARs for each activity was calculated for each half-hour. The day was then divided into five phases as shown in Fig. 14, which includes the means of the percentages calculated for the half-hour periods within each of the phases.

Table 7 The number and percentage of individual activity records (IARs) for feeding on identified plant species

a. Lemur fulvus rufus at Antserananomby

Plant Species	*Number of IARs*	*Percentage of IARs*
Tamarindus indica	1802	75.68%
Leaves	1793	75.30
Fruit	9	0.38
Acacia sp.	156	6.55
Ficus soroceoides Bak	141	5.92
Acacia rovumae Olia	74	3.10
Terminalia mantaly	21	0.88
Quisivianthe papinae	18	0.75
Other	169	7.06
Total	2381	99.94%

b. Lemur fulvus rufus at Tongobato

Plant Species	*Number of IARs*	*Percentage of IARs*
Tamarindus indica	276	48.85
Leaves	237	41.95
Flowers	26	4.60
Fruit	10	1.77
Bark	3	0.53
Terminalia mantaly	127	22.47
Flacourtia ramontchi	69	12.21
Acacia rovumae Olia	38	6.72
Vitex beravensis	18	3.18
Other	37	6.54
Total	565	99.97%

The most striking differences between *L. f. rufus* and *L. catta* are seen in the data from Antserananomby (Figs. 14*a* and 14*b*). In this forest, *L. f. rufus* rested throughout the afternoon, whereas *L. catta* rested only during the midday phase (phase 3). Data from all three forests indicate that *L. f. rufus* rests more than *L. catta* during the day. The activity/rest ratios for the hours from 6.00 hrs. to 18.25 hrs. were calculated from the data in Table 9. The ratios are as follows:

Lemur fulvus rufus

Tongobato	50/50 = 1.00
Antserananomby	44/56 = 0.79

Lemur catta

Antserananomby	59/41 = 1.44
Berenty	61/39 = 1.56

c. *Lemur catta* at Antserananomby

Plant Species	*Number of IARs*	*Percentage of IARs*
Tamarindus indica	374	24.36%
[Fruit	[183	[11.92
Leaves	174	11.33
Flowers]	17]	1.11]
Small trees:	320	20.84
[*Alchornea* sp.		
Flacourtia ramontchi		
Grevia sp.		
Poupartia caffra H. Perr]		
Ground plants (all species)	225	14.65
Ficus soroceoides Bak	194	12.63
Vines (all species)	140	9.12
Quisivianthe papinae	94	6.12
Vitex beravensis	82	5.34
Ficus cocculifolia	40	2.60
Acacia rovumae Olia	18	1.17
Other	48	3.12
Total	1535	99.95%

d. *Lemur catta* at Berenty

Plant Species	*Number of IARs*	*Percentage of IARs*
Tamarindus indica	519	22.97%
[Leaves	[274	[12.13
Fruit	225	9.96
Bark]	20]	.88]
Rinorea greveana H. Bn.	474	20.98
Pithecelobium dulce Benth	433	19.16
Phyllanthus sp.	137	6.06
Melia azedarach L.	132	5.84
Ehritia sp.	130	5.74
Ground plants (all species)	124	5.48
Opuntia vulgaris Mill	84	3.71
Annona sp.	38	1.68
Vines (all species)	27	1.19
Other	161	7.10
Total	2259	99.91%

L. f. rufus fed very early in the morning and late in the afternoon and travelled little to obtain its food. *L. catta* began to feed later in the morning and stopped earlier in the evening than *L. f. rufus.* Groups of *L. catta* travelled greater distances to obtain food, foraging throughout the afternoon. At Antserananomby, *L. f. rufus* fed mainly during the first and fifth phases and *L. catta* during the second and fourth. In all phases but the third (12.00 to 14.55 hrs), when both species were resting, the patterns of activity were different.

Table 8 Number and percentage of individual activity records for feeding on identified parts of plants

a. *Lemur fulvus rufus* at Antserananomby

Part of plant eaten	*Number of IARs*	*Percentage of IARs*
Leaves	2123	89.16%
Fruit	161	6.76
Flowers	90	3.77
Bark	7	.29
Total	2381	99.98%

b. *Lemur fulvus rufus* at Tongobato

Part of plant eaten	*Number of IARs*	*Percentage of IARs*
Leaves	275	52.08%
Fruit	224	42.43
Flowers	26	4.92
Bark	3	.56
Total	528	99.99%

c. *Lemur catta* at Antserananomby

Part of plant eaten	*Number of IARs*	*Percentage of IARs*
Leaves	670	43.64%
Fruit	516	33.61
Herbs	225	14.65
Flowers	124	8.07
Total	1535	99.97%

d. *Lemur catta* at Berenty

Part of plant eaten	*Number of IARs*	*Percentage of IARs*
Fruit	1335	59.30%
Leaves	550	24.43
Flowers	137	6.08
Herbs	124	5.50
Bark, sap, cactus	105	4.66
Total	2251	99.97%

At Berenty, groups of *L. catta* began to feed earlier in the morning and the peak of their resting activity was earlier than at Antserananomby. They also began to feed earlier in the afternoon and stretched out their afternoon feeding activity over a longer period of time. These differences, which can be related to the early sunrise and high early morning temperatures at Berenty, make the behaviour pattern of *L. catta* at this forest superficially similar to that of *L. f.*

Table 9 Mean percentages of individual activity records for each activity each half hour

Species and forest	*Activity*					
	Feeding	*Grooming*	*Resting*	*Moving*	*Travel*	*Other*
Lemur fulvus rufus						
Tongobato	16.59	11.30	49.73	7.77	12.21	2.41
Antserananomby	26.22	5.25	56.58	3.09	2.47	6.40
Lemur catta						
Antserananomby	24.94	4.67	41.42	7.45	11.34	10.19
Berenty	31.12	10.97	38.63	6.30	6.91	6.07

rufus in the other study areas. Even though the temperature and time of sunrise at Tongobato were more similar to those at Berenty than to those at Antserananomby, the pattern of behaviour of *L. f. rufus* was similar in both of the forests in which it was studied. The high percentage of movement and travel of *L. f. rufus* at Tongobato resulted from the difficulty in habituating the animals to the observer, because local inhabitants frequently hunted the lemurs in this forest. Seasonal differences (in this case, time of sunrise and sunset and time of the maximum temperature during the day) seemed to affect the daily activity cycle of *L. catta* more than that of *L. f. rufus.*

L. f. rufus and *L. catta* both exhibited 'sunning' behaviour early in the morning at Antserananomby. This is reflected in the high percentage of 'other' activities during the first phase of the day. *L. catta* sunned twice as much as *L. f. rufus* during this phase, and sunning may be an important thermo-regulating mechanism in this species. Very little sunning was observed for either species in the other two forests. The mean minimum temperatures at night at Berenty and Tongobato were much higher than those at Antserananomby during the months of observation in these forests.

Significance tests were run on the resting and feeding activities of *L. f. rufus* and *L. catta* at Antserananomby. Since the samples for different five-minute periods are not independent, Z scores were computed for each five-minute interval (Figs. 15 and 16). The methods used for these calculations are described by Sussman[2] and Sussman, O'Fallon and Buettner-Janusch (in prep.).

Fig. 15 illustrates the fact that *L. f. rufus* fed significantly more in the early morning and late evening, and *L. catta* fed significantly more in the late morning and late afternoon. Feeding behaviour in the midday phase was not significantly different; at this time, both species were mainly resting. Throughout the day there are very few times (a total of 25 minutes) when *L. catta* rested significantly more than *L. f. rufus* (Fig. 16). There are only 26 five-minute periods in which a higher percentage of *L. catta* than *L. f. rufus* were observed resting.

Figure 14. Mean percentage of individual activity records for each of six activities during five phases of the day. F = Feeding; G = Grooming; R = Resting; M = Moving; T = Travel; O = Other.

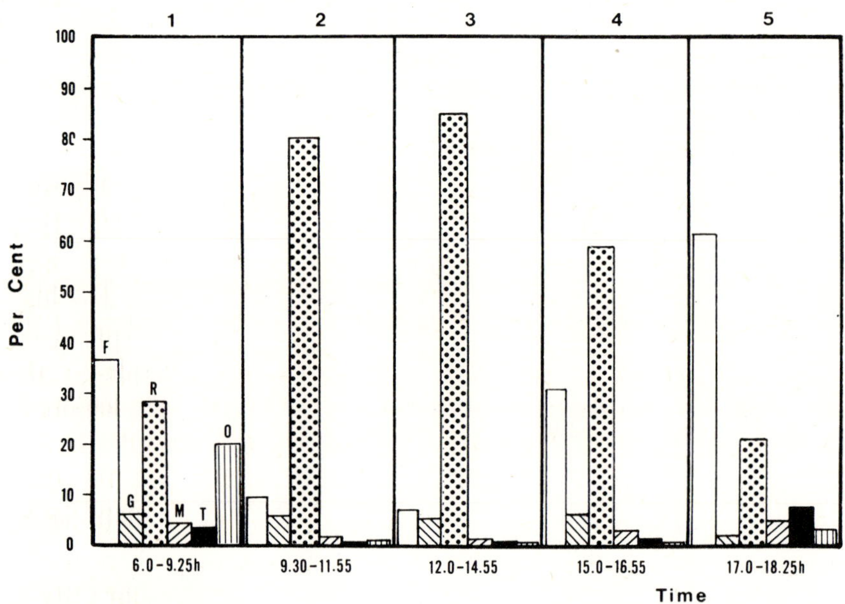

(*a*) *Lemur fulvus rufus* at Antserananomby

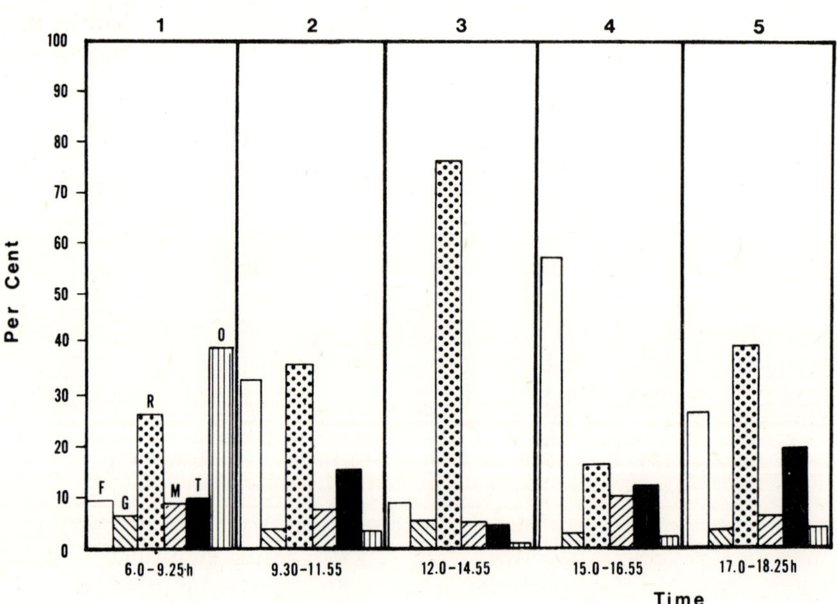

(*b*) *Lemur catta* at Antserananomby

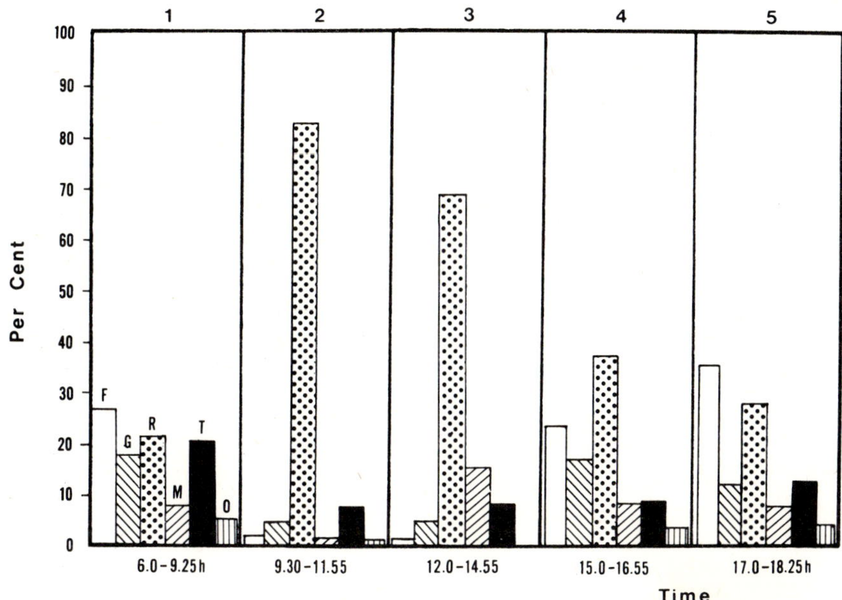

(*c*) *Lemur fulvus rufus* at Tongobato

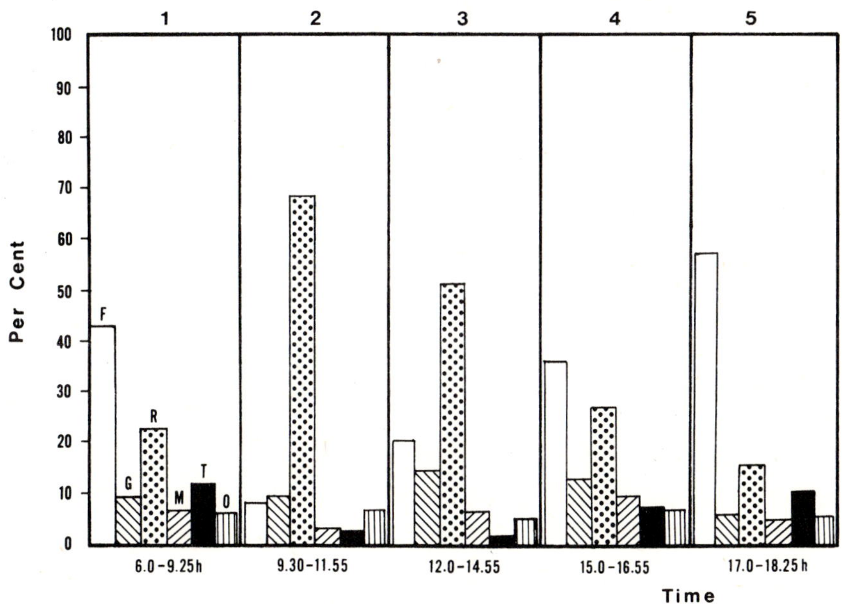

(*d*) *Lemur catta* at Berenty

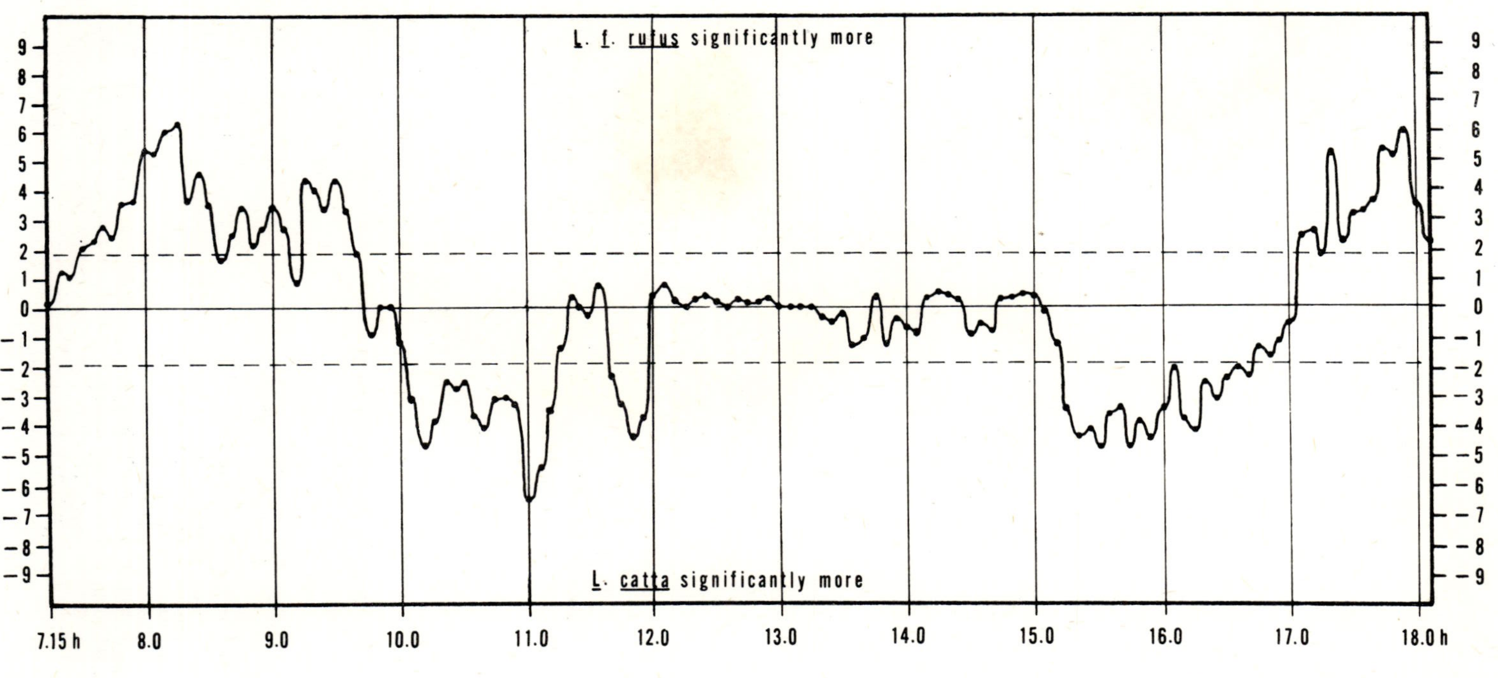

Figure 15. Z scores for the feeding activity of *Lemur fulvus rufus* and *Lemur catta* at Antserananomby computed for each five-minute observation period. A score equal to or more than 1.96 means that *L. f. rufus* feeds significantly more than *L. catta* ($p \leqslant .05$). A score equal to or less than -1.96 means that *L. catta* feeds significantly more than *L. f. rufus* ($p \leqslant .05$).

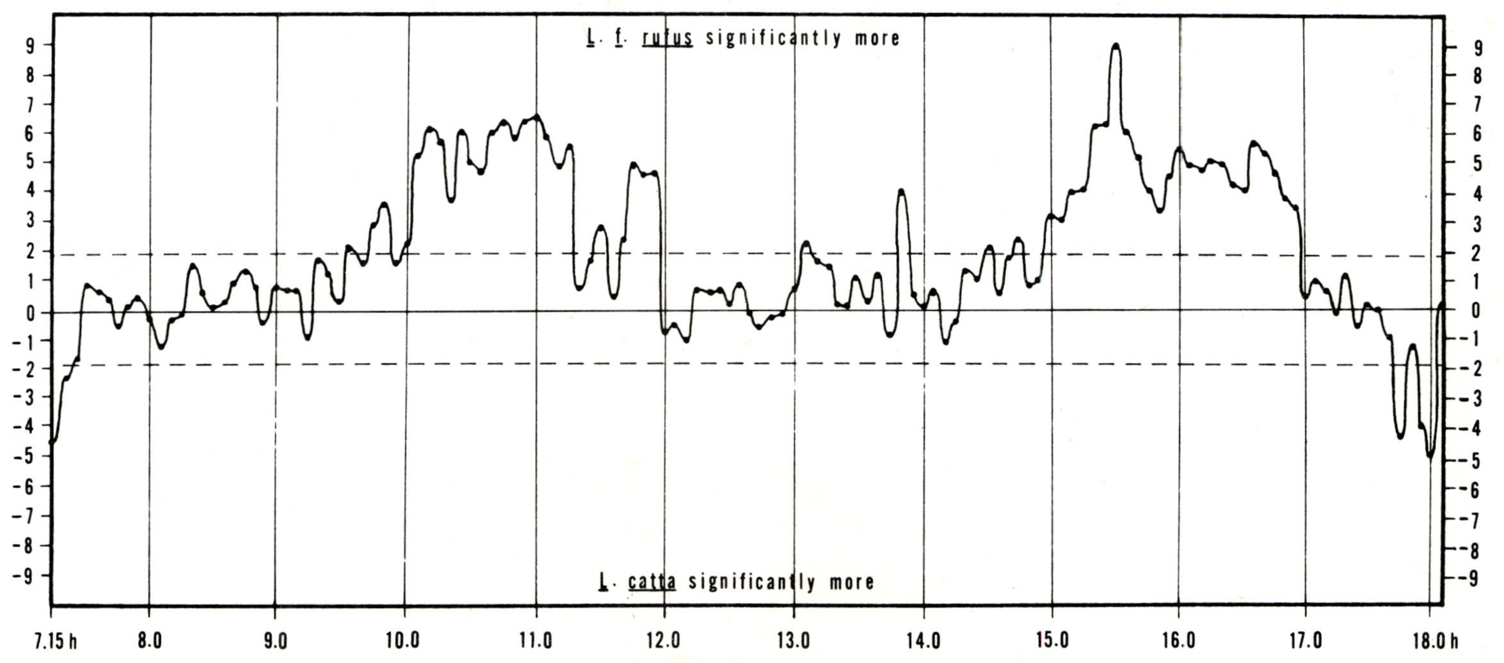

Figure 16. Z scores for the resting activity of *Lemur fulvus rufus* and *Lemur catta* at Antseranamomby computed for each five-minute observation period. A score equal to or more than 1.96 means that *L. f. rufus* rests significantly more than *L. catta* ($p \leqslant .05$). A score equal to or less than -1.96 means that *L. catta* rests significantly more than *L. f. rufus* ($p \leqslant .05$).

Statistical analyses comparing data from different forests were difficult to interpret because of the differing forest conditions. Significance tests on the behaviour of populations in different locations only indicate whether or not the behaviour is significantly different. They do not give us any insight into whether differences are due to characteristics of the populations or to environmental conditions.

5. *Group size and composition*

Groups were counted throughout the study period at each forest, and census data on each group were checked and rechecked continually. In all three forests, census data were collected by more than one observer. In most cases, two observers worked together.*

The following criteria were used to distinguish between sex and age classes. There are conspicuous differences between the sexes in *Lemur fulvus rufus.* The male is grey and has a pronounced white face mask with a bright red-orange tuft of hair on the top of the head (Fig. 1). The female is red-orange, with white patches over her eyes and a black bar across the forehead and between the eyes (Fig. 2). These distinctions are obvious even in the juvenile animals. The coloration of infants, however, is often misleading, and sex determination of these young animals is not possible. In *Lemur catta* there are no differences in pelage between the sexes. In order to sex individuals of this species, it was necessary to get into strategic positions to observe the genitalia of the animals. It was not possible to determine the sex of juvenile or infant *L. catta.*

Each individual was placed into one of three age categories – infant, juvenile or adult. Infants were those animals that were still being carried by their mothers. This included animals up to the age of about four to five months. Juveniles were those animals that were no longer being carried, but had not yet reached full size. However, in the field it is often difficult to distinguish older juveniles from adults. Adult size is attained at about two years of age. In both *L. catta* and *L. f. rufus*, births occur once a year within a short period of time (usually a two-week period). The young all mature at approximately the same rate and the achievement of different stages of maturity is synchronous.

Tables 10 and 11 include the census data collected on *L. f. rufus* and *L. catta.* The 17 groups of *L. f. rufus* which I censused ranged in size from 4 to 17 animals. The average size of the groups was 9.5 animals. The average size of the three groups of *L. catta* which I censused was 18.0 animals. Including the counts of Jolly, and Klopfer

* Linda Sussman, Alain Schilling, and field guides Folo Emmanuel and Bernard Tsiefatao assisted me in collecting census data.

Table 10 Composition of groups of *Lemur fulvus rufus*

Name of group	Adult Male	Adult Female	Juvenile Male	Juvenile Female	Infant	Total	Adult Sex ratio M:F
Antserananomby*							
AF-1	4	6	1	1	0	12	1:1.50
AF-2	4	5	1	0	0	10	1:1.25
AF-3	4	3	0	1	0	8	1.33:1
AF-4	2	5	2	1	0	10	1:2.50
AF-5	4	6	0	2	0	12	1:1.50
AF-6	3	2	0	0	0	5	1.50:1
AF-7	3	4	1	1	0	9	1:1.33
AF-8	2	2	0	0	0	4	1:1.00
AF-9	2	2	1	0	0	5	1:1.00
AF-10	4	5	2	0	0	11	1:1.25
AF-11	5	7	1	2	0	15	1:1.40
AF-12	4	4	1	0	0	9	1:1.00
Totals	41	51	10	8	0	110*	41:51=1:1.24
Means	3.42	4.25	.83	.66	0	9.17	
Tongobato							
TF-1	5	8	0	0	4	17	1:1.60
TF-2	2	3	1	0	1	7	1:1.50
TF-3	4	5	0	1	2	12	1:1.25
TF-4	3	4	0	0	1	8	1:1.33
TF-5	3	2	0	1	1	7	1.50:1
Totals	17	22	1	2	9	51	17:22=1:1.29
Means	3.40	4.40	.20	.40	1.80	10.20	
Overall totals	58	73	11	10	9	161	58:73=1:1.26
Overall means	3.41	4.29	.64	.59	.53	9.47	

* There was also an extra male group consisting of two individuals. Therefore, the total number of animals censused at Antserananomby was 112.

Table 11 Composition of groups of *Lemur catta*

Name of Group	Adult Male	Adult Female	Juvenile	Infant	Total	Adult Sex ratio M:F
Antserananomby						
AC-1	7	8	4	0	19	1:1.14
Berenty						
BC-1	4	5	2	4	15	1:1.25
BC-2	7	6	3	4	20	1.17:1
Totals	11	11	5	8	35	11:11=1:1.00
Means	5.50	5.50	2.50	4.00	17.50	
Overall totals	18	19	9	8	54	18:19=1:1.05
Overall means	6.00	6.30	3.00	2.70	18.00	

and Jolly,[6,8] all of the groups of *L. catta* which have been censused consist of 12 to 24 individuals and the average size of the groups is 18.8 animals. Thus, the sizes of groups of *L. catta*, on the average, are approximately twice those of groups of *L. f. rufus.*

Adult sex ratios were similar in groups of both *L. catta* and *L. f. rufus.* There were slightly more adult females than adult males in groups of both species. It is possible that groups of *L. catta* contain proportionately more infants per year than do those of *L. f. rufus.* The ratio of infants to non-infants for *L. f. rufus* at Tongobato was 1/4.66, while that for *L. catta* at Berenty was 1/3.38. Since there were no infants during the time of the census at Antserananomby, the ratio of juveniles (mainly one-year olds) to adults was calculated. This ratio was 1/5.11 for *L. f. rufus* and 1/3.75 for *L. catta.*

A characteristic which is not revealed by the census data is the tendency towards peripheralisation of subordinate males in groups of *L. catta.* This characteristic was associated with a well-defined dominance hierarchy. There was no noticeable hierarchy in groups of *L. f. rufus.* Jolly reported that peripheralisation of subordinate males occurred in the group of *L. catta* she studied intensively at Berenty.[6] The groups I studied also showed a tendency towards the peripheralisation of subordinate males. In one of the groups at Berenty, two adult males were actively kept away from the centre of the group. These two peripheral males were always together and sometimes intermingled with juveniles of the group, but were chased whenever they approached the centre of the group too closely.

Conclusions

Differences were found in the utilisation of space and time, the diet and the social structure of *Lemur fulvus rufus* and *Lemur catta.* These differences were related to the habitat preferences of the two species, and in each case were independent of the presence or absence of the other species. This indicates that the differences are not caused by the interaction between the populations of *L. f. rufus* and *L. catta*, but by adaptations to basically different environmental conditions which occur where the species are allopatric.

The quantitative data on the utilisation of the vertical habitat indicate that *L. f. rufus* and *L. catta* have distinctive preferences for different forest strata. *L. f. rufus* is very specific in its choice of vertical habitat; *L. catta*, on the other hand, is frequently found in all forest layers. In all three forests studied, vertical habitat preferences do not seem to be altered by the presence or absence of the other species. This indicates that the two species do not compete directly for the use of a particular vertical forest stratum. It also suggests that the differences in the choice of vertical habitat are not the result of interaction between the two species.

The day-ranges and home-ranges of *L. f. rufus* are much smaller than those of *L. catta*, and the population density is much higher. The home-ranges of groups of *L. f. rufus* at Antserananomby and Tongobato are restricted to portions of the forests with a continuous canopy. The home-ranges are not rigidly defended and have overlapping boundaries. The high density of the population and the small, overlapping ranges of the groups allow *L. f. rufus* to completely fill these restricted forest areas. This pattern of group distribution may be adaptive for the exploitation of small areas which are relatively rich in food supply and in which the food is evenly distributed.[9]

Populations of *L. catta* are found in the west and south of Madagascar in many areas in which there are no closed canopy forests but only brush and scrub forests. Many of these scrub forests are very dry and the vegetation is patchy and likely to be sparse for at least part of the year. The large home-ranges and day-ranges of groups of *L. catta* may be genetically fixed adaptive responses to these arid environments. Trends in population density and range size similar to those found for *L. f. rufus* and *L. catta* have been reported for forest monkeys and savannah or grassland monkeys, respectively.[9, 10, 11, 12, 13]

Differences in the diet of *L. f. rufus* and *L. catta* can also be related to differences in the habitat preferences of the two species. Although both seem to be opportunistic in their choice of food, the less restricted use of the vertical and horizontal habitat by *L. catta* allows this species a greater opportunity to obtain a more varied food supply. *L. f. rufus*, having a more limited vertical and horizontal range, has a more specialised relationship to the environment.

The foraging habits of *L. f. rufus* and *L. catta* seem to be adaptations to different econiches. *L. f. rufus* inhabits closed canopy forests, which in most cases are found in small, relatively uniform, circumscribed areas. In these forests, the population density of *L. f. rufus* is extremely high. The ability to exist almost exclusively on a diet of kily leaves, supplemented with the fruit of those trees which happen to be within the range of any one group, allows *L. f. rufus* to exploit efficiently this type of environment. A group of *L. catta* continuously forages within its range and exploits a number of plants in a large area. Within a period of seven to ten days, the group visits most of its total range. It is likely that this continuous foraging behaviour may be adaptive to arid environments in which the food supply is sparse. The constant surveillance of a large (or relatively large) range allows groups of *L. catta* to exploit a number of different plants which may have a patchy distribution over a wide area.

The utilisation of time by the two species further indicates that behavioural adaptations of *L. f. rufus* and *L. catta* have developed in response to different environmental conditions. *L. f. rufus* feeds early in the morning and late in the afternoon; it rests throughout the day and spends almost all of the warmest portion of the day in the closed canopy. The temperature of the micro-environments in which *L. catta* is found throughout the day is more likely to vary. At night, *L. catta* sleeps in the closed canopy portion of the forest. However, during the day it spends much of the afternoon in unshaded portions of the forest. In these areas, the ambient temperature is usually quite high. At Antserananomby, where the evening temperatures were low, *L. catta* sunned for long periods during the morning.

The distribution of *L. catta* within the forests in which it was studied, and the fact that it is found in the hot, arid regions of southern Madagascar, indicate that this species is well adapted to warm and dry climates. *L. f. rufus* is not found in the south of Madagascar, and it remains in the shaded portion of the forest throughout the day. Furthermore, *L. f. rufus* is active only during the coolest hours of the day. It seems likely that *L. catta* uses behavioural thermo-regulation (e.g. sunning) to help maintain its body temperature and in this way is able to utilise habitats with large ranges of ambient temperature.

The sizes of groups of *L. catta* are, on average, approximately twice those of groups of *L. f. rufus*. Terrestrially adapted species have generally been considered to form larger groups than arboreally adapted species.[9,10,11] There are, however, several exceptions, and phylogenetic relationships also play a role in the determination of group size.[12,14] Before relationships can be established between ecology and group size or social structure, further studies must be made on arboreally adapted species and their use of the environment.

There is a tendency in groups of *L. catta* towards peripheralisation of subordinate males. In a very dry environment, where food is sparse, the exclusion of excess males could be very adaptive for the survival of the group.[15,16] A study of *L. catta* in arid regions may reveal groups containing many more adult females than males.

In summary, *L. f. rufus* seems to be a very specialised, arboreally adapted species. *L. catta* is able to exploit a number of different habitats. However, the behavioural adaptations which have been found in *L. catta* seem to have developed in response to the more arid, warm environments in which this species is found (where resources are likely to be limited). *L. f. rufus* shows a configuration of behavioural adaptations which conform to those found in many arboreal species of New and Old World monkeys, whereas some behavioural adaptations of *L. catta* parallel those of many terrestrial primates living in savannah or grassland econiches.

Acknowledgments

I would like to thank Dr Brygoo of l'Institut Pasteur de Madagascar, Président M. Manambelona, Secrétaire Général du Comité National de la Recherche Scientifique, M. Ramanantsoavina, Directeur des Eaux et Forêts, and the staff of ORSTOM, Tananarive for their assistance and cooperation while I was in the field. I am also indebted to many people of the villages of Manja and Vondrove, whose kind hospitality is deeply appreciated. I am especially grateful for the continued assistance (both academic and otherwise) offered by Dr J. Buettner-Janusch throughout the period of this study. My wife, Linda, aided me in all aspects of the research and I am, of course, especially indebted to her. The study was supported in part by Research Fellowship MH46268-01 of the National Institute of Mental Health, United States Public Health Service and by a Duke University Graduate Fellowship.

NOTES

1 Brown, W.L. and Wilson, E.O. (1956), 'Character displacement', *Systematic Zool.* 5, 49-64.

2 Sussman, R.W. (1972), 'An ecological study of two Madagascan primates: *Lemur fulvus rufus* and *Lemur catta*', PhD thesis, Duke University.

3 Perrier de la Bathie, H. (1921), 'La végétation malgache', *Ann. Mus. Colon. marseille*, 3rd ser., 9, 1-268; Humbert, H. (1954), 'Les territoires phytogéographiques de Madagascar: leur cartographie', *LIX° Colloque International du Centre National de la Recherche Scientifique*, Paris, pp. 439-48; Humbert, H. and Darne, G.C. (1965), *Notice de la Carte de Madagascar*, Toulouse.

4 Crook, J.H. and Aldrich-Blake, P. (1968), 'Ecological and behavioural contrasts between sympatric ground-dwelling primates in Ethiopia', *Folia primat.* 8, 192-227.

5 Morton, D.J. (1924), 'Evolution of the human foot', *Amer. J. Phys. Anthrop.* 7, 1-52.

6 Jolly, A. (1966), *Lemur Behavior*, Chicago.

7 Charles-Dominique, P. and Hladik, C.M. (1971), 'Le *Lepilemur* du sud de Madagascar: écologie, alimentation et vie sociale', *Terre et Vie* 25, 3-66.

8 Klopfer, P.H. and Jolly, A. (1970), 'The stability of territorial boundaries in a lemur troop', *Folia primat.* 12, 199-208.

9 Crook, J.H. and Gartlan, J.S. (1966), 'Evolution of primate societies', *Nature* 210, 1200-3.

10 DeVore, I. (1953), 'A comparison of the ecology and behaviour of monkeys and apes, in Washburn, S.L. (ed.), *Classification and Human Evolution*, Chicago, 301-19.

11 Aldrich-Blake, F.P.G. (1970), 'Problems of social structure in forest monkeys', in Crook, J.H. (ed), *Social Behaviour in Birds and Mammals*, New York, 79-101.

12 Crook, J.H. (1970), 'The socio-ecology of primates', in Crook, J.H. (ed), *Social Behavior in Birds and Mammals*, New York, 103-66.

13 Denham, W.W. (1971), 'Energy relations and some basic properties of primate social organisation', *Amer. Anthrop.* 73, 77-95.

14 Struhsaker, T.T. (1969), 'Correlates of ecology and social organisation

among African cercopithecines', *Folia primat.*, 11, 80-118.

15 Kummer, H. (1967), 'Dimensions of a comparative biology of primate groups', *Amer. J. Phys. Anthrop.* 27, 357-66.

16 Kummer, H. (1971), *Primate Societies*, Chicago.

S.K. BEARDER and G.A. DOYLE

Ecology of bushbabies Galago senegalensis *and* Galago crassicaudatus, *with some notes on their behaviour in the field*

Introduction

The lorisoid primates of Africa and south-east Asia, unlike the lemurs of Madagascar, cohabit with monkeys, but – being exclusively nocturnal – do not compete ecologically with them. They have retained many characteristics of their primitive mammalian ancestors and remain similar to the most ancient primates. The two sub-families are best characterised by essential differences in the mode of locomotion of their representatives; the Lorisinae being slow-moving climbers, while the Galaginae are described as active leapers. The pattern of locomotion of at least one species of the Galaginae, *Galago crassicaudatus*, is midway between the two types.

In Africa, the family Lorisidae comprises seven species, five of which may be found occupying the same habitat in tropical West Africa (Gabon).[1] Ecological localisation results in virtual absence of interspecific competition. The remaining two species, which are the subjects of the present study, are found over a large area of Africa south of the Sahara, and may also occasionally be found together, in regions where their characteristic habitats overlap.

Field research was begun in South Africa and Rhodesia in 1968, using red torchlight to make continuous direct observations of bushbabies at all hours of the night. Over 1,200 observation hours have been made to date, in a number of geographical regions and habitat types. The animals soon became habituated to the presence of the observer and allowed his close approach. Separate and detailed studies of each species during 12-month periods were made in two main study areas. Further information was collected in areas where both species occurred together. A method of trapping was devised which made it possible to mark and release a number of animals for long-term studies of population dynamics, individual movement, use of space and social behaviour.

Habitat and distribution

1. Galago senegalensis

The lesser bushbaby (maximum weight 300 gm.) is to be found in a wide range of habitats from sea level to 1,500 m. These include semi-arid, steep-sided valleys, open woodland, orchard bush or scrub and isolated thickets with grassland. This species also occurs in the primary forests of mountain regions, river valleys and coastal areas. Its distribution is restricted by open, or relatively treeless, areas and is often associated with *Acacia* thorn trees, particularly *A. karoo.*

Observations of a single sub-species were made in the northern Transvaal, including the Springbok Flats and the Waterberg mountains; the north-eastern Transvaal, comprising the lowveld in the east and the Drakensberg escarpment in the west; and the eastern Highlands of Rhodesia.

Mean temperature figures varied considerably over this entire range, with a minimum just below freezing and a maximum of around 37°C. The rainfall showed a marked fluctuation from year to year, with mean figures per annum of 599, 722 and 1463 mm. for the three areas, respectively. Rainfall was mainly confined to the summer months between October and April, although rain was often recorded in any month. The actual subspecies studied was *G.s. moholi* A. Smith (Moholi Galago): a large-eared race, mainly grey in colour, but with the lower back washed with otter-brown which distinguishes it from typical *senegalensis* and other grey-bodied northern races. It has a thin tail, showing no tendency to bushiness.[2] It is found throughout the wooded and bushveld areas of South West Africa, Angola, Zambia, Botswana, extending east through southern Transvaal, and north through Rhodesia as far as Tanzania.[3]

2. Galago crassicaudatus

The thick-tailed bushbaby (maximum weight 1,800 gm.) has a more restricted distribution than the former species. It occurs between sea level and 1,800 m., but is mainly confined to dense evergreen indigenous forest and riparian bush, where there is an abundance of trees bearing edible fruits. The species may also be found in stands of plantation timber such as blue gum, black wattle or pine, where these adjoin natural bush. Such vegetation occurs in well-watered coastal regions, or inland as montane forest in steep-sided valleys and on hillsides watered by run-off from the mountain catchment areas. Much of this habitat has been exploited by man in the past and replaced by comparatively sterile plantations, but a considerable number of Government-protected forest areas still remain. As with the former species, *G. crassicaudatus* has a wide range of temperature tolerance.

Comparative information was collected on three subspecies in three different habitat types: (*a*) *G.c. umbrosus*, (*b*) *G.c. garnetti* and (*c*) *G.c. lönnbergi*, as described by Osman Hill.[2]

(*a*) *G.c. umbrosus* (Thos.), Transvaal Dusky Galago.
This subspecies may be distinguished by its brown colouration, particularly dark on the dorsal surfaces of the hands and feet. A certain proportion of the study population (8%) was melanistic, having a slightly darker overall hue and dark tips to the tails. One dark female gave birth to light-coloured offspring. Limits of distribution of this race are undetermined. It inhabits the forest areas and riparian bush of the Drakensberg foothills in the north-east Transvaal, where long term observations were carried out.

(*b*) *G.c. garnetti* (Ogilby), Garnett's or Black Galago.
A brown race, only the digits being blackish and with 80% of the study population having dark tips to their tails to a varying extent. The subspecies is found in Natal and Zululand, but limits of the range remain undetermined. Observations were made in dense coastal Dune Forest with a heavy understorey and many climbers, and also in Temperate forest with a canopy approximately 20-30 m. high.

(*c*) *G.c. lönnbergi* (Schwarz), Lönnberg's Galago.
A light-coloured variety having no dark tip to the tail and being somewhat smaller in size. There is no dark colouration on the dorsal surfaces of the hands and feet. It occurs in south-east Africa, from the Zambesi in the north as far south as the Limpopo. The exact western limit of its range is undefined. This subspecies was observed in relatively open bush in and around Umtali, Rhodesia.

The three most southern forms of *G. crassicaudatus*, together with four other races, are larger in size than the northern group, which comprises four subspecies.[2]

Other primates which occur in the regions described are: the Vervet monkey, *Cercopithecus aethiops*; the Samango monkey, *Cercopithecus mitis labiatus*; and the baboon, *Papio ursinus*.

Population density

On the basis of direct counts made over several consecutive nights in each of the different habitats, calculations have been made on the population densities of the two species (Table 1). In open bush it was possible to count practically all the members of a population within a set area with little chance of error, due to the extremely reflective nature of the tapetum lucidum of the bushbaby eye. This could be seen at a distance of more than 100 m. Counts of the lesser bushbaby were also made during the day by finding their sleeping places. In forest habitats, counting was done along transect lines coinciding with forest paths. In these cases, allowances have been made for the effective scanning area of the torch beam in relation to the density and height of the vegetation.

Table 1 Population densities of the two bushbaby species

G. senegalensis

Area	*Habitat*	*Density / square km.*
N. Transvaal	*Acacia* thornveld	200
N. Transvaal	*A. karoo* thickets	500
N. Transvaal	Wooded valley	275
Rhodesia	Mixed woodland and plantations	87
N.E. Transvaal	Lowveld: riparian bush and savannah	103
N.E. Transvaal	Escarpment: riparian bush and scrub	95

G. crassicaudatus

Area	*Habitat*	*Density / square km.*
Zululand	Dune forest	125
Zululand	Temperate forest	112
Rhodesia	Mixed woodland and plantations	110
N.E. Transvaal	Lowveld: riparian bush and savannah	72
N.E. Transvaal	Escarpment: riparian bush and scrub	88

The density calculations are indications of the situation in areas where the vegetation was of a more or less uniform type, in which the animals were distributed fairly evenly. However, due to the uneven or interrupted nature of the bush, the bushbabies were frequently separated into small aggregations. Where the change of vegetation was even more marked, populations of bushbabies were virtually or completely isolated from other populations, and this situation may account for the great variety of subspecific characters which have been described.

There is a marked disparity between the population densities calculated here and those of the remaining five African lorisids calculated by Charles-Dominique in Gabon, using similar techniques.[4] This may be attributed to the fact that far fewer mammal species live sympatrically in South Africa and Rhodesia than in tropical West Africa.

In the north-east Transvaal, where the two species lived sympatrically their population densities were lower than for those regions where they occurred alone. This may be a result of interspecific competition or of the fact that the vegetation did not represent the optimum habitat for either species. In fact, competition was minimal since each species showed preference for different parts of the habitat. *G. senegalensis* spent more time in open 'orchard' bush, or at the periphery of the riparian bush and forest which was most utilised by *G. crassicaudatus.* No direct competition through antagonism or fear was shown by either species during interspecific encounters. Only in Rhodesia, where *G. crassicaudatus* had adapted to a much more open environment than elsewhere, were there indications of competition for food, and here the population density of *G. senegalensis* was surprisingly low.

The highest population densities of *G. crassicaudatus* were found in humid, forest regions, while *G. senegalensis* was most abundant in arid thornveld. Some of the factors which bring about ecological separation are outlined below.

Social organisation

The majority of nocturnal prosimians so far studied, in contrast to other primates, are predominantly solitary in their habits. Their social behaviour may therefore be considered as being less complex than in animals which live together in larger groups forming social units. Both *G. senegalensis* and *G. crassicaudatus*, however, formed relatively stable associations throughout the year and familiar individuals often slept together during the day. Social interactions included all grooming, play, sexual behaviour, and quite complex patterns of communication. Each species made approximately 20 calls of social significance.[5,6]

Both species of bushbaby slept either alone or in small family groups of 2-6 animals, which often included an adult male, an adult female, and the offspring of one or more generations. It was found that the males and maturing offspring would often sleep alone, while the females and youngsters would usually sleep together.

Occurrence of groups in: (*a*) *G. senegalensis* (N = 119), and (*b*) *G. crassicaudatus* (N = 148) was as follows:

Number of animals in group:		1	2	3	4	5	6
% frequency of occurrence:	(*a*)	40	30	23	5	1	1
	(*b*)	46	26	18	6	3	1

At night, the members of sleeping groups behaved differently in the two species:

1. G. senegalensis

Individuals which had spent the day together generally split up at night and sometimes did not come together again until just before dawn. Social contact, including that between a mother and her offspring, was typically brief, with exceptions during courtship and when juveniles had developed sufficient agility to be able to follow their mothers before their movements became independent. Records show that bushbabies of this species were alone on average for 70% of each night.[7]

2. *G. crassicaudatus*

This species was somewhat gregarious in that the female and her offspring moved together at night as a cohesive group. Even when the young were too small to follow the mother, she carried them from place to place, either transporting them in her mouth one at a time or with them clinging to the fur of her back. The adult male generally remained alone, but he occasionally joined the family group and moved with them. Even larger groupings could be formed by the addition of previous offspring or members of other groups, the latter usually being the result of attraction to a particular food source.

The social systems of *G. senegalensis* and *G. crassicaudatus* are comparable to those described for *Lepilemur mustelinus* by Charles-Dominique and Hladik,[8] and for *G. demidovii, G. alleni, Euoticus elegantulus, Perodicticus potto* and *Microcebus murinus* by Charles-Dominique and Martin.[9] The animals are loosely organised into 'family groups', which may be defined as comprising those bushbabies whose individual ranges coincided or overlapped to a large extent and who would share the same sleeping place, at least on occasions.

It was found that the size, shape and number of home-ranges changed gradually during the year. It appears that as the young males matured their ranges became increasingly separate from those of the parents and they began to show territorial behaviour in the form of olfactory marking, vocal advertisement and antagonism towards rival conspecifics. Pair bonds developed with females, which also showed territorial behaviour, and the larger and stronger males were usually able to establish their home-ranges over those of one or several females. In this way, social groups were formed, having communal home-ranges, while the smaller males were rejected to the periphery. Within an expanding population, such a social system would cause a species to become spread throughout the available habitat and prevent overcrowding in any one place. It is suggested by Charles-Dominique[10] that the social structures of monkeys and other social mammals may be derived from this primitive type.

Use of space

Measurements were made of the home-range sizes of 'family groups' in the two main study areas. *G. senegalensis* had an average range size of 2.8 hectares while that of *G. crassicaudatus* was 7 hectares. The

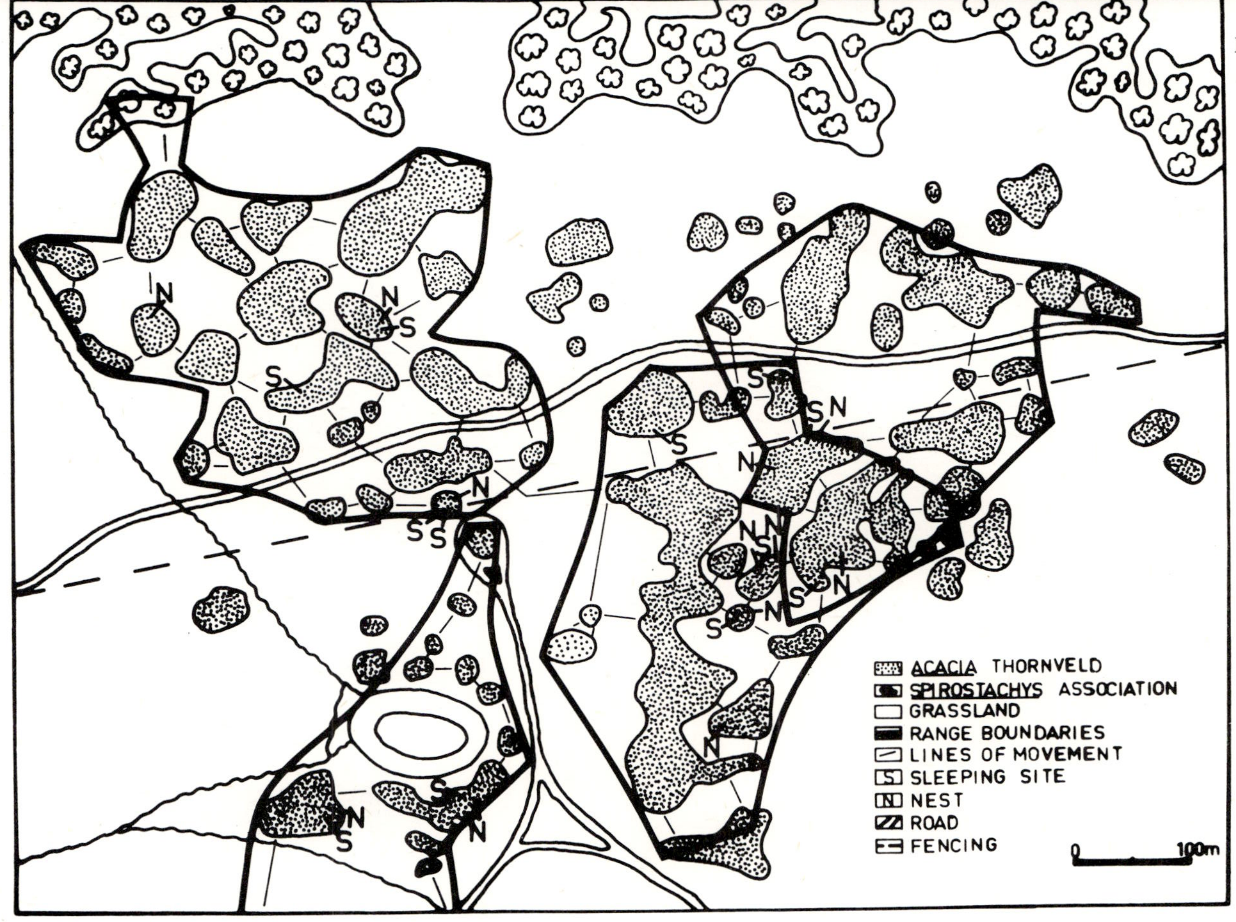

Figure 1. Home-ranges of four *G. senegalensis* groups, showing nests and sleeping places over a six-month period.

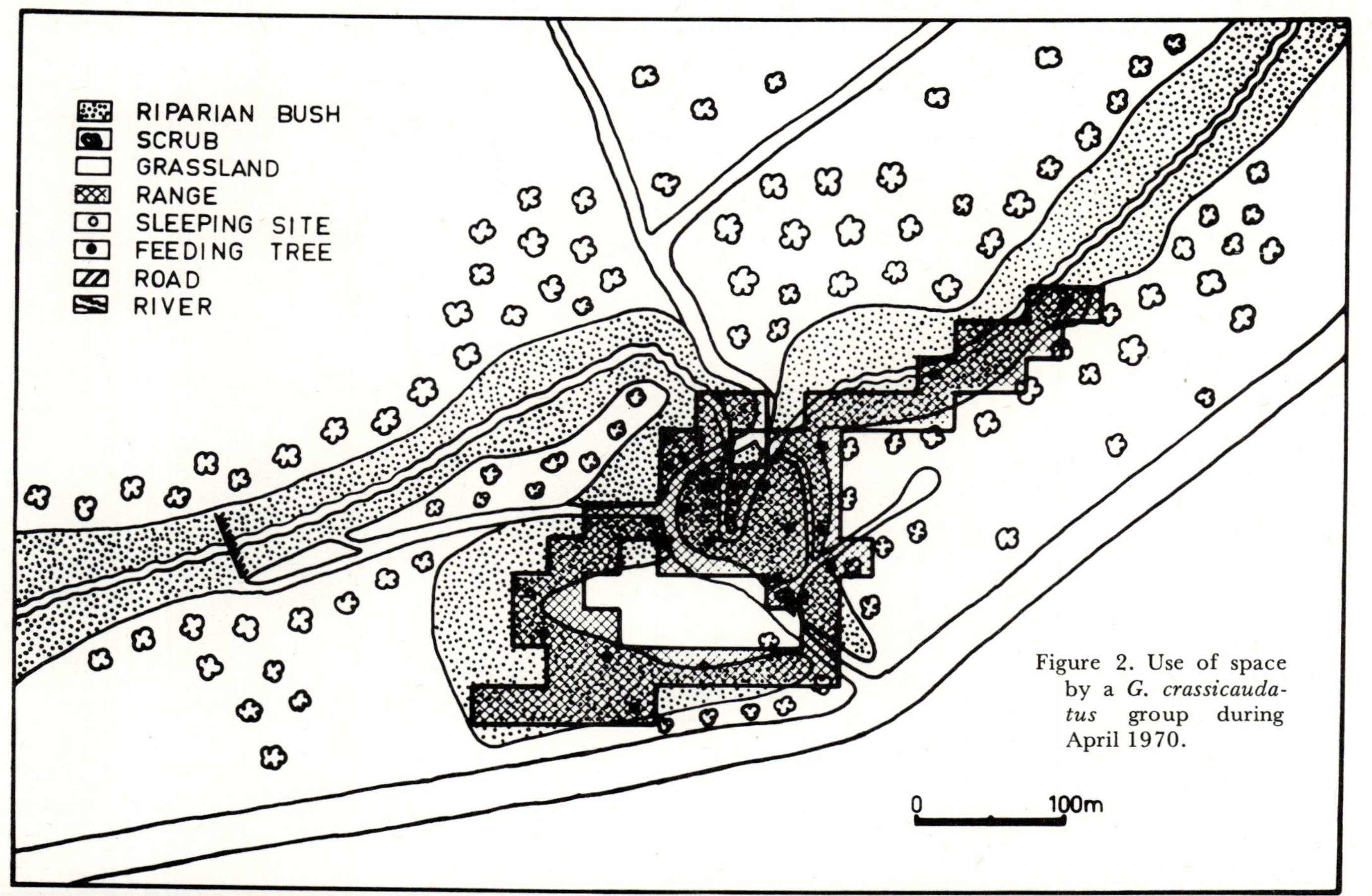

Figure 2. Use of space by a *G. crassicaudatus* group during April 1970.

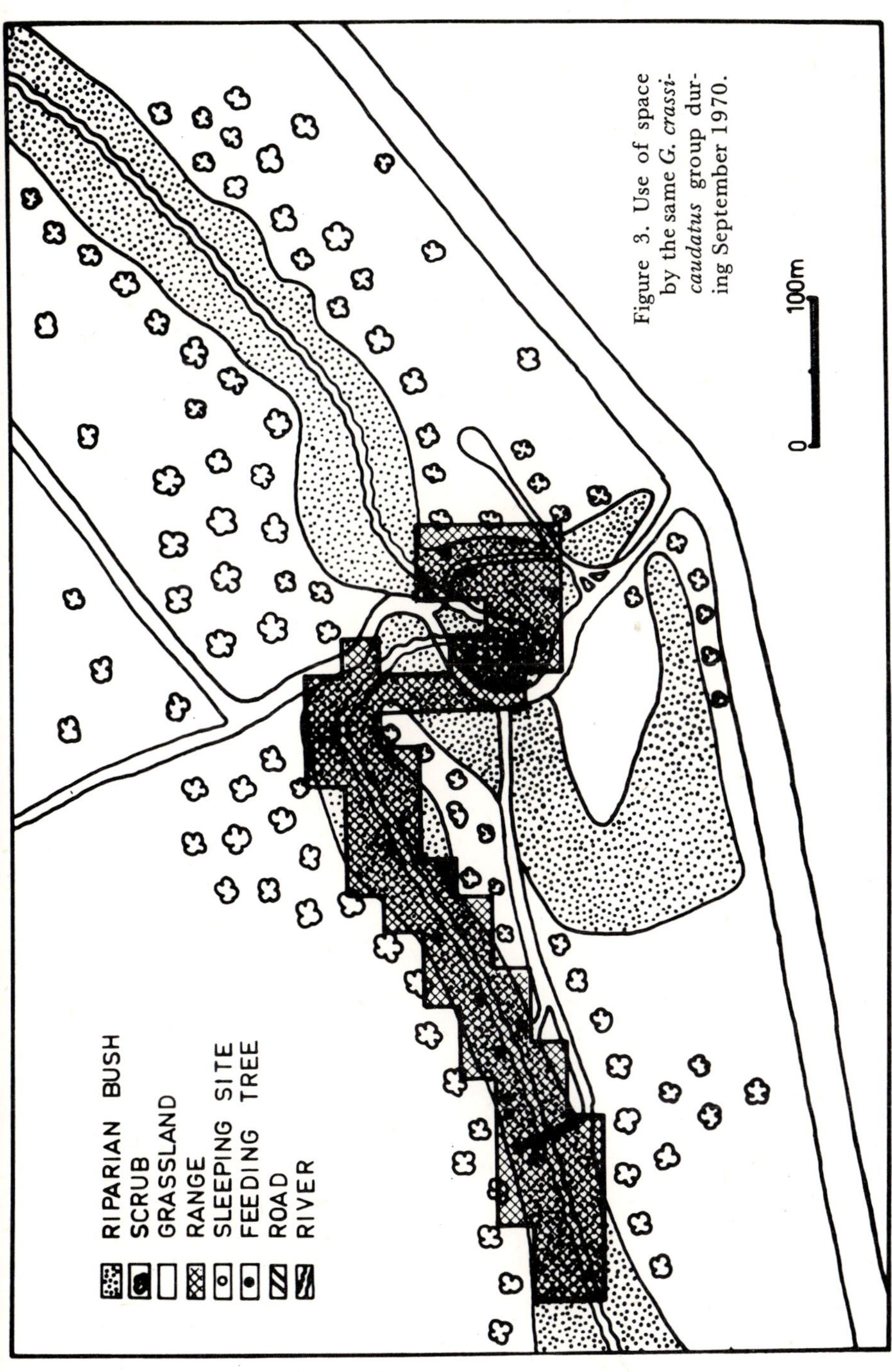

Figure 3. Use of space by the same *G. crassicaudatus* group during September 1970.

home-ranges of neighbouring groups were separate or overlapped to a small extent (Fig. 1).

The use of space by each animal may be analysed using a grid system (Figs. 2 and 3). The length of time which was spent in each square of the grid may be calculated for a single night and for consecutive nights at different times of the year. Activity is found to be centred around particular trees or groups of trees which provided sleeping sites, food sources or resting places, the use of which varied during the year depending upon their suitability. Such trees were not uniformly distributed throughout the range, and the centres of activity were therefore found at various unrelated points.

Bushbabies were active at all levels within their habitat, but the amount of activity at different levels varied between one habitat and another, depending on the spacing of trees and the availability of food. Those animals of both species which lived in open woodland descended to the ground readily in order to cross open spaces of up to 100 m. or to forage. In forest environments they only descended occasionally and with greater reluctance.

Each species used a variety of supports, but *G. senegalensis* – owing to its small size – could move along fine, closely-spaced and often very thorny branches without difficulty. It was thus able to exploit a large number of habitat types, and to a greater extent, than was *G. crassicaudatus*.

Movement and locomotion

Movements within the home-range were in accordance with the observations of Brown on other small mammals.[11] They were directional from one point to another, following habitual 'pathways'; but the routes taken and the exact movements made were often varied depending upon circumstances, indicating that the animals had a thorough knowledge of their entire range. One or more circular excursions were sometimes made during a night, starting at one point and returning to that point from the opposite direction some hours later. Alternatively, the same route was used to return as that used on the outward journey.

The movements of *G. crassicaudatus* differed from those of *G. senegalensis* in being generally more stereotyped. The pathways were followed more strictly, with similar movements being made on consecutive occasions. The more haphazard movements of *G. senegalensis* may be an adaptive feature related to its insectivorous diet. The food sources of *G. crassicaudatus*, which is predominantly frugivorous, were at fixed points within the home-range, and movements were made directly from one point to another. However, this did not prevent fairly continuous exploration of the home-range, so that food trees which were only productive at certain times of the

Table 2 A comparison of locomotion and diet in *Galago senegalensis*, *G. crassicaudatus* and *Perodicticus potto* [1 2]

Species	*Primary method of locomotion*	*Secondary method of locomotion*	*Movement on ground*	*Diet*
Galago senegalensis	Rapid leaping, using long hind legs as propulsive force	Slower 'stepping' movement within a tree	'Kangaroo-like', forelimbs not used	Insects and gum
Galago crassicaudatus	Quadrupedal running with occasional short jumps, or longer ones between trees, using hind legs as propulsive force	Slow, stealthy and quiet movement, sometimes beneath a branch. Weight transferred evenly forwards	1. 'Kangaroo-like', forelimbs not used 2. 'Galloping' action with both fore and hind feet together 3. Walk or run. Hindquarters and tail in the air	Fruit and gum Occasional slow-moving insects and birds
Perodicticus potto	Slow, stealthy and quiet movement, sometimes beneath a branch. Weight transferred evenly forward	Occasional running. No leaping	Walk. As along a branch	Fruit, gum, slow-moving insects and occasional small vertebrates

Figure 4. Locomotor postures in *G. senegalensis* (sketches 1 to 8) and *G. crassicaudatus* (sketches 9 to 14).

Figure 5. Potto-like movement in *G. crassicaudatus.*

year were not missed. Differences between the locomotor patterns of the two species may also be related to their diets.

If a comparison is made between the extreme form of active leaping locomotion shown by *G. senegalensis* and its associated methods of obtaining food with that of *G. crassicaudatus* and the slow moving *Perodicticus potto*.[12] it can be seen that the type of locomotion and feeding behaviour of *G. crassicaudatus* lies somewhere between that of the other two species, showing characteristics of both (Table 2; Figs. 4 and 5).

The most common method of locomotion in this species was not the rapid leaping gait characteristic of other galagos, but a 'monkey-like' running or walking along the tops of branches, with only short jumps when moving rapidly. Longer jumps were made when the occasion demanded, but these were generally in a downward direction with little forward propulsion and they differed from those of *G. senegalensis* in their execution.

It has been shown from the fossil remains of lorisoids from the Miocene of east Africa by Walker[12] that the leaping form of locomotion is a primitive state in the African prosimians, while stealthy locomotion is assumed to be a post-Miocene development, which makes the lorises dependent upon true forest conditions. It is of interest to note that while *G. crassicaudatus* may be found in relatively open areas it is more typically confined to dense bush and forest habitats, at least in the southern parts of its range.

Marking and rubbing behaviour

Six types of marking or rubbing actions have been observed which may carry olfactory clues of importance in territorial and social behaviour (Table 3). Some field notes are recorded here on the more conspicuous patterns, while the possible functions of marking actions are discussed by Doyle, and Andrew and Klopman.[13,14]

1. *G. senegalensis*

(*a*) *Urine-washing.*
This is a habit common to several nocturnal prosimian species and is described in detail elsewhere.[13,14] It was seen in all age classes and occurred several times in quick succession at the beginning of the active period, but thereafter only a few times during each hour of the night. Urine-washing at particular spots, apart from around the sleeping place, was not observed, and the animals showed a lack of obvious 'sniffing' when moving quickly through their home range. The idea of scent 'trails' has been discounted.[7,13] The deposit of urine on the branches may inform conspecifics of the presence, status and sexual condition of the marker, or it may be of significance to the animal itself. Adult males sometimes urine-washed in contact with conspecifics and in this way may have transferred their scent onto members of their 'family group'.

(*b*) *Chest rubbing.*
Chest rubbing was only performed by adult males, which had a triangular bald patch on their chests. It has yet to be established whether this patch incorporates glandular tissue. Rubbing, which was sometimes accompanied by biting of the branch, occurred in a similar fashion and context to that described for *G. crassicaudatus*. Rubbing of the head and mouth was also observed, but the significance of these movements is not known.

2. *G. crassicaudatus*

(*a*) *Urine-washing.*
Urine-washing was similar to that of *G. senegalensis*. It was performed by all age classes and presumably has the same significance. Both species were observed to urinate without urine-washing or rhythmically micturating.

(*b*) *Chest rubbing.*
Both adult males and females have a chest gland which may be seen as a longitudinal bald patch in the middle of the chest. The patch is larger and more conspicuous in the males. Chest rubbing was performed in a variety of postures, depending on the size and position of the branch or trunk which was being rubbed. The pattern was often executed beneath a branch in an upward-sliding movement. Particular places within the home-range were rubbed on repeated occasions, with a preference being shown for small upright projections and stumps of branches in large trees. Rubbing was frequently accompanied by sniffing of the branch, leg-rubbing, ano-genital rubbing and biting. It occurred during social encounters, when several places within a small area were sometimes marked one after another. Most chest rubbing observed was done by males, and it would appear to be one aspect of territorial behaviour in the species.

(*c*) *Ano-genital rubbing.*
Ano-genital rubbing was observed infrequently. The chest and whole underside of the body, including the genital region and sometimes the anus was rubbed along a branch in a forward sliding motion. It was sometimes performed beneath a branch with the legs and feet free on either side and the ano-genital region pressed against the branch by the pull of the arms. It was observed for both adult males and females and presumably has territorial or sexual significance.

Table 3 Marking and rubbing behaviour shown by age/sex classes of *G. senegalensis* and *G. crassicaudatus*

No.	*Behaviour*	*Species*	*Age class*	*Sex*
1	Urine-washing	*G. s.*	Ad. Juv. Inf.	Male and female
		G. c.	Ad. Juv. Inf.	Male and female
2	Rhythmic micturition	*G. s.**	Ad.	Male and female
		*G. c.**	Ad.	Male and female
3	Chest rubbing	*G. s.*+	Ad.	Male
	(median chest gland)	*G. c.*	Ad.	Male and female
4	Head and mouth rubbing	*G. s.*	Ad.	Male and female
5	Ano-genital rubbing	*G. s.**	Ad.	Male and female
		G. c.	Ad.	Male and female
6	Leg rubbing	*G. c.*	Ad. Juv.	Male and female

* Not noticed under conditions of observation in the field, but observed in the laboratory.[5,6]

\+ Existence of glandular tissue not definitely established.

(*d*) *Leg rubbing.*
Leg rubbing was a common behaviour pattern for which it is difficult to find an explanation. It involved a rapid scraping of the basal part of the hind limb and sole of the foot against a branch or sloping trunk. This action was then often repeated with the other leg. The leg was moved noisily back and forward several times, sometimes with the foot clasped gently round the branch. Leg rubbing was done by juveniles as well as adults, but was seen most often with adult males.

Despite the similarity of this action with that of scent-marking, and its possible significance in transferring urine from the soles of the feet onto the support, it did not show any temporal relationship with scent-marking. It was common during social encounters, together with chest rubbing, and occurred when the subject was confronted by the observer or a potential predator, indicating that it may be a displacement activity. The sound of scraping may serve a communicatory function.

Sleeping habits

At the end of each night's activity, the members of a group returned to particular sleeping trees which were used at various times during the year, with some of them being favoured more than others. Some trees were used by several generations.

Use of sleeping places by (*a*) *G. senegalensis* (N = 168), and (*b*) *G. crassicaudatus* (N = 35), were as follows:

Type of sleeping site:		*Nest*	*Branch/Fork*	*Building*	*Hollow*
% frequency of use:	(*a*)	52	40	2	6
	(*b*)	8	91	1	0

1. *G. senegalensis*

Members of this species were found to sleep in nests or on branches and forks within a single habitat with equal regularity. Buildings and hollow trees were also used. Sleeping places were usually between 4 m. and 6 m. above the ground in dense, thorny trees (*Acacia tortilis; A. nilotica*), and the nests (which were of a size corresponding to the number of occupants) consisted of platforms constructed of flat leaves taken from neighbouring trees. The nests differed from those of *G. senegalensis* in South-West Africa, which were globular in shape and covered on top.[15]

Up to 13 sleeping places were used in any one home-range during a year. Nests were often renewed within the same tree or close by and the animals frequently slept in the nest tree but not on the nest, so that records of the positions of sleeping sites showed that they occurred in clusters.

During the year, with changes in temperature and leaf-cover, particular trees became more or less suitable for sleeping purposes, and this was reflected by a change in their use. It was not uncommon for more than one sleeping place to be used on consecutive days or by different individuals within one home-range.

The choice of sleeping place seems to depend on several factors, some of which may be sacrificed at the expense of others. These included: direct protection from predators and extremes of temperature; indirect protection from predators by concealment; and comfort.

Although activity at night was unaffected by weather conditions, voluntary movement during the day appeared to occur in response to changes in temperature. Cover was sought during the heat of the day, particularly in the summer, while the animals often slept in exposed conditions in winter.

Laboratory data indicates that nests are built by females during a period just before giving birth and for some time afterwards.[16] Nests were found in the field during the spring and summer at the beginning of the birth season when suitable nesting material became available. A few infants, which were born just before suitable nesting leaves could be obtained, were found in old birds' nests and hollow trees. The nests were probably used as platforms on which to give birth and to provide support and camouflage for the young, and thereafter they continued to be used even though they were no longer essential for support of the infants. Haddow and Ellice note that in east Africa nesting is rare and appears to be a vestigial habit.[17]

2. *G. crassicaudatus*

This species also used more than one type of sleeping place. The use of hollow trees was not observed, but has been recorded by Astley-Maberly.[18] In most cases, bushbabies concealed themselves in dense tangles of branches and creepers at a height of between 5 m. and 12 m. above the ground. A number of different trees were rendered suitable by the fact that they supported dense growths of climbers such as *Dalbergia armata* and *Pteralobium* sp. Only occasionally did the animals sleep on branches which left them exposed. Nests were only seen when a female had young infants, and consisted of inaccessibly leafy platforms, somewhat depressed in the centre and sheltered from above by foliage.

Up to 12 sleeping places were found within each home-range and, as with those of *G. senegalensis*, they occurred in clusters and their use followed a similar pattern. Movement away from the sleeping place during the day was not observed.

Field observations indicate that *G. senegalensis* relies on its ability to move rapidly away from danger should it be disturbed during the day. *G. crassicaudatus*, on the other hand, generally sleeps in confined places, relying on concealment to avoid detection by potential predators.

Birth periodicity

Climatic factors

Birth records of *G.s. moholi* were made in an area (Lat. 24°35′, Long. 28°47′) only 240 km. from the recording area for *G.c. umbrosus*, which lies a few degrees north of the Tropic of Capricorn. These areas have a seasonally arid climate (Table 4). The winters (June to September) are dry and sunny with a large daily temperature range (up to 17°C). The hot summer rainy season occurs between October and April, with thunderstorms contributing a large proportion of the total precipitation, although spells of continuous rain are also recorded. The distribution and annual total of rainfall is highly variable from year to year. The climate is affected by the varied local topography and aspect. All valleys are subject to winter temperature inversions and the daily temperature range may cause marked changes in the relative humidity, dew occurring in autumn following the summer rains. There is considerable dying-back of the vegetation during the winter and leaf growth begins again before the first good rains, with sometimes severe consequences if they are late. Long dry spells may occur between rains.

1. *G. senegalensis*

In the study area for this species, temperatures may range between -6°C and +38°C with an annual absolute mean of 15.5°C. Only 464 mm. of rain was recorded during the year of study (135 mm. below average), with 449 mm. falling between the months of October and April. Nesting material was not available until November, which may have been an important factor influencing the survival of infants.

The occurrence of births in the laboratory has been described by Lowther, Sauer and Sauer and Doyle et al.[15,16,19] The gestation period has been established as between 120 and 126 days, and twins are usually produced after the first birth, which is often a singleton. Post-partum oestrus is common. In the main study area, 13 sets of twins and 7 singletons were observed.

Restricted breeding has often been claimed or suggested for *G. senegalensis*, while Manley has shown the potentiality of this species for recurrent oestrus cycles.[20] Butler notes that *G. senegalensis* appears to have an oestrus cycle lasting four to six weeks recurring

Table 4 Mean temperatures (t, °C), precipitation (p,mm) and relative humidity (h,%) for the study areas of (a) *G. senegalensis* and (b) *G. crassicaudatus*

(a) *Month*	*t,°C* *Max.*	*Min.*	*p,mm.*	*h,%* *0800 hr.*
J	29.4	17.0	120	67
F	29.0	16.6	101	70
M	27.7	14.8	73	72
A	26.4	11.2	35	72
M	23.5	6.4	14	72
J	20.9	3.1	6	70
J	21.0	3.0	5	66
A	24.5	5.8	5	61
S	27.5	10.1	17	53
O	29.7	13.8	46	55
N	29.8	15.4	84	61
D	30.3	16.7	108	63
YEAR	26.7	11.2	614	65

(b) *Month*	*t,°C* *Max.*	*Min.*	*p,mm.*	*h,%* *0800 hr.*
J	29.1	17.7	352	82
F	28.6	17.7	307	85
M	27.7	16.4	280	88
A	26.8	13.4	95	87
M	25.1	8.4	44	86
J	23.2	5.1	20	85
J	23.1	5.0	29	83
A	24.8	6.8	28	79
S	27.0	10.0	49	73
O	28.8	13.4	109	73
N	29.0	15.4	189	78
D	29.8	16.8	266	80
YEAR	26.8	12.2	1768	80

throughout the year.[21] The intensity of sexual activity varies during the year, so as to give the appearance of one or more restricted breeding seasons.

The present data suggest that there are two separate birth and mating periods for this species in South Africa. New-born infants were found in October and early November and again at the end of January until March. The majority of infants were born during February. It can be deduced that matings occurred during mid-winter (June and July) and they were subsequently witnessed at the end of October and the beginning of November, which is in accordance with the timing of the second birth peak.

It has already been noted that the timing of births and nest-building activity coincided with the onset of the rainy season, when food supply, nest trees and nesting materials became abundant. It is possible that the onset of the first birth period varies from one geographical area to another. The existence of a double birth-peak may be partly explained by the fact that female bushbabies do not become sexually mature until approximately 200 days old, so that those which are born in February are not receptive until September, which precludes them from giving birth in October or November. On the other hand, those females born in November can come into oestrus by May, and with a post-partum oestrus are capable of giving birth twice during the following summer.

2. *G. crassicaudatus*

Within the study area, temperatures range between -2.8°C and 41.9°C with an annual absolute mean of 19.5°C. 81 mm. of rain was recorded between April and September 1970, while 854 mm. fell before the following April. During this time the evergreen bush became lush and almost impenetrable.

G. crassicaudatus had only a single birth season. All infants were born at the beginning of November and of 16 infants there were 6 sets of twins, one set of triplets and one singleton.

Buettner-Janusch records the gestation period as being approximately 18 weeks with no defined birth season in the laboratory.[22] Seasonal births in the wild may be largely the result of environmental factors. The period of birth and lactation again coincided with the summer months, when rainfall and temperature were high, while food and dense cover were readily available. It may be assumed that mating took place at the beginning of July.

Discussion

Those factors which seem to be important in determining the number and type of bushbabies which may be supported in a particular habitat, as with many other mammal species, include: the availability and suitability of cover, sleeping sites and food; the presence and abundance of predators; and competition with other species. Social behaviour, particularly territoriality, may affect the spacing of members of a population, while the number and frequency of births has a direct effect on population numbers. Local events, such as disease epidemics, veld fires and bush destruction for development schemes, may have a disastrous effect on the species, which lack the ability to move over very long distances.

G. senegalensis and *G. crassicaudatus* showed tolerance to a wide range of temperatures; they had a basically similar social structure, with family groups in both species showing territorial behaviour in the form of olfactory marking, vocal advertisement and aggression towards rival conspecifics. They both bred during the summer months when food and cover were readily available; each utilised all levels within the habitat, and their feeding habits overlapped to some extent. The adaptability of both species is illustrated by their occurrence in and around human habitations.

Comparison of the ecology and behaviour of allopatric populations of the two species indicate that they are primarily adapted to different habitat types for which they showed a distinct preference. The smaller bushbaby, *G. senegalensis*, was found in a greater variety of habitats and at higher population densities than *G. crassicaudatus*. It bred more rapidly, required less food of a type which was widespread in its occurrence, and had a smaller home range. In addition it had greater agility, less stereotyped movements, fewer requirements for dense cover and was able to make use of smaller branches than *G. crassicaudatus*.

The low population densities of one or both species in regions where they lived sympatrically may be accounted for by the fact that these areas did not represent the optimum habitat for either species. Yet *G. senegalensis* was absent from some apparently suitable forest areas where *G. crassicaudatus* was abundant, and direct competition cannot be discounted.

Acknowledgments

This research was supported by a grant from the National Geographic Society, Washington, D.C., as well as by grants from the University Development Foundation and the University Council, University of the Witwatersrand, the Human Sciences Research Council and the Department of Cultural Affairs of the Republic of South Africa.

NOTES

1 Charles-Dominique, P. (1971*a*), 'Eco-éthologie des prosimiens du Gabon', *Biol. Gabon.* 7, 121-228.

2 Osman Hill, W.C. (1953), *Primates: Comparative Anatomy and Taxonomy. Vol. I: Strepsirhini*, Edinburgh.

3 Shortridge, G.C. (1934), *The Mammals of South West Africa.* London.

4 Charles-Dominique, P., this volume.

5 Andersson, A. (1969), 'Communication in the Lesser Bushbaby (*Galago senegalensis moholi*)', unpublished MSc thesis, University of the Witwatersrand.

6 Bearder, S.K. (in prep.), 'Aspects of the ecology and behaviour of the Thick-tailed Bushbaby, *Galago crassicaudatus*', unpublished PhD thesis, University of the Witwatersrand.

7 Bearder, S.K. (1969), 'Territorial and intergroup behaviour of the Lesser Bushbaby (*Galago senegalensis moholi*, A. Smith), in semi-natural conditions and in the field', unpublished MSc thesis, University of the Witwatersrand.

8 Charles-Dominique, P. and Hladik, C.M. (1971), 'Le lepilemur du Sud de Madagascar: écologie, alimentation et vie sociale', *Terre et Vie* 25, 3-66.

9 Charles-Dominique, P. (1972), 'Ecologie et vie sociale de *Galago demidovii* (Fischer 1808; Prosimii)', *Z.f. Tierpsychol*, Beiheft 9, 7-41; Martin, R.D. (1972), 'A preliminary field study of the Lesser Mouse Lemur (*Microcebus murinus*, J.F. Miller 1777)', *Z.f. Tierpsychol.*, Beiheft 9, 43-89.

10 Charles-Dominique, P. (1971*b*), 'Sociologie chez les lémuriens', *La Recherche* 15, 780-1.

11 Brown, L.E. (1966), 'Home range and movement of small mammals', in P.A. Jewell and C. Loizos (eds.), *Play, Exploration and Territory in Mammals*, London.

12 Walker, A. (1969), 'The locomotion of lorises, with special reference to the potto', *E. Afr. Wild. J.*, 7, 1-5.

13 Doyle, G.A., this volume.

14 Andrew, R.J. and Klopman, R.B., this volume.

15 Sauer, E.G.F. and Sauer, E.N. (1963), 'The South West African Bushbaby of the *Galago senegalensis* group', *J. S. W. Africa scient. Soc.*, 16, 5-36.

16 Doyle, G.A., Pelletier, A. and Bekker, T. (1967), 'Courtship, mating and parturition in the Lesser Bushbaby (*Galago senegalensis moholi*) under semi-natural conditions', *Folia primat.* 7, 169-97.

17 Haddow, A.J. and Ellice, J.M. (1964), 'Studies on Bushbabies (*Galago* spp.) with special reference to the epidemiology of yellow fever', *Trans. Roy. Soc. trap. Med. Hyg.* 58, 521-38.

18 Astley-Maberly, C.T. (1967), *The Game Animals of South Africa*, 2nd ed, Cape Town.

19 Lowther, F. de L. (1940), 'A study of the activities of a pair of *Galago senegalensis moholi* in captivity, including the birth and post natal development of twins', *Zoologica*, N.Y., 25, 433-59.

20 Manley, G.H. (1965), 'Reproduction in lorisoid primates', *J. Reprod. Fert.* 9, 390-1.

21 Butler, H. (1960), 'Some notes on the breeding cycle of the Senegal galago, *Galago senegalensis senegalensis*, in the Sudan', *Proc. zool. Soc. Lond.* 135, 423-4.

22 Buettner-Janusch, J. (1964), 'The breeding of galagos in captivity and some notes on their behaviour', *Folia primat.* 2, 93-110.

P. CHARLES - DOMINIQUE

Ecology and feeding behaviour of five sympatric lorisids in Gabon

Introduction

This study was carried out in Makokou (Gabon) in the period 1965-1969, on the basis of four study visits of 7, 8, 14 and 3 months' duration, respectively. Field-work was carried out from the CNRS Laboratory in Gabon (originally known as the Mission Biologique au Gabon, under the direction of Professor P.P. Grassé), which is now referred to as the Laboratoire de Primatologie et d'Écologie Equatoriale (Director: A. Brosset).

The study region is located in the heart of the Congolese block of rain-forest (Ogoue-Ivindo basin). It lies 550 km. from the Atlantic coast and 0.4° latitude north of the equator. Along the access routes

Figure 1. A potto (*Perodicticus potto*) moving around on large-diameter lianes in the forest canopy. Note the strongly reflecting tapetum of each eye, and the reduction of the tail (characteristic of Lorisinae).

Figure 2. An angwantibo (*Arctocebus calabarensis*) moving around in dense undergrowth, using fine lianes attached to smaller trees. Note the reduction of the tail and the use of all four grasping extremities in locomotion.

Figure 3. Demidoff's Bushbaby (*Galago demidovii*) moving through dense vegetation composed of inter-twined branches and lianes.

(roads, certain water-courses, etc.), the forest has been degraded for cultivation which has been subsequently abandoned at various times, thus giving rise to areas of secondary forest at different stages of reconstitution. Primary forest is found a few km. away from these inhabited zones.

Five prosimian species live sympatrically in Gabon; two lorisines:

1. *Perodicticus potto edwardsi* (Bouvier, 1879). Body-weight 1100 gm. Head + body length 327 mm.; tail-length 52 mm. (Fig. 1).
2. *Arctocebus calabarensis aureus* (De Winton, 1902). Body-weight 200 gm. Head + body length 244 mm.; tail-length 15 mm. (Fig. 2).

and three galagines:

1. *Galago demidovii* (Fischer, 1808). Body-weight 61 gm. Head + body length 123 mm.; tail-length 172 mm. (Fig. 3).
2. *Galago alleni* (Waterhouse, 1837). Body-weight 260 gm. Head + body length 200 mm.; tail-length 255 mm. (Fig. 4).
3. *Euoticus elegantulus elegantulus* (Le Conte, 1857). Body-weight 300 gm. Head + body length 200 mm.; tail-length 290 mm. (Fig. 5).

(These figures represent averages based on 33, 30, 66, 17 and 39 specimens respectively.)

The Lorisinae and the Galaginae represent two quite distinct subfamilies: the former are exclusively slow climbers which never leap and always move around slowly and cautiously, whilst the latter are rapid and vigorous leapers.* These behavioural differences are paralleled by numerous anatomical adaptations, primarily affecting the limbs and the tail. In particular, the tarsus is extremely developed in the Galaginae, whilst the tail is very reduced in Lorisinae. By contrast, the skull, the dentition, the digestive tract and the reproductive organs exhibit only very slight differences which would not, in themselves, justify a separation into two subfamilies.

The leaping specialisation of the bushbabies is generally regarded as providing a means of rapid escape, at the same time permitting exploitation of an extensive home-range. On the other hand, the slow and deliberate gait of the lorises has been given different interpretations by different authors: '. . . high-grade specialisations for an arboreal existence' (Hill);[1] '. . . directly related to catching prey

* Many authors have exaggerated the locomotor performance of the bushbabies. Sanderson,[5] in particular, writes of a leap of over 10 m. made by *E. elegantulus*, with the animal purportedly gaining in height! We have measured leaps of 2 m. for *G. demidovii*, and of 2.5 m. for *G. alleni* and *E. elegantulus*, with take-off and landing at the same level. The present record is a leap of 5.5 m. (orthogonal projection of horizontal displacement) made by *E. elegantulus*, with a loss in height of 3.1 m.

Figure 4. Allen's Bushbaby (*Galago alleni*) clinging to a small-diameter, vertical support in the undergrowth. Note the long tail, which is used to effect minor corrections in leaps from one vertical trunk to another.

Figure 5. Needle-clawed Bushbaby (*Euoticus elegantulus*) in the process of eating gums on a branch in the forest canopy. Note the crouched body-posture and the application of the snout to the branch surface.

such as insects and roosting birds' (Walker).[2] Some authors have, in fact, expressed astonishment that these animals, which are apparently so vulnerable, have not been decimated by predators. From the author's observations in the forest,[3] it seems very likely that some kind of cryptic mechanism is involved, though this would of course operate only in the natural habitat of these species. (The eye, particularly that of raptors, is more sensitive to rapid movement than to slow progression, and many nocturnal arboreal predators are guided primarily by the auditory sense in localising their prey.) A cryptic mechanism would not, in any case, be the sole means of defence. As a last resort, when an encounter occurs, the lorises utilise active defence mechanisms which vary from species to species. The behaviour of the potto and the angwantibo towards predators has already been described, along with the morphological adaptations involved.[3]

In sum, the two lorisid groups have developed two radically different methods of escaping from predators. Whereas the bushbabies flee rapidly once detected by a predator, the lorises avoid detection by means of an elaborate pattern of slow locomotion. It will be shown that such adaptation has been associated with extensive modification of the feeding behaviour and the diet of the lorisines.

Ecology

The lorisids represent a tiny fraction of the forest mammalian fauna; there are 5 species among 120 mammal species living sympatrically at Makokou. Over and above this, these lorisid species live at relatively low population densities. Systematic counting along pathways and the results of trapping[3] indicate the following average densities per square km.:

Perodicticus potto	–	8 per square km.
Arctocebus calabarensis	–	2 per square km.
Galago demidovii	–	50 per square km.
Galago alleni	–	15 per square km.
Euoticus elegantulus	–	15 per square km.

Using the same techniques, we have calculated far higher densities for lemurs in Madagascar, where the prosimians represent the bulk of the mammalian fauna.[4] However, one must make allowance for heterogeneous distribution of populations, which occur as 'nuclei' ('noyaux') which can be separated to varying degrees. The extreme case seems to be that of the angwantibo, which is abundant in certain parts of the forest (7 per square km.) and virtually lacking over large areas. Nevertheless, we have observed the sympatric occurrence of three, four and even five of these prosimian species in some areas. In fact, numerous interspecific encounters were observed in the forest, without any sign of attack or escape behaviour. In general, any two lorisids of different species which encounter one another exhibit a brief bout of mutual observation and then continue on their way. Thus, any hypothesis of direct interspecific competition based on aggression must be discarded.

The lorisids are all nocturnal. Certain authors, on the basis of observations made in captivity, have suggested that they are partially diurnal or crepuscular.[5,6,7] Such observations must arise from artefacts of captivity, since all of the individuals followed in the forest showed themselves to be strictly nocturnal. In fact, although the lorises only move around during the night, the bushbabies exhibit some initial activity in twilight. But night falls so rapidly at the equator that the bushbabies are only moving around for ten minutes or so before the twilight has faded. The following values for the luminosity of the sky were measured at the times when activity began:

Euoticus elegantulus	–	300 to 100 lux
Galago demidovii	–	150 to 20 lux
Galago alleni	–	50 to 20 ux

Even then, it must be remembered that the light penetrating into the undergrowth of primary forest represents only one hundredth of the values measured above the canopy.

Quite erroneously – again as a result of observations in captivity – numerous authors have reached the conclusion that all lorisids use tree-holes as retreats. Of course, when placed in cages deprived of foliage, the animals do actually retreat into the only boxes placed at their disposition. However, of the five species studied, *Galago alleni* is the only one which sleeps in tree-holes under natural conditions. Usually, such tree cavities are in split hollow trunks, which can be entered from above. The bushbaby can spend the whole day clinging to the internal face of such a 'chimney'. Sometimes a rudimentary nest is built at the base with a few collected leaves. Interspecific competition for daytime retreats can be excluded; the observer has to examine a large number before finding one which is occupied. In addition, *Galago alleni* seems to be quite tolerant of other mammal species. We observed one individual sleeping a few metres away from an arboreal rodent (*Anomalurus erythronotus*) and a group of five bats in a hollow trunk of *Scyphocephalium ochocoa*. The two other bushbaby species sleep singly or in small groups on a small, leafy branch or in a tangle of lianes. *Galago demidovii* will also sleep in spherical nests constructed with green leaves.[3,8,9] The two lorisine species sleep singly on branches or lianes, protected by foliage.

The modern concept of the 'ecological niche' involves a large number of factors. Here, we will only examine the spatial localisation of the animals and their respective diets. These two factors cannot be dissociated, since food-seeking occupies the major part of the activity period of these lorisid species.

Spatial localisation

In order to define in an objective manner the localisation and nature of the supports utilised, we considered the following criteria:

height relative to the ground
orientation of the support
diameter of the support
nature of the support (ground, small trunk, liane base, large trunk, large branch, foliage, foliage mixed with lianes, lianes)

Every time a lorisid was encountered in the forest, we noted the various characteristics of the support on which the animal was *first* seen. (The presence of the observer could modify the selection of any subsequent pathway taken by the animal.) A large number of such observations (642 sightings of the five species) permitted statistical consideration of the data, which are discussed in detail in a previous publication.[3] All that will be given here is a brief description of the forest biotopes most frequently visited by each lorisid species (see Fig. 6):

1. *Perodicticus potto*

The potto inhabits the canopy (10-30 m. in primary forest; 5-15 m. in secondary forest). It is an exclusive climber, using supports with a wide range of sizes (1-30 cm. diameter). This permits the potto to pass from tree to tree by successively utilising large forks (20%) and small branches which interlock from one branch to another (21%). Lianes which permit short-cuts are also utilised (20%). The support orientations are: horizontals 39%, obliques 35%, verticals 26%. In exceptional cases, the potto descends to the ground (escape from a conspecific, crossing a deforested area, etc.). When this occurs, the potto is extremely cautious and heads directly for the nearest tree.

2. *Arctocebus calabarensis*

The angwantibo lives at 0-5 m., both in primary forest and in secondary forest, though it may rarely flee up to 10-12 m. when greatly alarmed. When progressing, this slow climber generally utilises small lianes passing between the small bushes in the undergrowth (43%). Quite frequently, the angwantibo is sighted in the foliage of

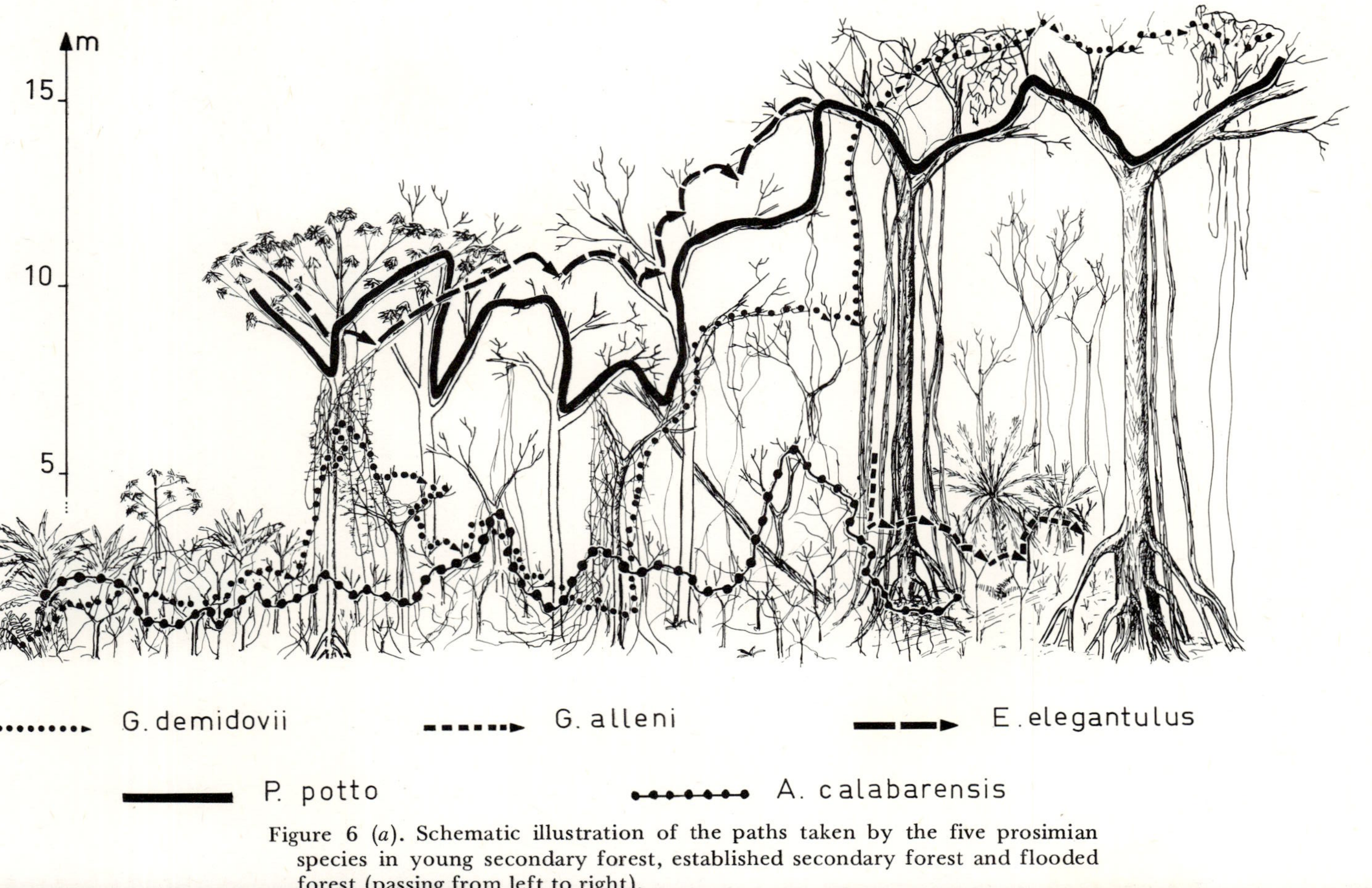

Figure 6 (*a*). Schematic illustration of the paths taken by the five prosimian species in young secondary forest, established secondary forest and flooded forest (passing from left to right).

Figure 6 (*b*). Similar illustration of the characteristic pathways taken by the five prosimian species in primary forest.

these small bushes (43%), where it hunts for insects. This species will descend quite readily to the ground to eat fallen fruits or to follow the tracks made by terrestrial mammals (4%). 40% of the supports utilised have a diameter of less than 1 cm.; 52% are between 1 and 10 cm. in diameter. Orientations of supports are: horizontals 20%, obliques 30%, verticals 50%.

3. *Galago demidovii*

This bushbaby lives primarily in dense vegetation invaded by small lianes (35%) and in foliage (25%). Accordingly, it is found at a great height in primary forest (10-30 m.) and low down in secondary forest (0-10 m.); the vertical distribution follows the distribution of the biotopes utilised. 64% of the supports used are less than 1 cm. in diameter; 25% are between 1 and 5 cm. in diameter. The support orientations are: horizontals 22%, obliques 30%, verticals 48%.

4. *Galago alleni*

This bushbaby species is more or less restricted to primary forest, living at a height of 0-2 m. It hunts for animal prey at ground-level, at the same time collecting fallen fruits; but locomotion consists primarily of leaping from one vertical support to another (small tree-trunks; bases of large lianes). The supports utilised have the following diameters: 9% less than 1 cm., 60% between 1 and 5 cm., 19% between 5 and 10 cm. The support orientations are: horizontals 6%, obliques 13%, verticals 81%.

5. *Euoticus elegantulus*

This bushbaby, which scarcely ever descends to the ground, lives in the canopy up to 50 m. However, it may descend to a height of 3-4 m. on certain large lianes which provide droplets of gum. Much progression is along large branches (62%) or large lianes (30%), and along the larger tree-trunks. The special structure of the nails, which terminate in small 'claws', permits movement head-downwards or head-upwards along large, smooth trunks where no other lorisid species would be able to maintain a grasp. It will be seen that this adaptation is related to the diet. The support orientations are: horizontals 22%, obliques 51%, verticals 27%.

To summarise the above information, it can be said that within each sub-family the closest species are ecologically separated by the height of the forest stratum exploited. In the Lorisinae, the potto exploits the canopy and the angwantibo lives in the undergrowth. Among the Galaginae, *Euoticus elegantulus* principally exploits the canopy, whilst *Galago alleni* inhabits the undergrowth. *Galago demidovii* provides a special case: this species depends upon dense vegetation, in which it can move easily thanks to its small size, and it follows the distribution of such vegetation (high up in primary forest and low down in secondary forest).

This latter example provides a good illustration of the fact that the height of the stratum exploited is not due to a 'height preference' exerted by each animal. It is an indirect consequence of choice of particular vegetation zones in each case. We were able to observe that the same applies to *Galago alleni*. After hand-rearing two young animals captured at the age of 1 week (at which stage they can only move a few centimetres), we found that, from the age of 1 month onwards, they sought out all supports with a vertical orientation (especially chair and table legs). Such selection, which has an apparent hereditary basis, is accompanied by a specific leaping 'style' which is adapted to this kind of support.[3] In addition, *Galago alleni* captured as adults and placed in secondary forest (where they scarcely ever occur naturally) moved around at heights of 10-15 m. They utilised vertical or oblique branches at this height when progressing, whereas the average height utilised in primary forest is 1-2 m. With *Galago alleni*, it seems very likely that the vertical orientation of supports of a certain size contributes greatly to the localisation of this species in the undergrowth zone of primary forest.

It will be seen that spatial localisation of these lorisid species is no more than the bare framework of ecological separation permitting closely related species to avoid dietary competition. In parallel with such spatial localisation, there has been dietary specialisation, which has particularly affected feeding behaviour.

Diet

Direct observation of lorisids under natural conditions is difficult. When they are surprised by the light-beam of a headlamp they remain immobile for a few seconds and then disappear rapidly into the vegetation. Under these conditions, analysis of the diet could only be carried out through examination of stomach contents. In Gabon, the prosimians are virtually untouched by hunters, and it was therefore possible to collect a large number of specimens without endangering the local prosimian populations. The animals were collected with the aid of a rifle, samples being taken at different times of the night and in all months of the year. This permitted us to study alimentary rhythms during the night as well as annual rhythms (examination of 174 digestive tracts for the 5 species). The food is quite finely chewed before swallowing; but it is still quite easy to separate the different constituents, which are mixed only to a slight extent in the stomach. We were thus able to separate animal prey from fruit and gums and to obtain fresh weights for these constituent fractions for each stomach examined. The fruits can be identified when the kernels are ingested, but it is often impossible to obtain a precise identification for animal prey. The analysis was therefore restricted to the level of large taxonomic groups: Coleoptera, Lepidoptera (both caterpillars and moths), Orthoptera, Hymenoptera (ants), Isoptera (termites), Myriapoda (centipedes and millipedes), Arachnida, Gasteropoda (slugs), Batracia.

After calculating for each species the percentage of the three principal dietary categories, we obtained the following results (Table 1):

Table 1 Stomach contents

	Animal prey	*Fruits*	*Gums*
Perodicticus potto	10%	65% (+ some leaves and fungi)	21%
Arctocebus calabarensis	85%	14% (+ some wood fibre)	–
Galago demidovii	70%	19% (+ some leaves and buds)	10%
Galago alleni	25%	73% (+ some leaves, buds and wood fibre)	– (small amounts only)
Euoticus elegantulus	20%	5% (+ some buds)	75%

If this first result were taken at face value, one might conclude that the bushbabies include one insectivore, one frugivore, and one gummivore, whilst the lorises represent one frugivore and one insectivore. Since the most closely related animals are separated by stratification of the vegetational zones utilised, it would seem that there can be no dietary competition. However, the problem is in reality more complex than this. The percentage figures do not take into account differences in body size of the five prosimian species concerned – that is, they do not reflect the real weight of food ingested. By calculating the average weight of each of the three dietary categories per stomach, the following results are obtained (Table 2).

Table 2 Weights of dietary components

	Animal prey	*Fruits*	*Gums*
Perodicticus potto	3.4 gm.	21 gm.	7 gm.
Arctocebus calabarensis	2.0 gm.	0.3 gm.	0 gm.
Galago demidovii	1.16 gm.	0.3 gm.	0.15 gm.
Galago alleni	2.2 gm.	9.2 gm.	(negligible)
Euoticus elegantulus	1.18 gm.	0.25 gm.	4.8 gm.

Although the differences for fruits and gums are very clear, this is not the case with animal prey, which are consumed in virtually equal quantities by all five species. Insects (which represent the bulk of the animal prey) are dispersed in the vegetation, and a hunting animal must explore large areas in order to capture large numbers of prey. A small bushbaby and a large bushbaby cover approximately the same distance in the course of the night, and thus each has approximately the same number of opportunities to encounter insects. The same applies to the potto and the angwantibo.

This mechanism has already been described by Hladik and Hladik[10] for the platyrrhines of Panama, which consume roughly the same absolute quantities of insects regardless of their own body-size. The smaller platyrrhines obtain almost all of their food by hunting animal prey, whilst the larger species have to supplement their diet with plant food.

Among the bushbabies, *Galago demidovii* (60 gm.) relies almost entirely on hunting, whereas the two larger species supplement their diets with fruits (*G. alleni*, 260 gm.) or with gums (*E. elegantulus*, 300 gm.). Among the lorises, the angwantibo (200 gm.) derives most of its food by hunting, whilst the potto (1100 gm.) augments its diet of insects with fruits and gums.

In captivity, lorisids are often seen to 'ignore' fruits when they have a large quantity of insects available. The same applies in the forest, where pottos which begin the night with a 'good hunting session' do not go on to eat fruits afterwards. All the individual pottos dissected which had 8-15 gm. of insect matter in the stomach had failed to eat fruits or gums. On the other hand, all individuals with less than 1 gm. of insect matter in the stomach had eaten 20-60 gm. of fruits and/or gums. In *Galago demidovii* the diet is observed to be slightly less insectivorous in the early part of the night, when the animals are still hungry, than in the second half of the night (35% fruits and gums devoured before midnight, and 20% thereafter). Accordingly, in the dry season (when insects are rarer) this species consumes 50% fruits and gums, as against 30% during the rest of the year.

Thus it seems probable that the availability of animal food is the major factor influencing the diet of these lorisid species. Naturally, secondary specialisations affect the feeding behaviour of the 'large' species oriented towards gums or fruits.

1. *Animal food*

In this analysis, we have considered the lorisines and the galagines separately. However, it is conceivable that the members of the two subfamilies may compete with one another in hunting insects. An examination of the categories of prey taken by each species shows that this is not the case. Table 3 shows, in order of importance, the different prey categories found in the stomach contents. The most common categories are followed by percentage figures based on the relative weights of animal food types in the stomachs.

Among the galagines, 78% of the prey are beetles, nocturnal moths and grasshoppers, whereas in the lorisines caterpillars and ants represent 70% of the prey. Pottos particularly prey upon ants (*Crematogaster* sp.) which release large quantities of formic acid, on centipedes (*Spirostreptus* sp.) which release large quantities of iodine, and on 'criquets puants' (malodorous orthopterans), which also emit repellent substances. The angwantibo feeds primarily on caterpillars, most of which are covered with stinging hairs. (The caterpillars eaten by the bushbabies never bear such hairs.) Thus, it would seem that the two lorisine species are specialised to tolerate 'noxious' prey left untouched by the galagines.

This special tolerance of the lorisines for 'noxious' prey permits the capture of a sufficient quantity with a minimum of movement from place to place. In fact, these efficiently protected 'noxious' organisms almost always display certain forms, colours or (especially)

Table 3 Animal prey in order of preference

	1	*2*	*3*	*4*	*5*
*Perodicticus potto** (N = 41)	ants (65%)	large beetles (10%)	slugs (10%)	cater-pillars (10%)	orthopterans (malodorous), centipedes, spiders, termites
Arctocebus calabarensis (N = 14)	caterpill-ars (65%)	beetles (25%)	ortho-pterans	—	dipterans, ants
Galago demidovii (N = 55)	small beetles (45%)	small nocturnal moths	cater-pillars (10%)	hemi-pterans	orthopterans, millipedes, homopterans, pupae
Galago alleni (N = 12)	medium-sized beetles (25%)	slugs	noc-turnal moths	frogs (8%) ants (8%)	grasshoppers, termites, millipedes, pupae caterpillars
Euoticus elegantulus † (N = 52)	grass-hoppers (40%)	medium-sized beetles (25%)	cater-pillars (20%)	nocturnal moths (12%)	ants, homopterans

* Examination of 41 stomachs from pottos did not reveal one case of vertebrate remains. However, whilst following a population of tame animals in the forest we were able to observe a female attempting to capture weaver-birds in a tree carrying a large number of nests. On another occasion, we surprised a female devouring a young frugivorous bat (*Epomops franquetti*). Capture of such prey must be relatively rare.

† No vertebrate remains were found in 52 *E. elegantulus* which were dissected; but we did once come across a tame individual eating a bird (*Camaroptera brevicauda*) in secondary forest.

odours, which normally signal their 'unpalatable quality' to potential predators. They are easily found, and their immobility permits easy capture. (In the case of ants of the genus *Crematogaster*, the potto follows moving columns, licking them up.) Over and above this, these prey organisms – which are ignored by many other predators – are abundant.

In this case, the lorisines are actually exhibiting a *tolerance* rather than a preference. If, in captivity, pottos or angwantibos are presented with their habitual prey alongside grasshoppers or moths normally eaten by galagos, they will eat the latter insects. In actual fact, angwantibos exhibit a particular form of behaviour prior to eating caterpillars. These animals are gripped by the head in the angwantibo's teeth, and the two hands 'massage' the prey for 10-20 seconds, with the result that many of the hairs are removed. Nevertheless, once it has eaten the caterpillar, the angwantibo spends some time wiping its snout and hands by rubbing them on a branch.

Despite their slow locomotion, which serves to protect them from predators, the lorisines succeed in capturing a sufficient quantity of animal prey to ensure a balanced diet. Proteins necessary for physiological equilibrium are found principally in animal prey or in the green parts of plants (leaves and buds). The primates are, in fact, either insectivorous/frugivorous (usually the more rapid species) or folivorous/frugivorous (usually the slower forms, with a few intermediate types).[10] Among the Lorisidae (insectivore-frugivores), the lorises exhibit an exceptional adaptation. These slow-moving animals have conserved the typical diet of the family Lorisidae by obtaining their proteins from a category of animal prey generally left untouched by insectivorous species.

One can therefore discount the idea of any competition between lorisines and galagines in predation on insects. A second contributory factor is important in this context: whereas the lorises capture resting, slow-moving prey, the bushbabies frequently capture rapid-moving prey, quite often when the latter are on the wing. In order to do this, the bushbaby projects its body forward with great rapidity, whilst maintaining a grasp on the branch with its hind-feet. The insect is trapped in flight with both hands, and the bushbaby returns to its starting position (immediately in the case of *Galago demidovii*, which returns like a spring, and only after a short time-lapse in the case of *Euoticus elegantulus*, which ends up suspended head-downwards after the capture).

On the one hand, the potto and the angwantibo – already separated by their height in the vegetational strata – hunt animal prey which are malodorous (potto) and irritating (angwantibo). On the other, the two large bushbaby species are separated by the heights at which they are active. *Galago demidovii* and *Euoticus elegantulus*, which both exploit the canopy in primary forest, are unlikely to compete with one another. The first species preys primarily upon small animals (beetles, moths), and the second feeds upon larger prey (grasshoppers, larger beetles).

2. *Fruits*

The dietary tables given above show that only the potto and Allen's bushbaby can be regarded as frugivorous. In fact, the other three species consume absolute quantities of fruits amounting to only 1/30th of that eaten by *Galago alleni* and 1/70th of that consumed by the potto. In general, all of the lorisids primarily select soft, sweet fruits (*Uapaca* sp., *Musanga ceroopioides, Ricinodendron africanus*). However, the potto can attack certain fruits with a hard exterior, and on two occasions we saw *Galago demidovii* profiting from the passage of a potto to eat the remains of a large fruit which the latter had opened. As far as *P. potto* and *G. alleni* are concerned, the former eats fruits in the canopy, whilst the second collects them on the ground. This excludes any competition between the two species. These field observations have been confirmed experimentally in captivity. In a cage containing a tree, the pottos preferred to eat fruits placed in the branches, whilst *G. alleni* preferentially ate those placed on the ground.

Trees which are in fruit not only attract animals which are seeking the fruit – they also attract predators. Frugivores are presented with an abundant source of food; but the more time they spend there, the more they expose themselves to predation. Thus, the optimal solution for them is to collect *rapidly* and *in large numbers* fruits which can be eaten in some protected place. These two necessities are met in different ways in different zoological groups: by the crop in birds, by the cheek-pouches in monkeys and rodents, by the rumen in ruminants, etc. The potto and Allen's bushbaby have particularly distensible stomachs which can contain up to 1/13th of the body-weight. With the other three species, by contrast, we have never found any individuals with more than 1/30th of their body-weights in the stomach. Whereas the species which are not specialised for a frugivorous diet primarily utilise their tooth-scraper for slow removal of small fruit morsels, the potto and Allen's bushbaby can swallow rapidly large pieces of fruit, often including the kernels. In 30-60 seconds, a potto can eat an entire banana, and Allen's bushbaby swallows cherry-sized pieces of fruit by pushing them into the mouth with both hands and keeping the head up. It is actually quite rare to find these two species in immediate proximity of trees in fruit. In general, they are found 30-50 m. away, digesting under cover.

In the forest, the sites of fruit production are diverse and continually fluctuating. Through systematic counts, we established that large trees with high productivity are roughly 50 times less numerous than small trees or lianes with medium or low productivity. This is of capital importance for the distribution of small territorial mammals with permanent, restricted home-ranges (in particular, murids and prosimians). When a large tree comes into fruit from time to time in such a home-range, it is exploited straight away; but for the rest of the year food is obtained from trees with low productivity. The latter suffice for mammals of small body-size, but they must be abundant enough to provide a permanent supply of ripe fruits in the existing home-ranges. At night, trees with high fruit productivity are rarely visited. Using traps, we never captured more than 2-3 individuals of any given mammal species in such a tree (i.e. just as many as around trees with low productivity). Conversely, trees with low fruit productivity are little exploited during the daytime – approximately ten times less, according to our counts. This is doubtless associated with the fact that diurnal animals detect fruits by sight (at long range, when large fruiting trees are involved), whereas most nocturnal mammals detect by smell fruit which is isolated and hidden in the vegetation.

In order to study the social life of the prosimians, we have conducted a great deal of trapping with banana as bait. The animals were marked and released, which permitted us – among other things – to investigate their natural feeding behaviour. A basket of lianes containing 10 bananas is discovered in 1-5 days by *Galago alleni* and in 1-10 days by *Perodicticus potto.* On subsequent nights, if the bananas are replenished as necessary, the animals will return regularly by direct routes, often immediately after waking. They spend a few hours close to the bait and then move on to other fruiting areas. If several baskets of bananas are placed in the home-range of one individual, they will all be discovered and 2-3 may be visited during one night. By dissecting numerous pottos and Allen's bushbabies which were collected at the end of the night, we were able to observe that they can eat 2-3 different types of fruit in one night, which must oblige them to visit several fruiting points every night.

From all of these observations, it would seem that frugivorous prosimians (and murids) exploit simultaneously several fruiting trees, and that they never cease to explore their home-ranges in search of new trees in fruit. This mechanism, which is based on memory and exploration, permits them to feed themselves even if habitually visited fruiting trees cease production, or if they have been visited and depleted by another animal. Through continuous exploration of the home-range, they can locate trees at the start of fructification which will replace those which are ceasing to bear fruit.

3. Gums

The tables show that the potto and the needle-clawed bushbaby (*E. elegantulus*) are the principal feeders on gums. (*G. demidovii* consumes only 1/50th of the quantity of gums eaten by *E. elegantulus*, whilst *G. alleni* and *Arctocebus calabarensis* scarcely eat gums at all.) Both the potto and *E. elegantulus* inhabit the canopy, so it might be expected that they compete for gums.

Gums form principally along trunks and large branches at the site of old wounds and holes made by the mouth-parts of homopterans. (Examination of stomach contents revealed the presence of certain Auchaenorhynch homopterans – Fulgorids, Membracids, Tibinicids, etc. – which were no doubt swallowed involuntarily along with the exudations of resins which they provoked.) In contrast to fruits, gums appear regularly at the same places throughout the year. However, their production – which is dependent upon the metabolism of the trees – may be diminished during the main dry season.

In equatorial West Africa, the main dry season (15 June-15 September) is characterised by almost complete absence of rain, continuous cloud cover during the daytime, and a reduction of 3-4°C in mean temperatures. During this period, we were able to identify a marked decrease in the biomass of insects and a reduction in fructification. In parallel, we have observed that the prosimians lose 1/10th to 1/14th of their body-weights during this critical period. The weight-loss is directly dependent upon food availability. For example, *G. demidovii* consumes an average of 0.65 gm. of insects per night during the main dry season, as against 1.28 gm. per night during the rest of the year (stomach contents taken between 20.00 hrs. and midnight). Under the same conditions, we found smaller quantities of insects and gums in the stomachs of *E. elegantulus* examined during the dry season than in those collected at other times of the year. Thus, any competition would be exaggerated during this critical period.

When the dietary habits of the lorisids are examined in greater detail, it emerges that the potto eats only fruits and insects during the dry season, whereas fruits, gums and insects are taken at other times of the year. Conversely, *E. elegantulus* examined during the dry season had not eaten any fruits, although they eat small quantities during the rest of the year. Throughout the critical period, apart from the animal prey consumed, the potto is thus strictly frugivorous and *E. elegantulus* is strictly gummivorous.

Observation of the feeding behaviour of these two species renders the mechanism of such competition easily comprehensible. Whereas the activity of the potto is primarily oriented towards searching for and visiting trees which are in fruit, *E. elegantulus* spend most of their time visiting trees which are gum-producers. Needle-clawed bushbabies have an excellent memory for gum-production sites, and they follow veritable 'rounds' which permit them to collect (with the aid of the tooth-scraper) tiny droplets of gum formed after their last visit. Visits are made almost every night, even though each animal must visit a very large number of production-sites (about 300) in order to collect sufficient quantities of gums. *E. elegantulus* rapidly covers its 'rounds' thanks to its powerful leaps, stopping only for a few minutes at each gum-exuding site. In addition, the 'claws' at the ends of the nails permit access to sites which are inaccessible to the other prosimian species, right along the largest trunks. During the dry season, when the gums accumulate more slowly, *Euoticus* must visit a larger number of production-sites, whereas smaller 'rounds' suffice at other times of the year. Under these latter conditions, large aggregations of gums collect on trees which are not often visited, and it is such gums which are eaten by the pottos. Indeed, the gums found in the stomachs of pottos are often harder and darker in colour than those found in stomachs of *E. elegantulus*, and they occur as large lumps.

Thus, one cannot really talk in terms of real competition for gums between these two species, since the gums eaten by the potto are those left untouched by *E. elegantulus.* Dietary specialisations are most evident during the critical period of the year, and it is doubtless during this time that natural selection for adaptive characters is most active.

Conclusions

In both the Lorisinae and the Galaginae, the different sympatric species are morphologically relatively similar, and their dietary requirements are quite comparable: insects, with a supplement of fruits and gums in the larger species. In captivity, the five lorisid species can be maintained easily without any provision of foods (e.g. gums, fruits or certain insects) of the kinds eaten under natural conditions. In our present captive colony, all five species have become perfectly adapted to the same diet: milk, banana, apple, pear and crickets. In the forest, it is primarily through exploitation of different vegetation strata that the various species avoid dietary competition. Their ecological delimitations (in particular, their stratification) follow from 'preferences' which orient each species towards a particular type of support. This orientation is complemented by certain behavioural and anatomical specialisations associated with utilisation of supports. These anatomical adaptations are relatively minor; it is primarily behavioural differences which permit the separation of ecological niches.

Acknowledgments

My thanks go to Professor P.P. Grassé and A. Brosset for their generous provision of facilities and assistance at the CNRS field laboratory in Makokou, Gabon, throughout this study. All the photographs were taken in primary forests by my colleague, G. Dubost.

I should also like to make a special note of thanks to my friend and colleague, R.D. Martin, who offered to translate the manuscript for this article, despite his heavy commitments with the organisation of the Research Seminar. Our discussions, which have been numerous and very rewarding, have permitted me to extract a number of valuable conclusions.

NOTES

1 Hill, W.C.O. (1953), *Primates: Comparative Anatomy and Taxonomy*, 1. *Strepsirhini*, Edinburgh.

2 Walker, A. (1969), 'The locomotion of the lorises with special reference to the potto', *E. Afr. Wild. J.* 7, 1-5.

3 Charles-Dominique, P. (1971), 'Eco-éthologie des prosimiens du Gabon', *Biol. Gabon.* 7 (2), 121-228.

4 Charles-Dominique, P. and Hladik, C.M. (1971), 'Le *Lepilemur* du Sud de Madagascar: écologie, alimentation et vie sociale', *Terre et Vie* 25, 3-66.

5 Sanderson, I.T. (1940), 'The mammals of the North Cameroons forest area', *Trans. Zool. Soc. Lond.* 24, 623-725.

6 Napier, J.R. and Napier, P.H. (1967), *A Handbook of Living Primates*, New York, 258.

7 Jones, C. (1969), 'Notes on ecological relationship of four species of lorisids in Rio Muni, West Africa', *Folia primat.* 11, 255-67.

8 Vincent, F. (1969), 'Contribution à l'étude des prosimiens africains: le Galago de Demidoff', Doctoral thesis, CNRS, AO 3575.

9 Charles-Dominique, P. (1971), 'Ecologie et vie sociale de *Galago demidovii*', *Z.f. Tierpsychol,*, Suppl. 9, 7-41.

10 Hladik, A. and Hladik, C.M. (1969), 'Rapports trophiques entre végétation et Primates dans la forêt de Barro Colorado (Panama)', *Terre et Vie* 23, 25-117.

M.P.L. FOGDEN

A preliminary field study of the western tarsier, Tarsius bancanus *Horsefield*

Introduction

Hill recognised three tarsier species, all of them closely related and referable to the single genus *Tarsius.*[1] *T. bancanus* occurs in eastern Sumatra and Borneo, *T. syrichta* in the Philippines, and *T. spectrum* in Celebes. Exceedingly little is known about the ecology and behaviour of any of the three, even though a knowledge of their biology, and that of other prosimians, is of great importance to a proper understanding of Primate evolution as a whole. There has been much recent research on the other prosimian groups – the Madagascar lemurs and loris/bushbaby group (see Charles-Dominique and Martin for references)[2,3] – and the tarsiers now remain as the most important of the unknowns in our knowledge of prosimian ecology and behaviour.

This paper is offered as a small contribution towards filling this gap and concerns the western or Horsefield's tarsier, *T. bancanus* (Fig. 1). The study came about by accident when a number of tarsiers were trapped in mist-nets during the course of an ornithological research programme. The opportunity was taken to mark them and many were subsequently retrapped or seen in the field. Other commitments prevented tarsiers becoming more than a subsidiary subject of research; nevertheless, in view of the general lack of information about tarsiers, enough data were accumulated on their ecology and behaviour to justify this present account.

Methods

The study was carried out between October 1964 and September 1966 in the Semengo Forest Reserve, 20 km. south of Kuching, the capital of Sarawak, Borneo. The study area has been described elsewhere (Fogden).[4,5] Briefly, it consisted of 25 hectares of

Figure 1. (*a*) Photograph taken under natural conditions at dusk, showing a western tarsier (*Tarsius bancanus*) clinging to a thin, vertical trunk in the forest understorey. Such vertical supports are abundant in relatively clear secondary forest zones.

Figure 1. (*b*) Close-up photograph showing the typical vertical clinging posture. Note the expanded, tactile discs at the finger-tips and the greatly elongated hind limbs (particularly extended in the tarsal region).

rainforest more or less equally divided between primary mixed dipterocarp forest, which is the climax vegetation type of lowland Sarawak, and secondary forest at a comparatively early stage in the succession, most of it being composed of trees 5-10 m. tall and perhaps 10-15 years old. Within this area, netting was carried out for about two weeks in every month of the study period except April 1965. Nets were set at ground level, apart from an occasional net set in the canopy, and cutting of vegetation was kept to a minimum by using natural sites wherever possible. Nets were 12 m. long and about 2½ m. high, and most sites were sufficiently long to take two nets in line. The trapping effort was about 500 net-days per month, and on any one day or night 30-40 nets were in use. The nets were rotated

Figure 1. (*c*) Close-up photograph showing an identification ring on the left hind limb (tarsus). Note the large eyes, with the pupil greatly reduced. There is no reflection from the rear of the eye, since the conspicuous tapetum of the lemurs and lorises is lacking in the tarsier.

around the 200 available sites so that the study area was more or less evenly covered every two months.

The 26 tarsiers that were trapped were marked with numbered metal bird rings placed on the tarsus, and were subsequently released at the spot at which they were trapped. Each was weighed and sexed, and particular note was taken of any physical peculiarities which might assist subsequent recognition in the field. For example, a female was noted to be missing the distal half of its tail, a male had lost the last joint of each of the three outer toes of its left foot, and several had characteristic notches in one or other of their ears. These natural recognition characters were supplemented by clipping away the tuft of hairs from the end of the tail of some individuals, and by noting whether the ring was placed on the left or right leg. By using various combinations of these characters it was possible to identify in the field under favourable conditions 19 of the 26 tarsiers which were trapped in the study area.

No special effort was made to observe tarsiers directly in the field, but in the course of the two years of the study they were seen on about 40 occasions (including some sightings by other observers), generally at dusk. Though this was an average of less than two sightings per month of the study, the information gained was considerable. It was often possible to approach to within 5-10 m. of a tarsier, and so establish its identity, and also to follow it for some time provided that the vegetation was not too dense.

The study of other prosimians has been greatly facilitated by the ease with which they can be located at night by the reflection of a light beam from the tapetum at the back of their eyes. It is a great disadvantage that the eyes of tarsiers lack a tapetum and that they cannot be located in this way. Many hours spent in tarsier habitat with torches at night failed to reveal any, even though civets and other animals were easily located. Tarsiers are comparatively silent so that they are not conveniently located by sound either. However, they do have a bird-like call (described by Hill, Porter and Southwick, and by Harrisson) which is probably associated with courtship.[6,7] On the few occasions that it was heard at Semengo it was usually possible to locate the animal that was calling. Tarsiers also have a strong and readily identifiable smell which occasionally led to the discovery of an individual close by.

The seasonality of insects and fruit in the study area

Tarsiers are largely insectivorous, so some knowledge of the seasonal fluctuation in the number of insects in the study area is essential. During the course of the study it became apparent that fluctuation in the amount of fruit in the study area also had an important effect on the behaviour of tarsiers (see below). The seasonality of both insects and fruit has been discussed elsewhere (Fogden).[4,5] Briefly, insect numbers are ultimately related to rainfall which, though heavy throughout the year, increases during the north-east monsoon lasting from December to February or March. The increased rainfall results in increased leaf production, and there is an associated flush of insects which is most marked from January to about June. This is the period during which insects were most abundant in both 1965 and 1966. In contrast to the insects, fruit is patchily distributed and erratic in its seasonality, particularly in secondary forest. There may be long periods when fruit is abundant, and others when there is little or no fruit over wide areas. There is no annually recurring period when fruit is always relatively abundant or scarce, as there is with insects. At Semengo, fruit was plentiful for the whole of the first year of the study, but generally scarce from October 1965 to August 1966.

Results

The basic trapping and observational data on which the following sections are based are recorded in Table 1 and Figs. 2, 3 and 4. In Table 1, a record refers to the trapping or sighting of an individually recognisable tarsier (identified by a number in the Table and Figures). The location of such records are plotted on a map in Figs. 3

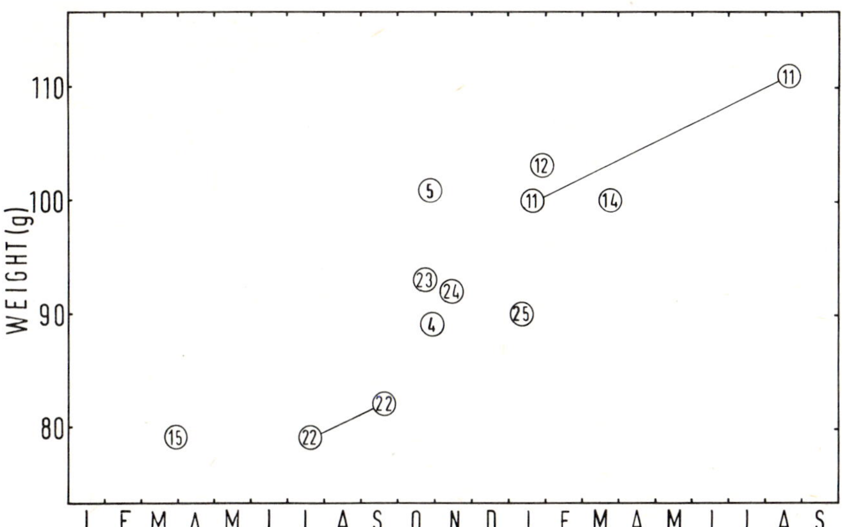

Figure 2. The pattern of weight increase of immature tarsiers during the 18-20 months after their birth in January or February. The records from two 20-month periods (January 1964-September 1965 and January 1965-September 1966) are plotted on the same scale. Each individual has an identity number which corresponds with those used in Table 1 and in the text.

and 4, except in the case of tarsiers recorded three or more times, in which case their range has been indicated by shading an area which just covers all points at which each was recorded. The 26 tarsiers involved were recorded 71 times (43 trapping records and 28 sight records), but the bulk of these records refer to only 12 individuals which were recorded 54 times. About a dozen sight records of

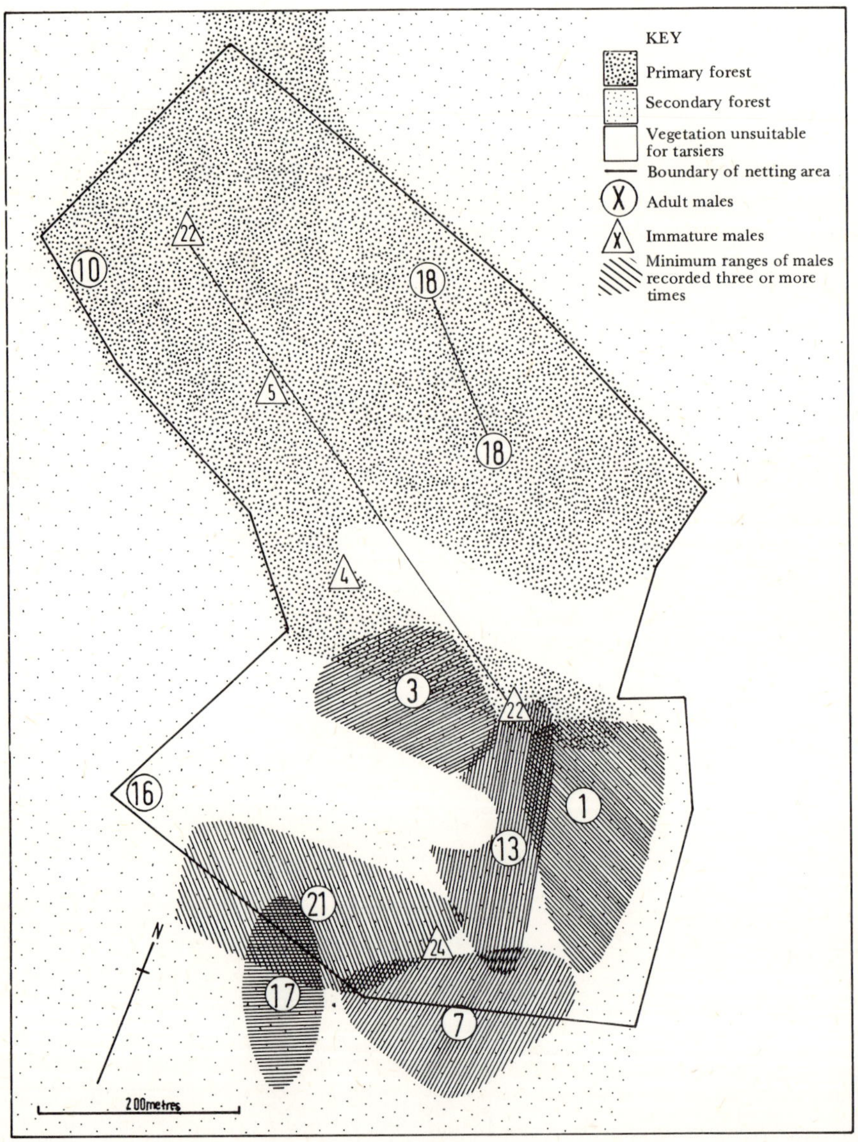

Figure 3. Map showing the distribution of male tarsiers in the Semengo study area. Each individual has an identity number which corresponds with those used in Table 1 and in the text.

unidentified tarsiers are included neither in Table 1 nor Figs. 3 and 4.

1. Sex ratio and body weights

The 26 trapped tarsiers consisted of 13 males and 13 females, so the sex ratio must be more or less equal.

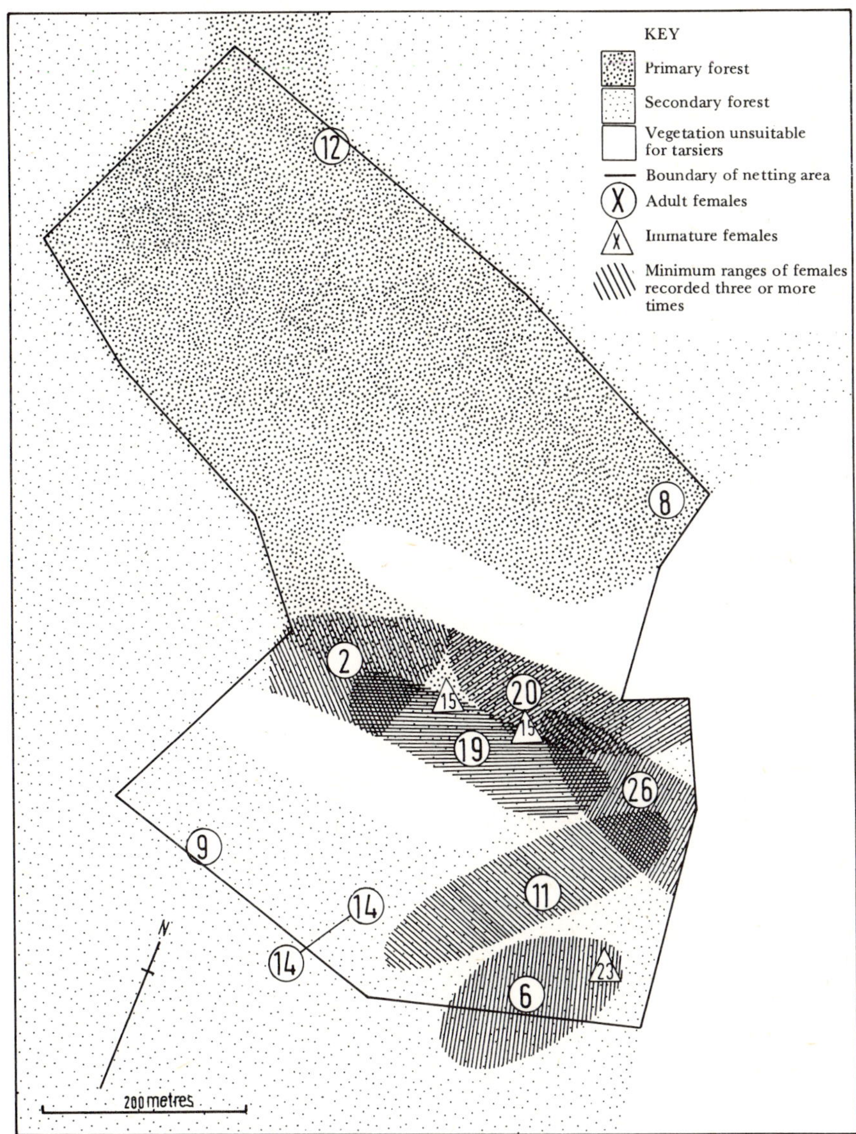

Figure 4. Map showing the distribution of female tarsiers in the Semengo study area. Each individual has an identity number which corresponds with those used in Table 1 and in the text.

Sexually mature males with large descended testes weighed 108-131 gm., with a mean of 120 gm. Such males had large testes throughout the year, though there may well have been some minor seasonal variation in their size. Individual body weights varied by up to 7 gm., but much of the variation could be attributed to animals being caught at different times of the night. There are insufficient data to show whether any of the variation in body weight was seasonal.

Mature females averaged lighter than males, weighing 100-119 gm. with a mean of 111 gm. However, the three lightest females (weighing 100 gm., 102 gm. and 103 gm.) might have been considered to be immature but for the fact that they included one

Table 1 Records of identified tarsiers

Identity No.	*Weight (gm.)*	*No. of records*	*Days between first and last record*
Males			
1	119	5	300
3	119-126	6	363
7	108-112	6	446
10	112	1	—
13	124-128	4	453
16	120	1	—
17	113	3	171
18	111-117	2	13
21	123-131	6	251
Immature males			
4	89	1	—
5	101	1	—
22	79-82	2	58
24	92	1	—
Females			
2	116	5	329
6	114	4	148
8	119	1	—
9	110	1	—
11	102-111	5	211
12	103	1	—
14	100	2	125
19	110	4	312
20	113-116	3	98
26	106	3	298
Immature females			
15	79	1	—
23	93	1	—
25	90	1	—

(no. 12) that gave birth while trapped in a net, and another (no. 14) that was lactating. It is unlikely that such females would be light because they were in poor condition, and the most likely explanation is that they were young females breeding when one year old and not yet fully grown. This view is compatible with the pattern of weight increase of young tarsiers (see below and Fig. 2), and is strengthened by the fact that the only one of the three to be re-weighed at a later date had increased in weight by 11 gm. in seven months (Fig. 2, no. 11). None of the females that were weighed was obviously pregnant, and it is probable that females attain much higher weights than the 119 gm. maximum recorded. New-born young are known to weigh 23-24 gm. (Le Gros Clark),[8] so female weights of 140 gm. or more are to be expected. In fact, Le Gros Clark recorded a weight of 142.5 gm. without giving the sex of the individual concerned.

Four males with very small testes weighing 79-101 gm. and three females weighing 79-93 gm. were considered to be definitely immature and not yet fully grown. The timing of their appearance in the trapped population is consistent with this view (see below and Fig. 1).

2. *Reproduction*

Tarsiers bear one young and have a sharply defined breeding season. In southwestern Sarawak, Le Gros Clark found that impregnation takes place from October to December and that the young are born in January and February, while Banks gave October to March as the breeding season.[8,9] Harrisson observed copulation between a pair of captive tarsiers in an outdoor enclosure in October,[7] while in this study a female was recorded to give birth in January and another to be lactating in March. All these observations are in good agreement. This sharply defined timing of breeding in south-western Sarawak results in coincidence of both the final stages of pregnancy and the period of maximum growth of the young with the time when insects are most abundant. There can be little doubt that the insect food supply is the ultimate factor determining the timing of the breeding season.

The timing of the appearance of light-weight and presumably immature tarsiers in the trapped population at Semengo is in good agreement with the timing of this restricted breeding season. The lightest individuals, weighing about 80 gm., were trapped in March and July. These probably represent early and late-born young. Thereafter immature tarsiers of gradually increasing weight were caught over a period of several months, adult weights being attained at an age of about 15 to 18 months. This progression of weights is illustrated in Fig. 1. It can be seen that the two females (nos. 12 and 14), weighing about 100 gm., which are presumed to have bred at an age of only one year, fit neatly into the pattern.

3. *Habitat and general behaviour*

Statements in the literature agree that *T. bancanus* is much more common in secondary than in primary forest (Le Gros Clark, Banks, Davis).[8,9,10] This is also borne out by this study, for 86% of records of tarsiers were either from secondary forest or from the narrow strip of primary forest cutting across the centre of the study area. Only seven tarsiers were caught in the block of primary forest in the northern half of the study area, and three of these were adults (male no. 10 and females nos. 8 and 12) trapped right on its edge, whose main ranges were probably in the adjoining secondary forest outside the study area. Of the four tarsiers that were trapped more than 50 m. inside primary forest, three were immature males, while the fourth (no. 18), which weighed 111-117 gm., was the only male tarsier weighing more than 110 gm. to have small testes. It seems probable that primary forest is a marginal habitat for tarsiers, and that it is utilised mainly by young males which are unable to compete with mature males for space in secondary forest.

Tarsiers have a greatly elongated tarsus and a low intermembral index (the ratio of forelimb length to hindlimb length) of about 0.58, and they are clearly specialised for vertical clinging and leaping. In accordance with this specialisation they were found to occur only in forest understorey with an abundance of vertical supports, particularly in the form of saplings 2-6 cm. in diameter and 1-2 m. apart (Fig. 1). They apparently avoided very dense tangled vegetation (many mist-nets were sited in such areas), the forest canopy in general, and areas of relatively open secondary vegetation with few saplings and a dense herbaceous cover.

Within the limits of the forest understorey, tarsiers have a wide vertical distribution, for they were seen from ground level up to heights of at least 8 m. There are indications that their vertical distribution varies seasonally (see below), and that at times they spend much of their time on or close to the ground. It is probable that most tarsiers were trapped while on the ground, for the majority were entangled in the very bottom of the net in positions which would have been impossible if they had been jumping between trees.

4. *Food and feeding behaviour*

T. bancanus is basically insectivorous, though given the opportunity it no doubt catches other small animals in the wild, particularly lizards, just as it does in captivity (Harrisson).[11] The stomach contents of seven specimens examined by Davis contained nothing recognisable apart from fragments of orthopteroid insects and a spider.[10]

At Semengo most of the locations at which tarsiers were seen and trapped were close to fruiting trees, or more specifically trees with ripe fruit lying on the ground beneath them. In fact, one of the best ways to see tarsiers was to wait quietly in the vicinity of such trees at dusk. Tarsiers were not attracted to the fruit itself, but by insects and other animals which were attracted to the fruit. These included large numbers of cockroaches, crickets, other orthopterans and large moths, as well as geckos and other lizards feeding on the host of smaller insects concentrated in the area. In these circumstances tarsiers captured most of their prey on the ground. They would scan the forest floor from a perch a metre or more above the ground, and having located prey (which was generally moving) leap directly on to it, killing it by biting with tight-shut eyes (see photograph in Harrisson).[7] Tarsiers were seen to catch prey in this way with leaps of up to 2 m., but most leaps were considerably shorter. The scanning phase of hunting sometimes lasted for up to ten minutes at a single perch, with the tarsier remaining more or less immobile; but more usually a failure to locate prey at one perch resulted in it moving on after only a minute or two. Some observations suggested that hearing as well as sight plays a part in locating prey.

On a number of occasions tarsiers were seen foraging on the ground by feeling in the leaf litter with their hands and apparently sniffing with their nose close to the ground. They appeared to be hunting by touch and/or smell, but the method also appeared effective in causing cryptic insects to move (this was observed in the case of stick insects and leaf-mimicking orthopterans), so that they could be seen and then pounced on in the normal way. Harrisson described captive tarsiers searching the floor of their large outdoor cage in an apparently similar way.[11]

5. *Seasonal variations in vertical distribution*

Tarsiers were neither trapped nor seen in the field at the same frequency throughout the two years of the study. They were trapped or seen 29 times (24 trapping records) in the first six months of the study, 34 times (13 trapping records) in the second six months, but only 16 times (seven trapping records) in the whole of the final year. This difference can probably be attributed to a change in the vertical distribution of tarsiers, resulting from the fact that fruit was abundant at Semengo for the whole of the first year, but very scarce for most of the second (see above). During the first year tarsiers were probably attracted to fruiting trees where they hunted low down and were easily trapped and observed, particularly as many mist-nets were sited, and much time was spent in visual observation, in such areas. During the second year there were no fruiting trees to act as focal points for either tarsiers or netting and observation. Consequently, tarsiers were neither easily trapped nor observed. Free-moving tarsiers were seen nine times during the second year and, though this sample is small, it is probably significant that they were at an average height of nearly 3 m., which is higher than the height of a mist-net.

The period during which tarsiers were only rarely recorded covered a time of insect abundance as well as scarcity, so it is unlikely that a proportion of the population left the area (in any case there was nowhere more suitable to go to), or went through a period of dormancy as do some lemurs when food is scarce (see Martin for references).[3]

It might also be argued that the decline in trapping records resulted from the tarsiers learning to avoid net sites. This may be true to a slight extent, but this explanation would not account for the fact that no tarsiers whatsoever were caught in areas of suitable habitat (to the south of the main study area) in which nets were sited for the first time during the second year of the study. Nor would this explanation account for the decline in sight records of tarsiers in the second year.

6. *Activity patterns*

Tarsiers are crepuscular and nocturnal. Most probably spend the day clinging to a suitable support in dense vegetation, though in the Philippines a small proportion of *T. syrichta* were found in hollow trees (Wharton).[12] Tarsiers were not systematically observed at night so nothing can be said about rhythms of nocturnal activity. However, it is clear that they become active at around dusk, for the vast majority of sight records were at that time, and a high proportion of tarsiers were trapped in the first two hours of the night. Several sight records were made a full hour before dusk on overcast evenings, and one as early as 17.15 hrs., nearly two hours before dusk (W. Corris, personal communication).

7. *Social organisation*

Table 1 and Figs. 3 and 4 show that adult tarsiers were sedentary, for the nine individuals for which most data are available (males nos. 1, 3, 7, 13 and 21, and females nos. 2, 11, 19 and 26) were recorded several times within relatively small areas over periods of 200 days or more. The ranges of individual tarsiers were about 2-3 hectares, and averaged larger in males than females. The ranges of individuals of the same sex appeared to be more or less exclusive, though there were small areas of overlap which were greater in females than in males, but the ranges of males and females overlapped extensively. Whether or not these ranges were territories in the sense of being defended is not clear. Marking with urine was frequent, but it is not known whether this had a territorial function.

Banks[9] stated that tarsiers usually occur in pairs, and statements by Harrisson[7,11] implied that established pairs are the rule in the wild. In this study two tarsiers were seen together on only eight occasions, two of which involved the same male with different females within the space of a few days. Pairings which were identified were male no. 3 with females nos. 2 and 19, male no. 13 with female no. 11, and male no. 21 with female no. 14. The known ranges of males generally overlapped with those of more than one female, though there is no evidence of any female actually associating with more than one. In the circumstances it is impossible to decide whether permanent associations between particular males and females were formed, but it does seem likely that males associate with more than one female. It is perhaps significant that the identified males that were seen with females were the three heaviest (Table 1). This might indicate that heavy males are dominant over lighter males in associating with females which overlap their ranges.

It has already been noted that immature male tarsiers differed from adults in being recorded mainly in primary forest. There is also a suggestion that they ranged more widely than adults, for one was re-trapped after moving nearly 600 m., twice the maximum recorded movement of an adult. It seems likely that immature males disperse into marginal habitat after becoming independent, and range widely until able to secure a permanent range in better habitat. It is possible that they only become sexually mature when they secure a permanent range, for one male (no. 18), which was caught twice in primary forest, had very small testes even though it weighed 111-117 gm. and must have been at least 15-18 months old.

Immature female tarsiers appear to secure a permanent range at an earlier age than males, for two females (nos. 12 and 14) bred when probably only one year old. Whether immature females disperse or remain in the area in which they were born is not clear, but it is noteworthy that no female tarsiers, adult or immature, were recorded far inside primary forest. Together with the fact that female ranges are smaller and overlap more than those of males, this might indicate that immature females establish ranges close to their parents.

Discussion

The restriction of tarsiers to secondary forest is a surprising feature of their ecology, for the understorey of primary forest has an abundance of suitable vertical supports and appears structurally more suitable than secondary forest for an animal that is specialised for vertical clinging and leaping. This apparent anomaly is emphasised by the fact that Allen's Bushbaby (*Galago alleni*) of West Africa, which resembles the tarsier in being specialised for vertical clinging and leaping, does occur in the understorey of primary forest and not in secondary forest (Charles-Dominique).[13] It is difficult to believe that the food supply in primary forest is unsuitable for tarsiers. The answer may lie in competition from other species, though there are no obvious competitors of the tarsier in primary forest, unless the various species of treeshrews *Tupaia* exploit the same food supply in spite of being diurnal. The only nocturnal treeshrew, *Ptilocercus lowii*, and the slow loris, *Nycticebus coucang*, occupy niches that are quite distinct from that of the tarsier, and neither can be regarded as a potential competitor.

The suggested social organisation of the tarsier has a number of possible points of similarity with the social organisation of Demidoff's bushbaby, *Galago demidovii* (Charles-Dominique) and the lesser mouse lemur, *Microcebus murinus* (Martin).[2,3] One can think in terms of heavy dominant males which pair with more than one female (the equivalent of the heavy 'A-central males' of *G. demidovii* and the 'central males' of *M. murinus*), lighter males which are in contact with females but subordinate to heavier males when it comes to pairing (the equivalent of 'B-central males' of *G. demidovii*), and immature males which inhabit marginal habitat and lack a permanent home range (the equivalent of the 'vagabond males' of *G. demidovii* and 'peripheral males' of *M. murinus*). On the other hand, there is no evidence that female tarsiers sleep socially in nests, as females do in the case of *G. demidovii* and *M. murinus.* Both the latter species are very vocal and apparently have calls which aid the reassembly of nesting groups each morning. The relative silence of tarsiers can probably be related to their not being highly sociable.

Acknowledgments

This study would not have been possible without the great help and encouragement of Tom Harrisson, Curator of the Sarawak Museum, and many members of the Museum staff, particularly Lian Labang, Gaun anak Sureng and Ambrose Achang. The mist-netting study was partly financed by an equipment grant which was kindly made available by Dr H. Elliot McClure in his capacity of Director of the United States Armed Forces Migratory Animal Pathological Survey. Thanks are also due to my wife who criticised the manuscript and drew the figures.

NOTES

1 Hill, W.C.O. (1955), *Primates: Comparative Anatomy and Taxonomy*, II: *Haplorhini: Tarsioidea*, Edinburgh.

2 Charles-Dominique, P. (1972), 'Ecologie et vie sociale de *Galago demidovii* (Fischer 1808)', *Z.f. Tierpsychol.*, Beiheft 9, 7-41.

3 Martin, R.D. (1972), 'A preliminary field-study of the Lesser Mouse Lemur (*Microcebus murinus* J.F. Miller 1777)', *Z.f. Tierpsychol.*, Beiheft 9, 43-89.

4 Fogden, M.P.L. (1970), 'Some aspects of the ecology of bird populations in Sarawak', D. Phil. thesis, University of Oxford.

5 Fogden, M.P.L. (1972), 'The seasonality and population dynamics of equatorial forest birds in Sarawak', *Ibis*, 114; 307-43.

6 Hill, W.C.O., Porter, A. and Southwick, M.D. (1952-3), 'The natural history, endoparasites and pseudoparasites of the tarsiers (*Tarsius carbonarius*) recently living in the Society's menagerie', *Proc. zool. Soc. Lond.* 122, 79.

7 Harrisson, B. (1963), 'Trying to breed *Tarsius*', *Malay. Nat. J.* 17, 218-31.

8 Le Gros Clark, W.E. (1924), 'Notes on the living tarsier *Tarsius spectrum*', *Proc. Zool. Soc. Lond.* 1924, 217-23.

9 Banks, E. (1949), *Bornean Mammals*, Kuching.

10 Davis, D.D. (1962), 'Mammals of the lowland rain-forest of North Borneo', *Bull. Singapore Nat. Mus* 31, 1-129.

11 Harrisson, B. (1962), 'Getting to know about *Tarsius*', *Malay. Nat. J.* 16, 197-204.

12 Wharton, C.H. (1950), 'The tarsier in captivity', *J. Mamm.* 31, 260-8.

13 Charles-Dominique, P. (1971), 'Eco-éthologie des prosimiens du Gabon', *Biol. Gabon.* 7, 121-228.

F. D´SOUZA

A preliminary field report on the lesser tree shrew Tupaia minor

Introduction

Tree shrews are small squirrel-like mammals occurring throughout South-East Asia. The family Tupaiidae is currently classified into two sub-families: the Tupaiinae, which includes five genera, and the Ptilocercinae having only one genus, the pen-tail tree shrew. Within the Tupaiinae, there is a range of species which are said to occupy different ecological niches. For example, *Tupaia minor* (Fig. 1) is reported as being almost wholly arboreal, *Lyonogale tana* (Fig. 2) as predominantly terrestrial, and *Tupaia glis* or *T. belangeri* as semi-arboreal. The morphological differences between these species, particularly in the skull region (Fig. 3), are assumed to be functionally related to the differing habits.[1-4]

The following table illustrates the broad 'extremes' within the sub-family Tupaiinae.

Species	*Assumed habit*	*Av. weight*	*Av. length of head and body*	*Tail length as % of head and body*[5]
T. minor	Arboreal	44 gms. Range 33-59	126 mm. Range 114-138	100-130%
T. glis	Semi-arboreal	145 gms. Range 88-173	177 mm. Range 156-197	75-95%
Lyonogale tana	Terres-trial	200 gms.	180-230 mm.	70-90%

As yet there have been no systematic field studies of tree shrew species; therefore the relationships between the morphological and behavioural differences of both Tupaiinae and Ptilocercinae and their precise ecological niches have yet to be established.

Figure 1. *Tupaia minor*; note the short snout.

Figure 2. *Lyonogale tana*; note the long snout.

Methods and study area

The following report is based on a short-term observational study carried out in a patch of primary forest, 17 km. north-west of Kuala Lumpur, Malaya. The forest contained, among other mammalian species, three tupaiid species: the diurnal *T. glis* and *T. minor*, and the only nocturnal tree shrew, *Ptilocercus lowii.* In addition to the field study data, records of the current Mark/Release Programme

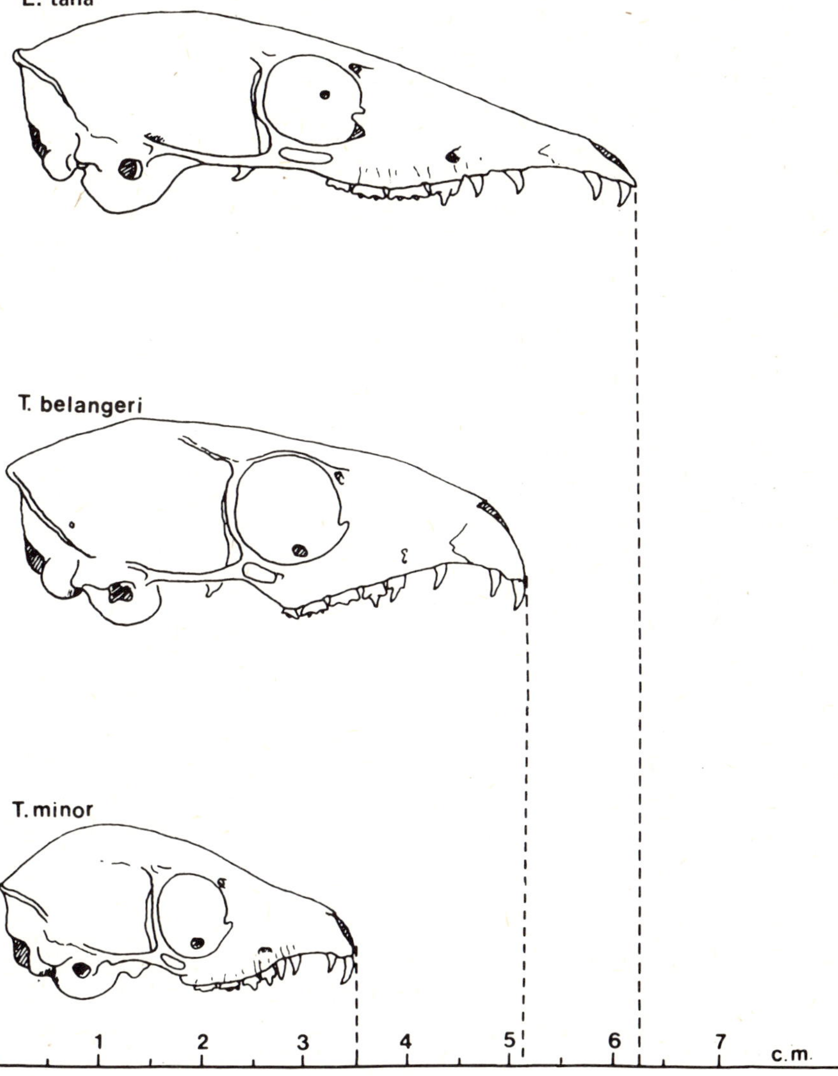

Figure 3. Scale drawings of the skulls of three tree-shrew species. (*Tupaia belangeri* is very similar to *Tupaia glis*).

were made available through the generosity of Dr Illar Muul of the United States Army Medical Research Unit, Institute for Medical Research, Kuala Lumpur.

Observations at canopy and sub-canopy level were made possible due to a system of ladders or transects built into the trees and interspersed with platforms. The transect system in the study area was 187 m. in length and had 7 viewing platforms varying in height above ground level from 9 to 26 m.[6] Binoculars were used to locate the animals initially and all observations were recorded on a cassette tape recorder and transferred the same day to data sheets supplied by the IMR. These data sheets incorporated maps of numbered trees surrounding the transect system. Thus it was possible to chart the progression of any one animal by tracing its movements from one numbered tree to another. Temperature and humidity were recorded on a thermohygrograph situated at a height of 8 m.

The broad habitat of tree shrews is that of tropical rain forest; however, within this category there are several relevant divisions. In Malaya, 5 vegetational zones can be distinguished: rice fields, lallang (long grasses), scrub, secondary and primary forest.[7,8] *T. minor* is said to inhabit primary forest predominantly, but it has been seen and trapped in both secondary forest and scrub. *T. glis* is a more wide-ranging species and has been recorded in primary and secondary forests as well as in scrub and cleared patches of land.

Bukit Lanjan, where this pilot study was carried out, is an area of approximately 25 square km. of predominantly primary forest bounded by villages, rubber plantations and roads. A good cross-section of both large and small mammals has been recorded in the forest, ranging from civets, leopards and bears to the minute bamboo mouse (*Chiropodomys gliroides*). Elevation ranges from sea level to approximately 1000′ (304 m.) and the area is defined as lowland rain forest. Temperatures in the lowlands are uniformly high, rarely falling below 21°C. The relative humidity ranges from 17%-96% during any one year. An analysis of variation in both temperature and humidity over a period of 18 months showed no significant fluctuations from one part of the year to another. The north-east monsoon and the south-west monsoon give rise to two slightly wetter periods during the year; however, this is not to imply that there are 'dry' periods or seasons. It is rare for rainfall to be less than 12 cm. in any one month.[9]

Primary forests are dominated by dipterocarp trees, which reach up to 30 m. in height, the crowns touching to form a continuous canopy. Above this, emergents rise to a height of 55-60 m. Below the canopy is a layer of saplings and smaller tree species which form a discontinuous under-canopy. The ground has a shrub and herbaceous layer which is generally sparse, although it can become dense where the canopy has been broken and the light penetrates to ground level. In the study area, clearings were not infrequent, as dead trees were uprooted by heavy rains and soil erosion. Thus, the pattern was of tall trees having a thick canopy and a lower interrupted layer of young trees and shrubs.

Intensive observation was concentrated within the area covered by the hill transect system. This observation station was situated well within the boundary of primary forest and at an elevation of 275 m. Occasional observations were carried out at the valley transect system, which again was just within the borders of primary forest, but located at sea level.

A third observation post, situated on the fringe of primary forest opening onto patches of secondary forest and lallang, was also selected. Only two *T. glis* were sighted in the secondary forest and only four in primary forest areas during the study period. Two *Ptilocercus lowii* were seen at night within the area of the hill transect, though the aborigines say it is mainly a primary forest creature. Thus the following report is mainly concerned with the lesser tree shrew *T. minor*, in a primary forest habitat. Apart from the lesser tree shrew, two other arboreal or semi-arboreal species were prevalent in the area; the black banded squirrel (*Callosciurus nigrovittatus*) and the slow loris (*Nycticebus coucang*), which is a nocturnal species. *T. minor* was distinguished from *C. nigrovittatus* and other small forest mammals (e.g. rodent species) by the following criteria:

(*a*) rapid 'jerking' type of locomotion, accompanied by tail-flicking
(*b*) light-coloured shoulder stripe
(*c*) pronounced length of tail in proportion to body size
(*d*) olivaceous colouring and ventral buff-coloured patches
(*e*) occasional distinctive vocalisations (recognised on the basis of recordings made in the laboratory).

C. nigrovittatus was recognised by its long lateral black band and *N. coucang* by the reflection from the tapetum.

Results

1. *Arboreality*

Since *T. minor* represents one of the 'extreme' species of the sub-family Tupaiinae, one of the main aims of the study was to determine the broad ecological niche and, in particular, the degree of arboreality. It is important to note that it is on the basis of structural similarities between the arboreal tree shrew species and some extant prosimian groups that the tree shrews have been included in the Order Primates. Thus, features such as the short rostrum, reduction of olfactory apparatus, relatively large eyes, forward rotation of the orbits, the development of a post-orbital bar and the structure of manus and pes, have been assumed to represent both arboreal adaptations and evidence of primate affinity.[1–2,10–14]

In order to assess the degree of arboreality, three observation points were selected from each of which there was a clear view of the ground, middle and canopy zones of the forest. Observation sessions of 15 minutes each were carried out three times daily for two weeks, between the hours of 09.00 and 10.30; 15.30 and 17.00; 20.00 and 21.30. Initial observations had confirmed that these times coincided with the main activity phases of the three most common species in the study area. Since the animals were not marked, it was not possible to assess the time spent at varying levels for *each individual.* Thus the animals were simply counted and the height above ground level recorded. During the 31.5 hours of observation, the sighting scores for the three species were as follows: *T. minor* = 35; *C. nigrovittatus* = 40; *N. coucang* = 21.

Of the 35 *T. minor* sightings, 11% (N = 4) were on the ground, 31% (N = 11) between 0.6 and 3 m., 34% (N = 12) between 3 and 7 m., and 23% (N = 8) between 7 and 12 m. Fig. 4 shows that both the squirrel and the loris appear to be restricted to heights above 3 m., and neither species was sighted on or near the ground.

These sighting scores may suggest that *T. minor* is less arboreal than usually supposed. If this is the case, it cannot be said that specialisations of the skull and hands and feet are all part of a complex adaptation to an exclusively arboreal way of life. This is

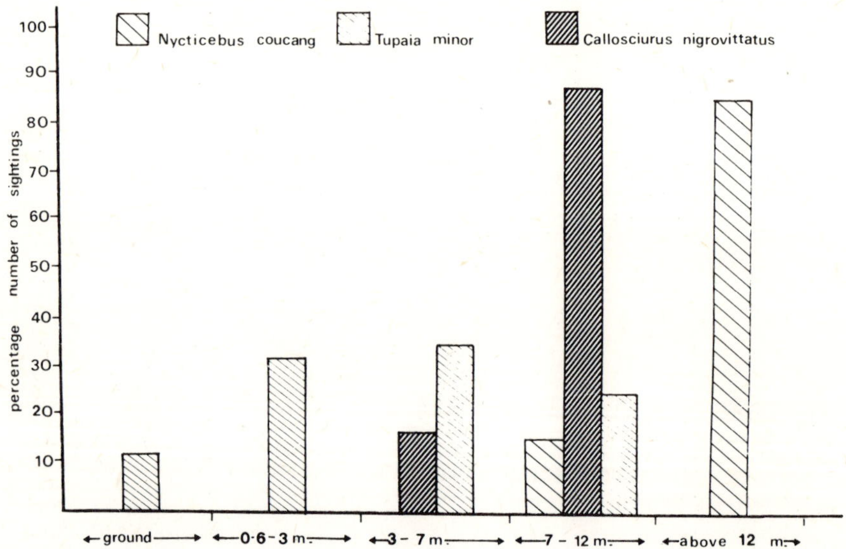

Figure 4. Sighting scores at varying heights above ground level for three primary forest mammal species.

particularly important if any comparison is to be made with the 'semi-arboreal' *T. glis*, which is presumed to show fewer morphological arboreal adaptations but which spends at least part of the time in the trees.

The structure of the primary forest in the study area was such that, in order to remain in the trees, an animal would be forced either to negotiate gaps of up to 6 m. or to exploit the 'fine branch' zone.

The slow loris, which has an average weight of 800 gm.,[12] moved through the canopy of the trees grasping extremely fine branches. The squirrel was able to negotiate gaps of up to 4 m., and one individual was seen to drop a vertical distance of 7 m. It would seem that *T. minor* shows neither of these types of locomotion and that it relies on scansorial adaptations in order to progress through the forest. The lack of extreme arboreal specialisation is correlated with the frequency of its occurrence on or near the ground (see Fig. 5).

Both the manus and pes of tupaiine species are narrow and elongated and have long, curved claws.[15,16] Palmar and plantar pads are well differentiated and the terminal digital pads have transverse ridges (except possibly in *T. javanica*).[14] *T. minor* has a rapid running and hopping type of locomotion; it balances on, rather than grasps, surfaces; though the curved claws and the frictional resistance of the palmar and plantar pads against the tree surfaces both contribute to the maintenance of balance. In climbing vertical branches, laboratory observations and tests show that use of the claws is more important than any grasping action of the digits.

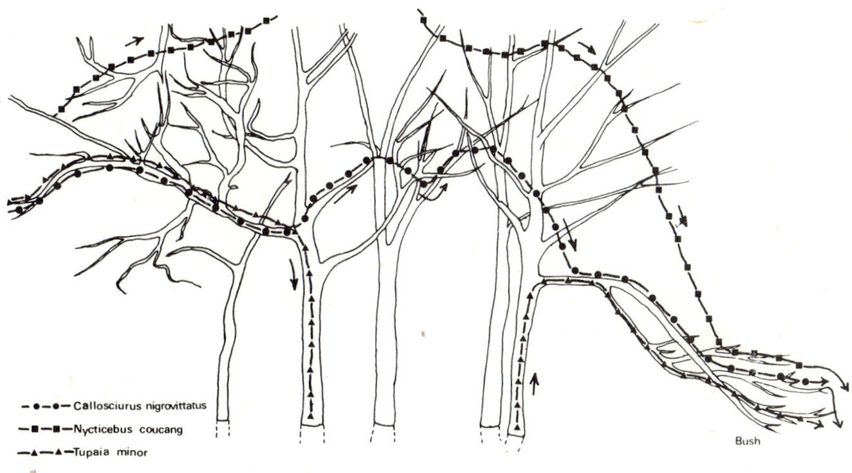

Figure 5. Characteristic routes through the forest taken by the three mammal species.

Tupaiine species do *not* show any marked divergence of either pollex or hallux, nor do they have truly grasping hands and feet.[4,12,17] Laboratory tests have shown that, when climbing, the limbs are well under the animal and not at an angle to the body; this is further evidence that the Tupaiinae, though excellent climbers, are not able to exploit the fine branch habitat, and thus their degree of arboreality is limited by the diameter of branches in the habitat.

In captivity, *T. minor* selects broad branches (i.e. 5 cm. or more in diameter) for activities such as eating, mating, grooming, defaecating and marking. Fine branches (i.e. approximately 1.5 cm. in diameter) are avoided, and this is markedly so for newly imported animals. If animals are given only fine branches, they will utilise the branches after some months as a means of getting from one cage area to another; however, all other activity in these cages takes place at ground level. There also seems to be a *preference* for height, both for general daily activity and for selection of nesting boxes. Given suitably broad branches at varying heights from ground level, individuals invariably select the highest nest boxes and topmost branches. Food pieces, for example, are taken in the mouth from the cage floor to preferred branches high up in the cage. This behaviour is markedly different from that observed in *Lyonogale*, which does not exhibit a *preference* for height, either for nesting or daily activities.

The ability to negotiate gaps between the trees depends to a large extent on the accuracy of visual perception of depth. Polyak (pp. 901-2) suggests that this ability is developed in Tupaiinae species (*Urogale everetti* and *T. glis*), but less highly than in some sciurid species.[18] Perhaps one of the limiting factors is the extremely small body size of *T. minor*. The type of locomotion exhibited by *T. glis* in the forest has not yet been studied, but its larger body size may indicate that it is able to leap across gaps more frequently and would thus be less scansorial than *T. minor*.

2. *Activity patterns*

Within the intensive study area, there was one main 'pathway system', consisting of a series of interconnecting broad branches. Another pathway, which was clearly visible from the ground, was located 150 m. from the transect system. These two pathways were used regularly by *T. minor* and *C. nigrovittatus*, and on two occasions by *T. glis*. The importance of these pathway systems lies in the fact that, once they had been located, peak activity periods could be established by counting the numbers of individuals crossing them. One such pathway system was within a few feet of the observer's platform, and it was possible to observe the animals from first light without causing any undue disturbance. Initial observations indicated that the earliest phase began at approximately 05.45 hrs. with the appearance of *C. nigrovittatus. T. minor* first appeared considerably later (approximately 07.45 hrs.), when the sun was well up.

Both pathway systems were observed for a total of 8 sessions. Each session began just before dawn at 05.20 hrs. and continued until just after sunset at 18.30 hrs. During two of the sessions, another observer, independently, scored the individuals sighted, in order to provide a check for accuracy. The species, exact time, and gross behavioural category (i.e. moving, feeding, grooming, mating, resting) were recorded. Laboratory observations on a captive colony have shown that maximum activity can be reliably estimated by both the total *amount* of movements made and the different *kinds* of activity engaged in at any·one time. Movement from one part of the cage to another in addition to feeding would seem to characterise the height of the activity phase. Accordingly, in the field, movement or feeding were used as behavioural indicators of activity.

Only those days during which temperature, humidity and rainfall were uniform were selected. Rainstorms or even heavy clouds tended to curtail the activity of *Tupaia minor* and some other forest species. The requirements for an 'observation' session were as follows:

(*a*) The previous afternoon and early evening had been free of rainfall.

(*b*) The hill was not obscured by heavy mist at 05.30 hrs. on the morning of the observation day.

Fig. 6 shows that *T. minor* have a bimodal pattern of activity, with peaks at 08.30-11.30 hrs. and 16.00-17.30 hrs. These data correspond well to the activity phases recorded in the laboratory. The decline in activity during the middle part of the day is characteristic for *T. minor* and *T. belangeri*,[3,19] and in the field these hours are the hottest. Temperatures may reach 28°C, and the minimum temperature recorded during the observation sessions for the period 12.00-15.30 hrs. was 21°C (cf. Figs. 6 and 7).

A preliminary analysis of 4-day periods following a previous day of either (*a*) frequent and heavy, or (*b*) continuous rainfall, suggests that these activity phases are extended for both species by up to 1½ hours during the day. *C. nigrovittatus* was sighted as early as 05.15, and *T. minor* at 06.45. On these days, the termination of activity was characterised by frequent observation of the two species together on the pathway systems, which does not otherwise occur.

The predominant cone-type structure of the tupaiid retina (*Urogale everetti* and *T. glis*)[18,20] suggests that it is adapted to diurnal vision. Polyak (p. 823) writes: 'Tree-shrews must be considered as specialised almost exclusively for photopic conditions.'[18] The sciurid retina, however, is composed of both rods and cones and this structural difference may account for the differential activity phases. Martin reports that in the laboratory *T. belangeri*

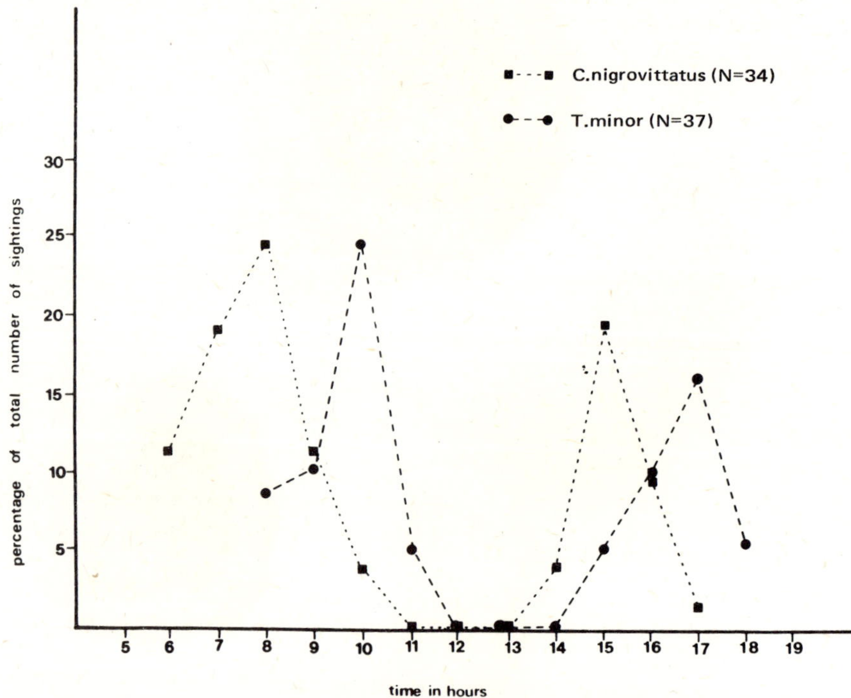

Figure 6. Bimodal activity periods for *T. minor* and *C. nigrovittatus.*

returned to the nest when heavy banks of cloud obscured the light.[3] On two occasions during the present field study, *T. minor* were seen disappearing into tree-hollows when the sun was only momentarily obscured by cloud, remaining there until the cloud dispersed. Captive *T. minor* appear to spend up to 5 minutes peering out of their nest boxes before emerging. This behaviour was also noted in the field with an individual *T. minor* whose nest entrance was within sight of the observer's platform. Observation of this nest entrance revealed that a period of simply looking out of the nest before emerging was a regular feature. The exact level of light intensity which promotes emergence and the onset of activity, or which determines return to the nest, has yet to be established, but it would seem that light intensity itself is an extremely relevant cue, as Pariente has shown for *Phaner* and *Lepilemur.*[21]

3. *Diet*

Laboratory observations show *T. minor* to be generally omnivorous, with a preference for live animal food (i.e. mealworms, insects and day-old mice). It is extremely difficult to observe the feeding habits

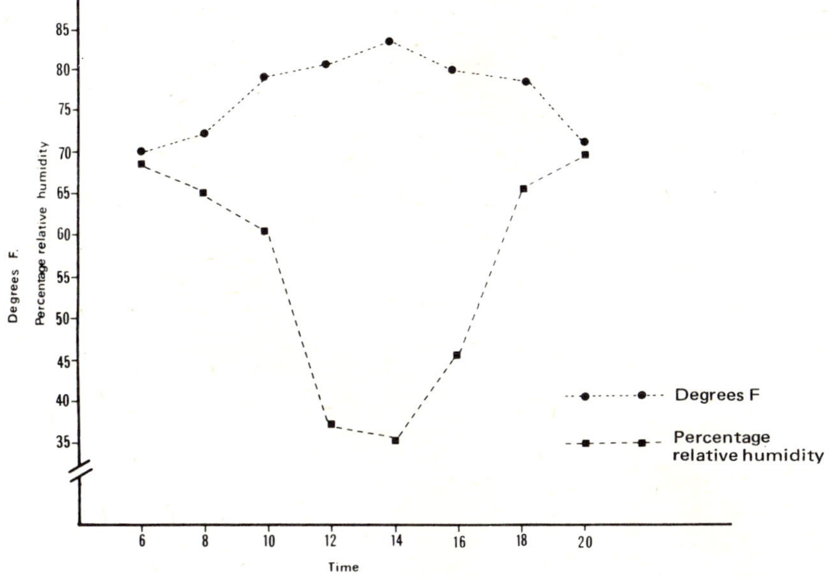

Figure 7. Average daily fluctuations of both temperature (°F) and % relative humidity during the study period.

of this species in the field. On two occasions, *T. minor* were seen carrying pieces of overripe bananas from the ground and feeding at a height of approximately 12 m. On another occasion, one individual was seen nosing and scratching the terminal branch of a dying tree, which was subsequently found to be infested with termites. Groups of *T. minor* have also been observed feeding on figs (*Ficus glabella*) (Dr I. Muul, personal communication). An analysis of the stomach contents of 10 shot *T. minor* showed fruit seeds, and the remains (e.g. wings and legs) of hymenopterans and other arthropods. The stomachs of two specimens contained several hairs, the structure of which indicated derivation from other mammal species. In the laboratory *T. minor* will tackle an adult mouse and kill it swiftly by grasping it in both hands and biting the head and neck. This method of killing appears to be typical for the genus *Tupaia*.[19] and it is quite possible that *T. minor* occasionally preys on small forest mammals.

4. *Nesting behaviour*

One *T. minor* nest was located and collected and two others inferred from the regular emergence and return of two individuals. The following table shows the location and height above ground level of the three nests:

Height	*Location*	*Zone*	*Remarks*
13 m.	In crevice of dried palm leaf	Middle of under canopy	Isolated tree surrounded by thick undergrowth. Nest constructed of tightly woven dried grass, leaves and strands from the nylon rope supporting the transect.[22]
19 m.	In hollow branch of 'medang' tree	Middle of under canopy	Entrance approximately 4 cm. in diameter. Isolated tree, not on pathway.
18-21 m.	'Medang' tree	Middle of under canopy	Isolated tree. Diameter of entrance approximately 7.5 cm., but situated on underside of branch and thus effectively protected from rain.

The first nest was only located, collected and examined on the last day of the study period, so it was not possible to observe the times of emergence and re-entry. This nest contained one female. The latter two nests were observed daily (when the entrances were not obscured by mists) at the beginning and end of both morning and afternoon activity phases. It appears that *T. minor* return to their nests either: (*a*) at the end of the afternoon activity phase at approximately 18.00 hrs., or (*b*) when the sun is obscured by cloud. They do not normally return to the nest during the midday decline in activity, which is also true of the captive colony. A fourth nest was inferred from the characteristic vocalisations of nestling *T. minor* and the regular disappearance of one adult into the tree hollow each morning at 09.00-09.15 hrs. (5 observations) and on one occasion at 18.00 hrs. If *T. minor* exhibit the 'absentee' system of maternal care reported for *T. belangeri*.[3,23,24] this latter behaviour pattern would suggest that the young had been deposited in a separate nest and were suckled at long intervals. Laboratory observations on the captive colony suggest that *T. minor* do in fact deposit their young in a separate nest box and that the inter-feeding interval is at least 24 hours and probably longer (D'Souza, in prep.).

Reproductive data

1. Seasonal weight changes

The Medical Ecology Division of the IMR initiated a field trapping programme in April, 1970, as part of the current research project into the transmission of diseases. Trapped individuals are taken back to the laboratory for examination, marked and then released the following day. In addition to examination for endoparasites and ectoparasites, any signs of pregnancy or lactation are noted for females, and testis descent and size are noted for males. Weights and body measurements are recorded, unless less than two weeks have elapsed since the last trapping date. During the course of this programme a total of 75 *Tupaia glis* (32 males and 43 females) and 36 *Tupaia minor* (17 males and 19 females) were trapped. Recapture dates for several *T. glis* are available, but none have been obtained for *T. minor.* Though the sample is as yet small, the data already collected permit some provisional statements as to the presence or absence of a breeding season in *T. glis.* Harrison demonstrated that male *T. minor* and *T. glis* show a significantly lower testicular weight during the period October-March than from April-September,[25] and this indicates a lower breeding potential during the former period. However, the number of pregnancies recorded for both species throughout the 4-year project was small (9), and did not show an absolute restriction to the April-September period. Martin suggests that the breeding potential of males as shown by a decline in testicular weight may vary according to environmental conditions and he concludes that pregnancy can generally occur throughout the year.[3]

In the IMR Mark/Release Programme there are only three recorded incidents of pregnancy, birth and suckling. One *T. glis* gave birth prematurely in the laboratory (26/3/71); a palpable foetus was noted for another individual (5/5/71); one *T. minor* was lactating (14/9/71). However, the recapture weights of several female *T. glis* show a greater incidence of fluctuation during the period March-August than from September to February. The average weight of wild-caught female *T. glis* is 140 gm. (43 individuals). Table 1 shows the weight changes for six females during the period May 1970-January 1972.

Rapid decreases in weight after capture may be a result of the stress involved in handling and anaesthetising the animal. However, there are four cases of substantial weight *increases* subsequent to capture and examination, one associated with birth and another with lactation (which occurs a few days prior to parturition in captive animals). It is thus reasonable to assume that major and rapid weight fluctuations in females are fair indications of pregnancy and birth. Furthermore, weight increases in males greater than 5 gm. are always associated with physical growth, as shown by gross changes in body measurements. Weight decreases in males appear to be rare; only one example is available, and the individual concerned died shortly after capture.

The greatest weight fluctuations, both in amount per individual and total recorded, occur in the period late March to the beginning of August, with only one exception (TG 33). The total number of animals trapped per month does not differ significantly. Weights of females captured during the period September-February do not generally show weight gains or losses greater than 7 gm. These data may support Harrison's evidence of a breeding peak, if not a well-defined breeding season.[25] It is unlikely that the weight changes

Table 1 Weight changes in female *Tupaia glis*

Animal number	*Trapping dates*	*Weight (grams)*	*Reproductive data*	*Weight change*
TG 11	19/5/70	129		
	12/4/71	179		+50
	21/4/71	147		-32
	14/5/71	144		-3
	6/6/71	153		+9
TG 23	11/12/70	158		-13
	30/12/70	145		
TG 27	17/12/70	149		
	11/1/71	137		-12
	26/3/71	179	Gave birth	+42
	13/4/71	162	3	-17
	5/5/71	153		-9
	19/7/71	180		+27
	6/8/71	103		-77
TG 33	29/12/70	155		
	20/1/71	113		-42
TG 62	3/6/71	170		
	7/6/71	130	Lactating	-40

are purely a result of variation in food sources, since males are not similarly affected.

2. *Testis retraction*

It has been stated that *Tupaia* exhibits non-seasonal permanent descent of the testes, whilst *Ptilocercus* exhibits seasonal descent of the testes;[2,11] however, the records from the Mark/Release Programme show that testis descent in *Tupaia* is extremely variable and is not restricted to any particular part of the year. Martin and von Holst have demonstrated that in captivity *Tupaia belangeri* can retract the testes into the abdominal cavity under conditions of social stress.[3,26] Martin (p. 421) writes: 'Dissection of the adult male *T. belangeri* demonstrated that the inguinal canal is open throughout life.'[3] The field records suggest that the incidence of testis-retraction is higher in animals confined in traps before examination than in those animals shot and examined immediately after death. Fig. 8 shows the percentage of testis descent and retraction in both shot and live-trapped individuals. It is interesting to note that 50% of the trapped *T. glis* exhibiting testis descent had extremely small testes, whereas the shot *T. minor* exhibited large testes. All individuals were adults, using body measurements as the criterion. Von Holst reports that under laboratory conditions, full testis-retraction occurs over a period of approximately 10 days.[26] It may be that in a stress situation as severe as confinement in a small trap for a period of some hours, testis-retraction is an extremely rapid stress response.

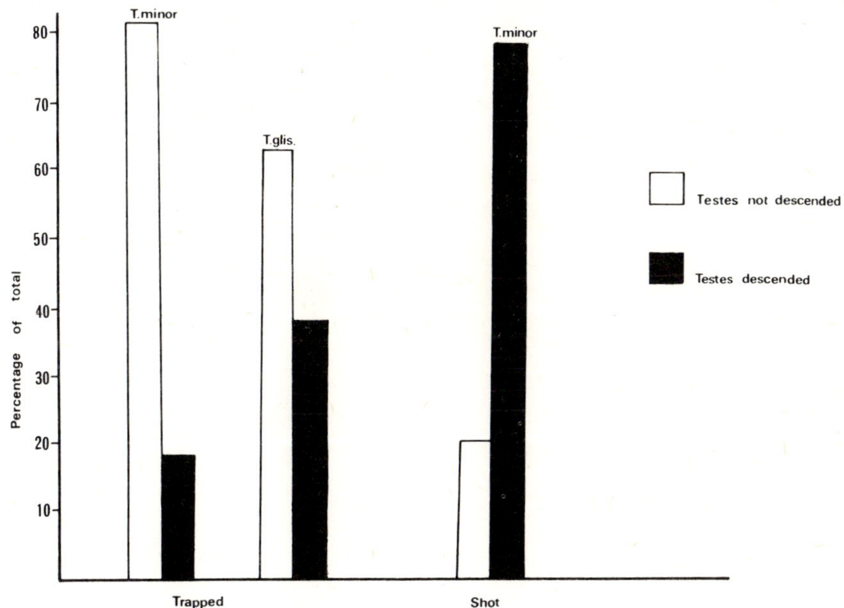

Figure 8. Histogram illustrating the percentages of testis-retraction in both shot and live-trapped individuals.

Acknowledgments

Many people have helped both in making the field study possible and in the preparation of the manuscript. In particular I should like to acknowledge the co-operation and generosity of the following persons during the field study period: Illar Muul, Lim Boo Liat, Michael Lee (IMR, Kuala Lumpur) and S.J. D'Souza, London. I am especially grateful to my supervisor Robert Martin, for his advice, criticism and encouragement, and to Dietrich von Holst and Alan Walker for their helpful discussions. Thanks are also due to Professor G.A. Doyle for his patience in awaiting the final draft of this paper. Finally, I should like to thank the following people for their help in preparation of the diagrams: Mrs Wendy Davies, Mrs Janet Maxwell, Ted Miller (University of Reading) and Miss Christine Roberts, London.

NOTES

1 Gregory, W.K. (1913), 'Relationship of the Tupaiidae and of Eocene lemurs, especially *Notharctus*', *Bull. geol. Soc. Amer.* 24, 247-52.

2 Le Gros Clark, W.E. (1962), *The Antecedents of Man*, 2nd ed., Edinburgh.

3 Martin, R.D. (1968*a*), 'Reproduction and ontogeny in tree-shrews (*T.*

belangeri) with reference to their general behaviour and taxonomic relationships', *Z.f. Tierpsychol.* 25, 409-532.

4 Martin, R.D. (1968*b*), 'Towards a new definition of primates', *Man*, (n.s.) 3, 377-401.

5 Data for *T. glis* and *T. minor* have been taken from the IMR field records. Data for *Lyonogale tana* are from Harrison (1964).[7]

6 Muul, I. and Lim, B.L. (1970), 'Vertical zonation in a tropical rain forest in Malaysia: method of study', *Science* 169, 788-9.

7 Harrison, J. (1964), *An Introduction to the Mammals of Sabah*, Jesselton.

8 Harrison, J. (1966), *An Introduction to the Mammals of Singapore and Malaya*, Singapore.

9 Banks, E. (1935), 'The change in climate and fauna with ascending altitude in Malaysia', *Sarawak Mus. J.* 4, 343-56.

10 Le Gros Clark, W.E. (1924), 'On the brain of the tree-shrew (*Tupaia minor*)', *Proc. zool. Soc. Lond.* 44, 1053-73.

11 Le Gros Clark, W.E. (1926), 'On the anatomy of the pen-tailed tree-shrew (*Ptilocercus lowii*)', *Proc. Zool. Soc. Lond.* 46, 1179-309.

12 Napier, J.R. and Napier, P.H. (1967), *A Handbook of Living Primates*, New York.

13 Wood Jones, F. (1917), *Arboreal Man*, London.

14 Wood Jones, F. (1929), *Man's Place Among the Mammals*, London.

15 Haines, R.W. (1955), 'The anatomy of the hand of certain insectivores', *Proc. zool. Soc. Lond.* 125, 761-77.

16 Haines, R.W. (1958), 'Arboreal or terrestrial ancestry of placental mammals', *Quart. Rev. Biol.* 33, 1-23.

17 Napier, J.R. (1961), 'Prehensility and opposability in the hands of primates', *Zool. Soc. Lond. Symp.* 5, 115-32.

18 Polyak, S. (1957), *The vertebrate visual system*, Chicago.

19 Sorenson, M.W. (1970), 'Behaviour of tree-shrews', in Rosenblum, L.A. (ed.), *Primate Behavior*, vol. 1, New York.

20 Wolin, L.R. and Massopust, L.C. (Jnr.) (1970), 'Morphology of the primate retina', in Noback, C.R. and Montagna, W. (eds.), *The Primate Brain*, vol. 1, New York.

21 Pariente, G., this volume.

22 Although the female inhabitant of this nest was seen taking nesting material to this nest, it could be that originally this 'woven' nest was inhabited by some other mammal or bird species. Laboratory data show that *T. minor* consistently builds a nest of loosely packed leaves and other fibrous material, but has never been noted 'weaving' the materials.

23 Martin, R.D. (1966), 'Tree-shrews: unique reproductive mechanism of systematic importance', *Science* 152, 1402-4.

24 Martin, R.D. (1969), 'The evolution of reproductive mechanisms in primates', *J. Reprod. Fert. Suppl.* 6, 49-66.

25 Harrison, J. (1955), 'Data on the reproduction of some Malayan mammals', *Proc. zool. Soc. Lond.* 125, 445-60.

26 von Holst, D. (1969), 'Sozialer Stress bei Tupajas (*Tupaia belangeri*). Die Aktivierung des sympathischen Nervensystems und ihre Beziehung zu hormonal ausgelösten ethologischen und physiologischen Veränderungen', *Z. vergl. Physiol.* 63, 1-58.

G. PARIENTE

Influence of light on the activity rhythms of two Malagasy lemurs: Phaner furcifer *and* Lepilemur mustelinus leucopus

Introduction

In the course of work on vision in Madagascar lemurs, it was considered necessary to obtain both ecological and ethological data for the species under investigation. The behaviour of the two species studied is known only from recent studies in which the author himself participated.[1,2] The author's study was concerned with the relationship between light and vision on the one hand and activity on the other.

The most striking feature of activity was its extreme regularity. Such a phenomenon requires either an 'internal clock' or a link with a regular physical parameter, like light or temperature. Since the meteorological conditions were particularly stable from one day to another at the time of the study, attempts were made to connect the activity of the animals with the cycles of light or temperature.

Description of species

1. Lepilemur mustelinus leucopus

The 'sportive lemur' (Fig. 1*a*), is a strictly nocturnal animal living in the south of Madagascar. By night its two big eyes strongly reflect the light of the observer's lamp, owing to the presence of a golden tapetum lucidum, which constitutes an efficient adaptation for nocturnal vision (Fig. 1*b*).[3] *Lepilemur* was observed in two different biotopes: the bush zone characterised by Didiereaceae, and the gallery forest along the Mandrare River. This study was carried out in

Figure 1 (*a*). *Lepilemur*: appearance of the animal in full light (note the vertical pupils in myosis).

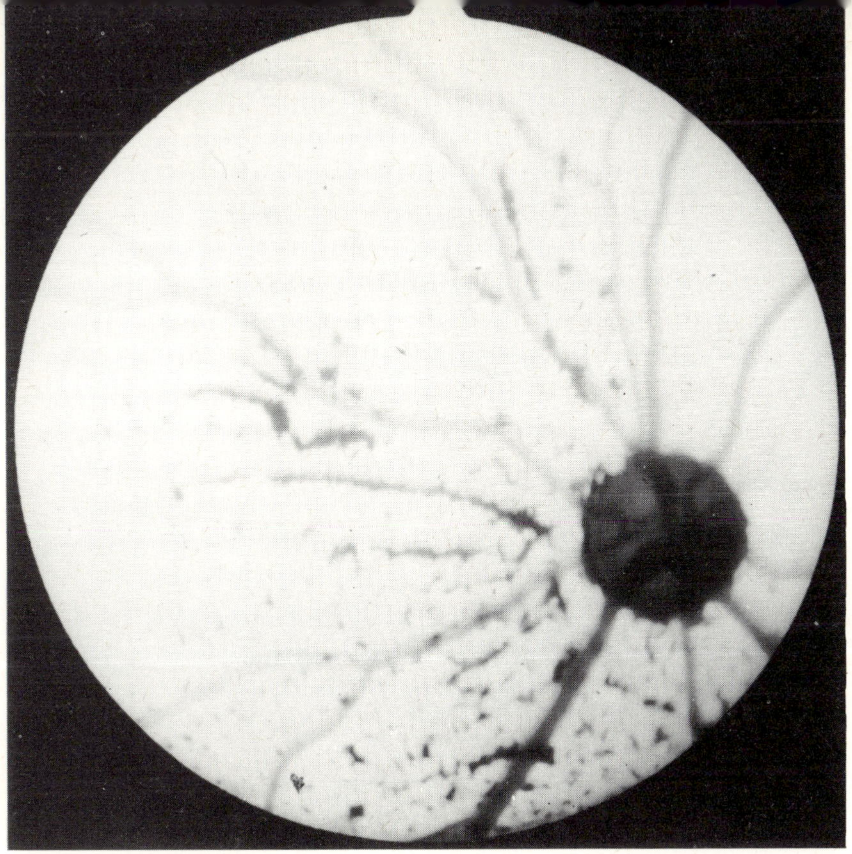

Figure 1 (*b*). Retinography of the papilla (blind spot) in *Lepilemur*. No foveal zone is seen on the golden tapetum.

Figure 1 (*c*). Pigmentation near the ora serrata (*Lepilemur*). No clear limit can be distinguished.

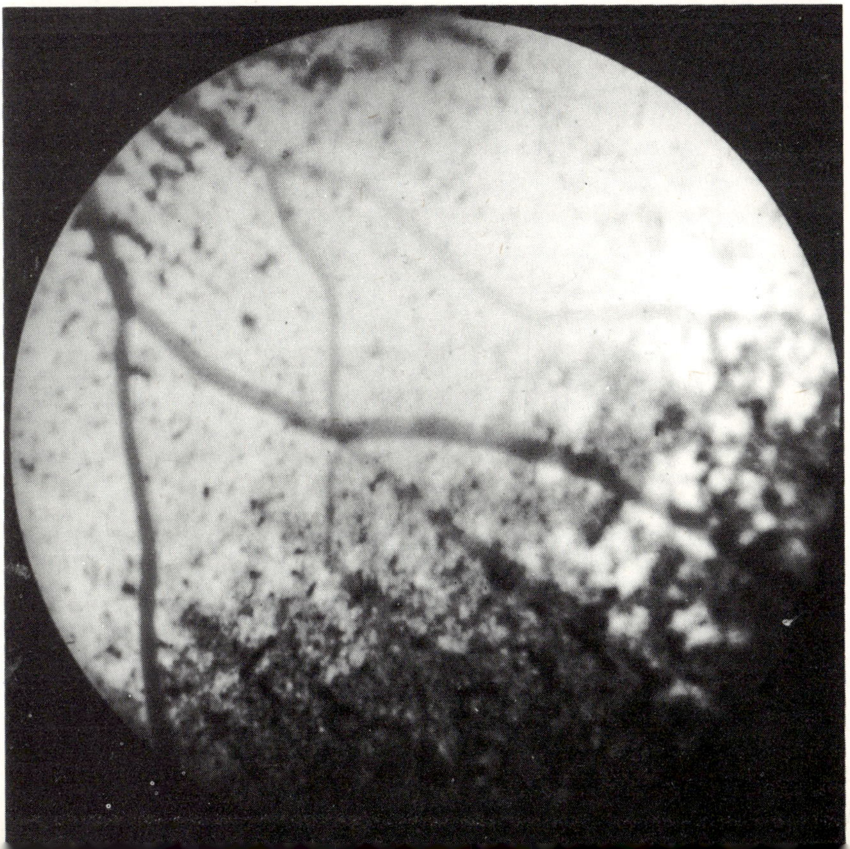

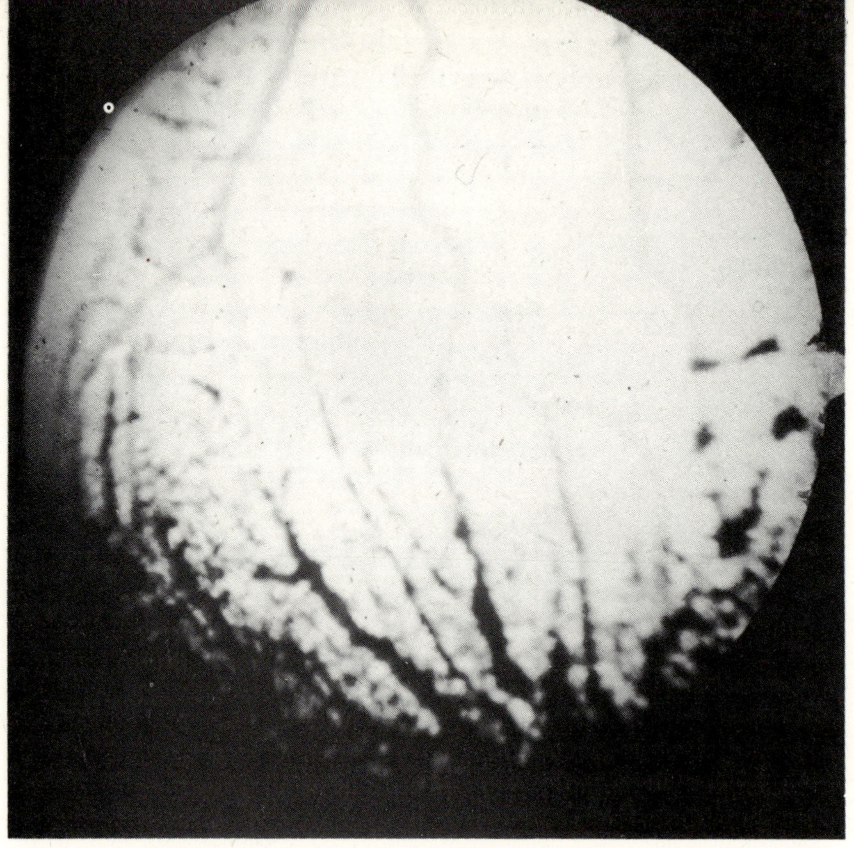

Figure 1 (*d*). Bright-gold appearance of the tapetum lucidum in *Lepilemur*.

Berenty (25°S, 46°24E), to the north-west of Amboasary, during the dry season when the temperature was 40°C at noon and about 14°C at 04.00 hrs. and the sky was cloudless most of the time.

2. *Phaner furcifer*

The 'fork-crowned lemur' (Fig. 2*a*) is also a strictly nocturnal animal, similarly possessing a tapetum lucidum.[3] It is as big as a squirrel with unbelievable agility in the trees. It was studied in Analabe, just North of Morondava (20°S, 44°33′E) in a typical forest zone where baobabs are very numerous. At the end of October and the beginning of November the weather was dry, very hot with a perfectly clear sky by night and by day with temperatures about 38°C at noon and 19°C at 05.00 hrs.

Apparatus and methodology

An autonomous and portable photometer IL 600 (International Light) was connected either to two different captors (PT 100 and PT 200 A) or to a photomultiplier system (Type PM 200 C).[4] Four filters were used with every captor (T_{max} at 350, 400, 520 and 640 nanometers). All measurements were taken at one metre above

Figure 2 (*a*). *Phaner furcifer*: the two eyeballs are very large in size.

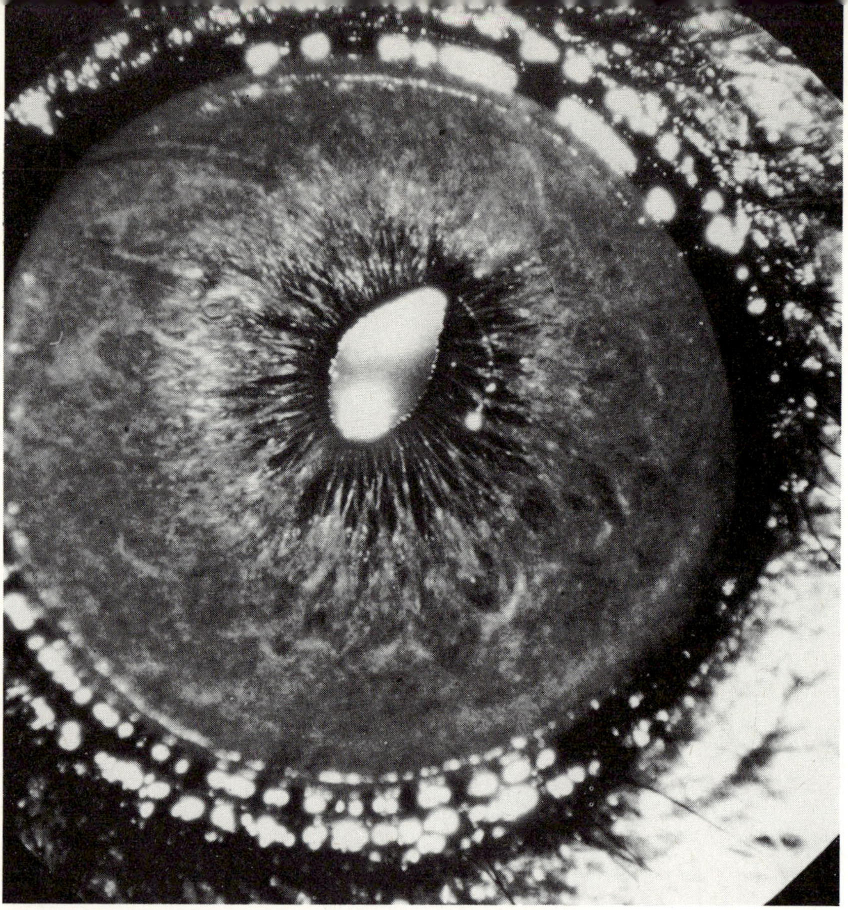

Figure 2 (*b*). Myosis showing the pear-shaped pupil of *Phaner*.

Figure 2 (*c*). *Phaner*: typical aspect of the ora serrata (it is quite different from *Lepilemur*, Fig. 1*a*).

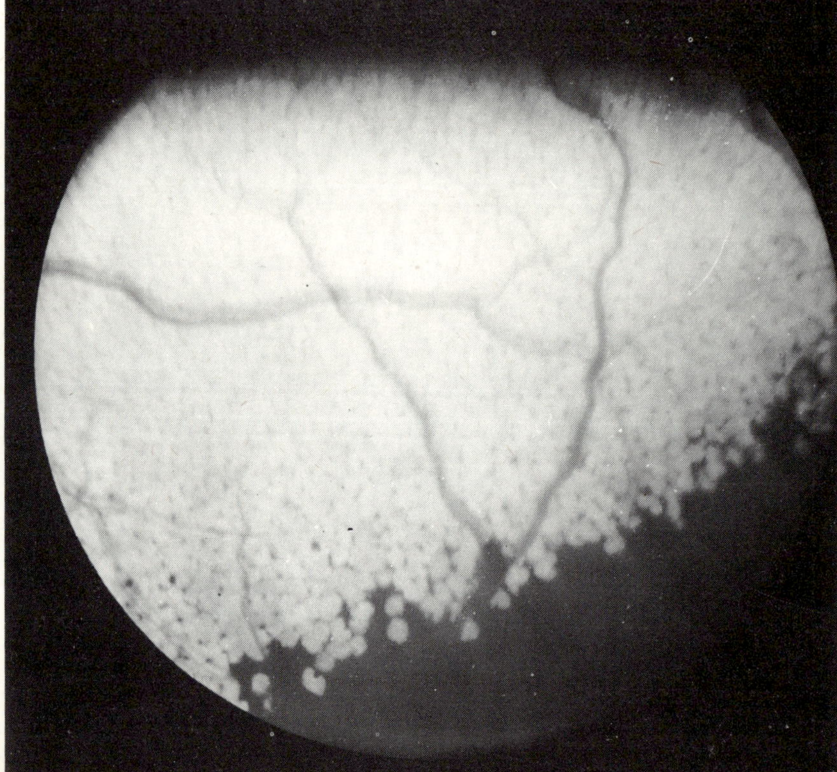

Figure 2 (*d*). Zenithal view showing the screening effect of vegetation on the sky in Analabe.

ground-level, every two hours and in five different spatial directions, north, south, east, west, azimuth (Fig. 2*d*). Orientation was always with reference to the geographic north (magnetic variation: W 18°32 in Analabe; W 13° in Berenty). At dawn and twilight, continuous recordings of light were carried out.

As a reference, a series of measures was taken during the same period away from any vegetation (at airfields one or two miles from the forest). All the data obtained in this way are energy values expressed in μwatt/square cm., which have in some cases been converted into illumination units considering:[4]

(*a*) the spectral sensitivity curves of the three captors;
(*b*) the spectral distribution of energy of the sun for three colour temperatures;
(*c*) the transmission curve as a function of wavelength of the four filters;
(*d*) the spectral response of a standard observer (CIE) under two different conditions of adaptation (daylight adaptation for human eye = photopic; nocturnal vision = scotopic).

It must be stressed that the values in lux (Figs. 6-9) are not related to the animals (for their spectral sensitivity response is still unknown), but to the human observer.

Most of the measurements used for the graphs were south measurements (with the back towards the sun in the southern hemisphere). For nocturnal measurements it was also necessary to distinguish between the very low energies among the trees (illumination used for vision in real conditions) and the light coming from the sky through the trees.

Finally, the times of sunset and sunrise were calculated for the upper edge of the sun, taking into account the effect of atmospheric refraction:

(*a*) *Berenty* (24°59′ S; 46°24′ E)
27/9/1970 sunrise 05h. 40m. 36s. local time
meridian crossing 11h. 45m. 47s. local time
sunset 17h. 50m. 58s. local time

(*b*) *Analabe* (20°44′ E)
28/10/1970 sunrise 05h. 25m. 12s. local time
meridian crossing 11h. 47m. 48s. local time
sunset 18h. 10m. 24s. local time

Results

1. Lepilemur mustelinus leucopus

These animals have been observed in the Didiereaceae bush by Charles-Dominique and Hladik[1] and in the gallery forest by the author. Sleeping in tree holes or between branches, the animals in this area are relatively well protected from the environment and provided with shaded surroundings. At sunset, they habitually extend their heads out of their shelters and watch to see whether the light conditions are favourable. The regularity of emergence is remarkable, taking place between 18.02 hrs. and 18.23 hrs. (Fig. 3). Despite the relatively small number of observations, a distinction

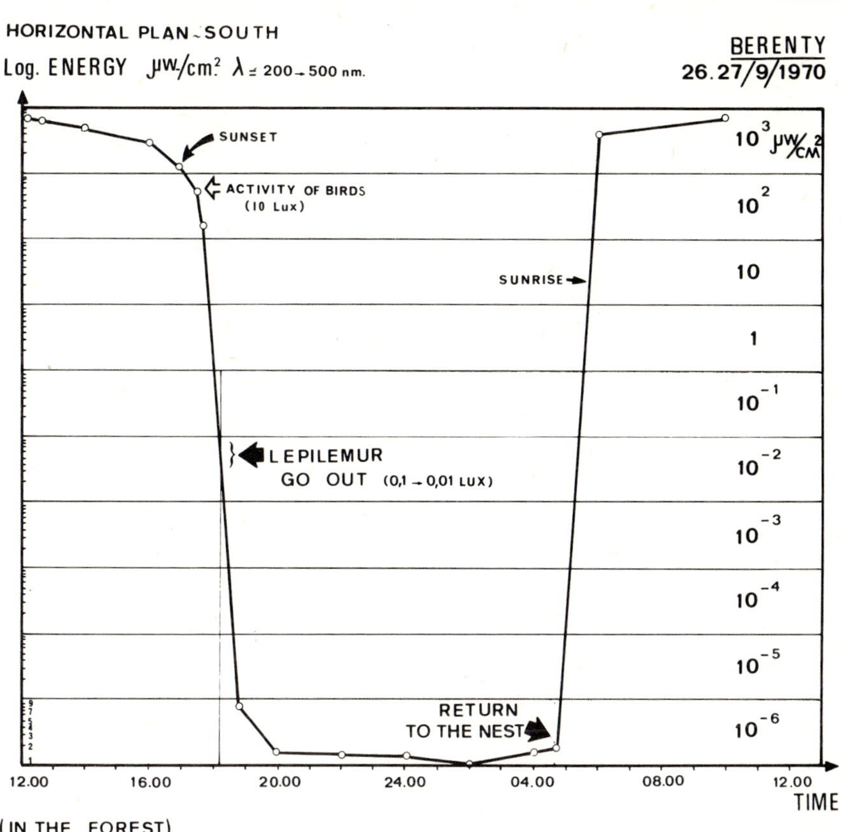

Figure 3. Diurnal variation in light energy on a south-horizontal direction during a 24-hour period at Berenty in September, showing the times *Lepilemur* emerges from and returns to the nest.

may be made between the animals living in the bush and those living in the gallery forest (Fig. 4). In terms of light penetration, the bush is much more open than the forest and is more similar to the open field. The average time of emergence is 18.08 hrs. in the forest and 18.16 hrs. in the bush.

This difference of 8 minutes permits all animals to leave their shelters when the light intensity is about 5 x 10^{-2} μw/square cm. When an intensity of 5 x 10^{-2} μw/square cm. is measured in the forest, the light is still about 1.5 μw/square cm. in the open field and 0.4 μw/square cm. in the bush. This illustrates the importance of light intensity for the activity rhythms of these animals. Temperatures and activity are plotted in Fig. 5; no obvious correlation between the two can be seen. Charles-Dominique and Hladik[1] consider that the rising of the moon during lunar periods induces the animals to vocalise; but no confirmation of this theory was sought in the present study.

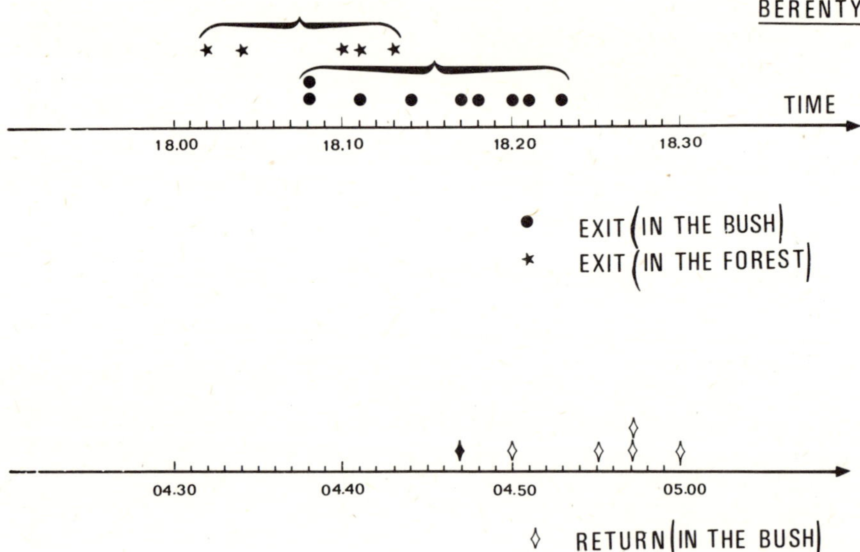

Figure 4. Comparison of nest-emergence time and nest-return time for *Lepilemur* in two different biotopes at Berenty, the bush and the forest.

During the night, the activity of *Lepilemur* is very variable. There is no doubt, however, that these lemurs have very good vision, since they are still able to leap several metres, even with the light level as low as 5 μlux (Fig. 6). The comparison of measurements in the open field and in the forest shows that the vegetation decreases the light almost 100 times.

The animals are less regular when returning to their retreat (cf. Fig. 4), but always do so when light is still at a very low level. The human observer is unable to detect a change in illumination at such times. In fact, the measurements show that the earlier returns correspond with an increase in light by at least 50% (100% in a number of observations). In all cases, the light conditions change extremely rapidly at about 04.45 hrs.

2. *Phaner furcifer*

In most cases, the first leap out of a spherical leaf-nest is accompanied by loud shrieks from the animal. In Analabe, this usually occurred with great regularity at 18.20 hrs. (only once did it happen at 18.10 hrs.). At 18.30 hrs. many animals started vocalising together. In fact, the timing of the first leap is different from that of true awakening; as with *Lepilemur* the animals wait for favourable environmental conditions before leaving the nest. It was thought at first that this mechanism was governed by temperature variations.

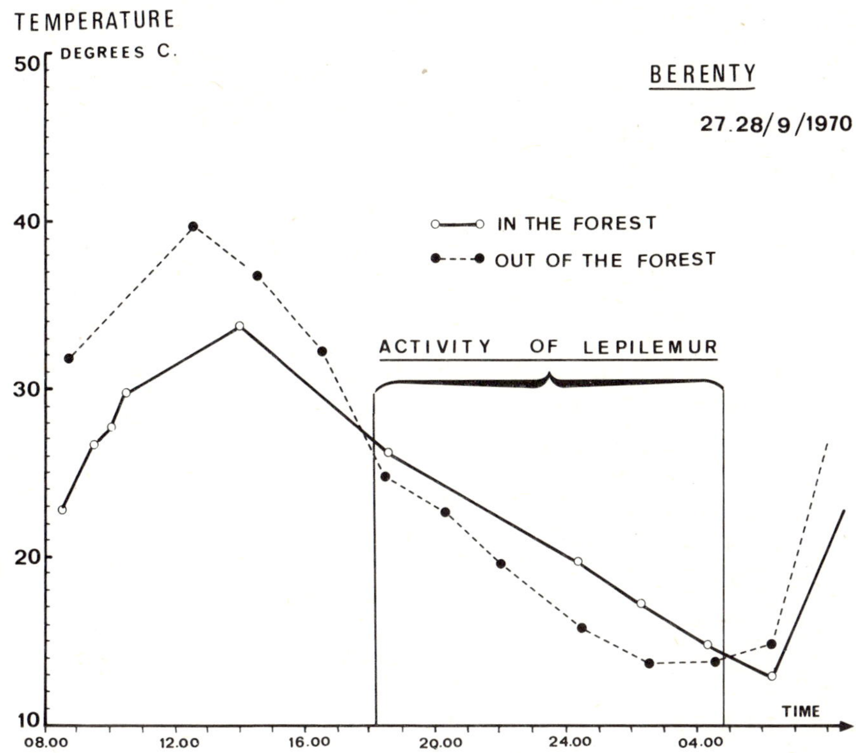

Figure 5. Diurnal variation in temperature within the forest and outside the forest at Berenty in September, showing the range of temperatures within which *Lepilemur* is active.

However, it can be seen from Fig. 7 that the temperature is 30°C at emergence and 18°C on return. Furthermore, temperature is decreasing progressively (1°C per hr.) when emergence takes place, whereas return to the nest precedes a rise in temperature by 30 to 40 minutes. Temperature variations do not therefore appear to be correlated with activity.

Fig. 8 shows that light variation is a very regular phenomenon. It can be seen from the same figure that activity starts as soon as the light falls below 2 lux. It is worth recalling here that in tropical zones (as in this study) the sun follows a half circle very high in the sky and sets along a line very similar to a perpendicular on the horizon. This is the reason why the change in luminosity is so pronounced at dawn and twilight.

For the observer's eyes, it is necessary to pass from photopic to scotopic conditions within a few minutes. Thus the time when the animals become highly active is a very characteristic one for human beings, as it corresponds to marked discomfort due to lack of visual adaptation.

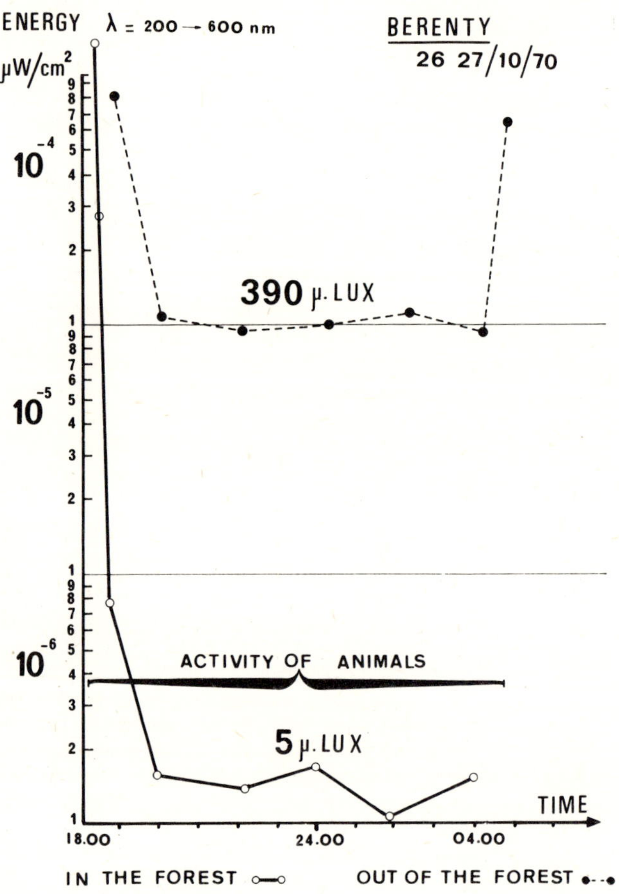

Figure 6. Nocturnal variation in light energy at Berenty in September, showing levels of light adequate for activity of *Lepilemur* both within the forest and outside the forest.

In the succeeding ten hours, *Phaner* shows evidence of intensive activity (running and leaping), while the light conditions remain at about 30 μlux, which represents very bad conditions for a human observer (Fig. 9). The time of returning to the nest is less regular. However, Petter et al.[2] never saw *Phaner* engaged in activity after 05.00 hrs. (a few lux on the curve). Most of the time, the returns were at 04.20 hrs. while luminosity was 0.01 lux, at which time the increase in light intensity is maximal (33 μlux at 04.00 hrs.; 30.000 μlux at 04.30 hrs.)

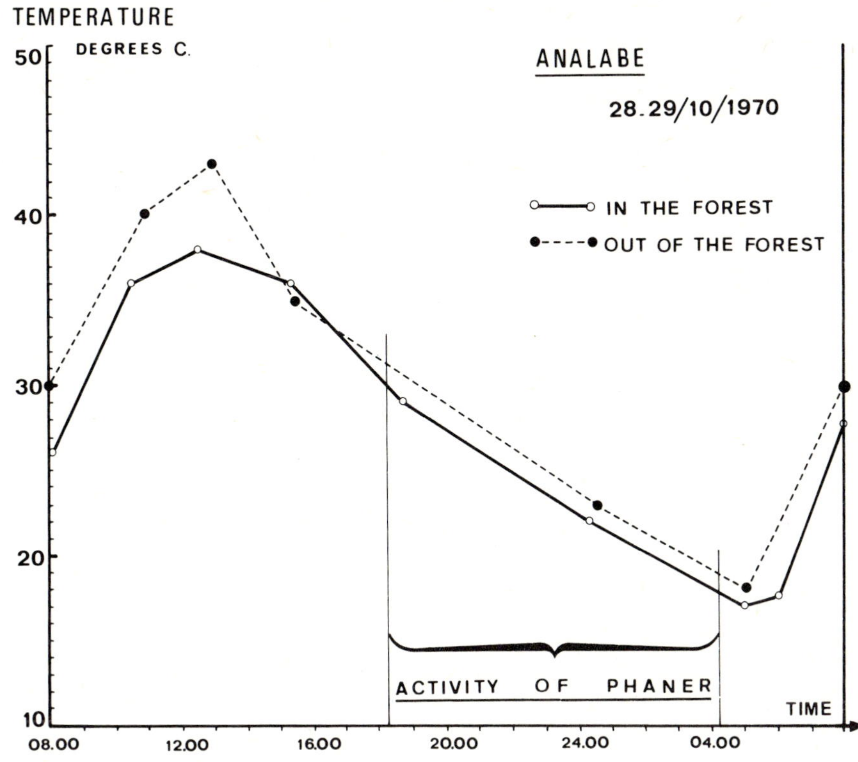

Figure 7. Diurnal variations in temperature within the forest and outside the forest at Analabe in October, showing the range of temperatures within which *Phaner* is active.

Conclusions

It can easily be seen that the activity rhythms of the two nocturnal Madagascan lemurs studied are perfectly adjusted to the daily cycle of the sun. Both species showed that they were able to leap several metres by night, which requires good judgment both of distance and of the strength of supports. Only vision can supply such information quickly and at a distance, since no echolocation seems to occur.

The regularity of emergence and return demonstrates that their eyes have the ability to discriminate, in decreasing and increasing light, a special level at a given moment. It is not known what mechanism is used, but it is conceivable that it would call for a duplex cone-rod system. Such a mechanism would be very surprising in nocturnal animals, which are reported to possess pure rod retinas.[5] The author is at present undertaking histological and behavioural

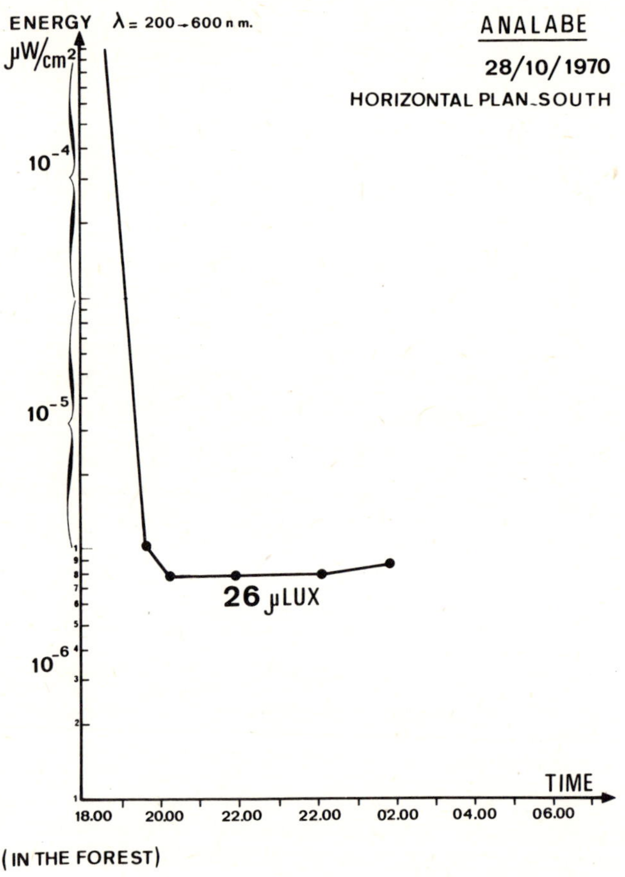

Figure 8. Nocturnal variations in light energy at Analabe in October. The level in μlux has been calculated as a reference.

investigations to determine whether colour vision is present in these two species.

If we consider the extremely low levels of radiation available by night in tropical forests, it is obvious that the vision of these animals presents very interesting and remarkable ecological/physiological adaptations. It should be added, as a final note, that these strictly nocturnal animals are still able to see quite well in full daylight. This fact demonstrates the perfect adaptation of their eyes, which can work right across the extraordinary scale of 10^{10} energetic units. Studies aimed at a better understanding of all these phenomena are at present underway in Brunoy, France (Laboratoire d'Ecologie Générale).

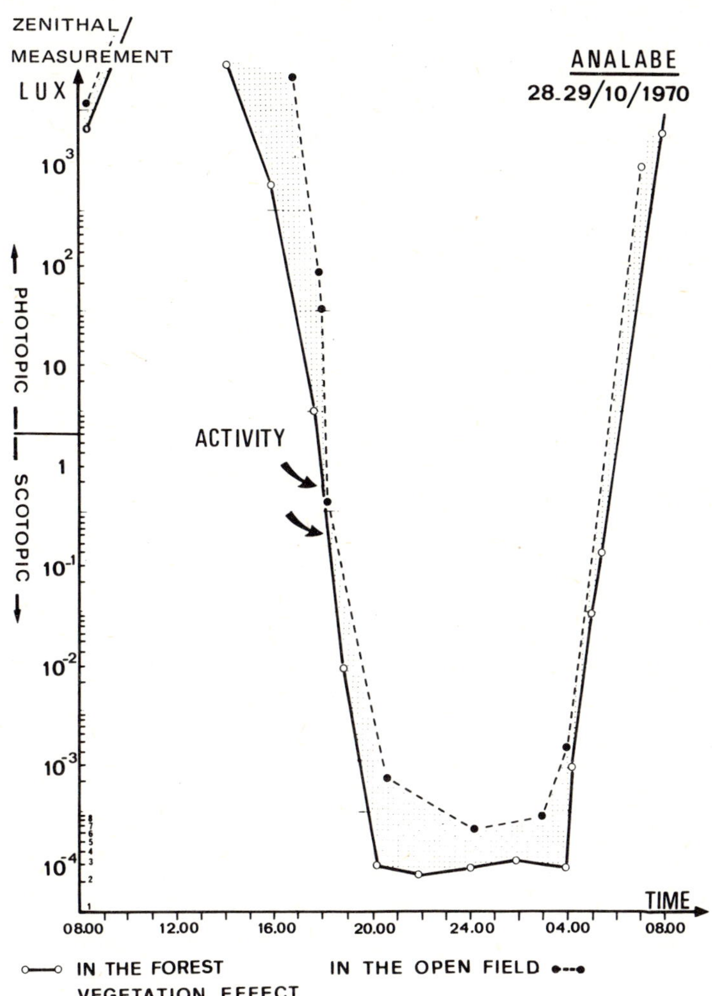

Figure 9. Visual conditions both inside and outside of the forest at Analabe, the difference between the two curves showing the screening effect of the vegetation.

Acknowledgments

This study was carried out with the help and generosity of Mr de Heaulme who made available the facilities both for living and working in the field at Analabe. The assistance of the Service des Eaux et Forêts of Madagascar is also gratefully acknowledged.

NOTES

1 Charles-Dominique, P. and Hladik, C.M. (1971), 'Le *Lepilemur* du sud de Madagascar: écologie, alimentation et vie sociale'. *Terre et Vie* 25, 3-66.

2 Petter, J.-J., Schilling, A. and Pariente, G. (1971), 'Observations éco-éthologiques sur deux lémuriens malgaches nocturnes: *Phaner furcifer* et *Microcebus coquereli*', *Terre et Vie* 25, 287-327.

3 Pariente, G. (1970), 'Retinographies comparées des lémuriens malgaches', *C.R. Acad. Sci.* 270, 1404-7.

4 Le Grand, Y. (1968), *Light, colour and vision*, London.

5 Wolin, L.R. and Massopust, L.C. (1970), 'Morphology of the primate retina', in Noback, C.R. and Montagna, W. (eds.), *Advances in primatology*. New York.

PART I SECTION B
Laboratory Studies of Behaviour

J. A. BERGERON

A prosimian research colony

Introduction

The Duke University Primate Facility contains about 180 prosimian primates used for research in anthropology, genetics, anatomy and animal behaviour. It is located in a secluded section of Duke Forest approximately two miles from the main campus of Duke University, adjacent to the Duke University Field Station for the Study of Animal Behaviour.

The Primate Facility staff consists of four faculty members and seven full-time non-academic employees. During the 1971-1972 academic year, seven graduate students and six undergraduate students conducted research at the Facility.

The Primate Facility was originally conceived by J. Buettner-Janusch while he was in the Department of Anthropology at Yale University. He founded the nucleus of the present prosimian colony in 1959, working with Professors William Montagna, then at Brown University, and Richard Andrew, then at Yale. On moving to Duke University in 1965, he brought with him the nucleus of the present colony. Peter Klopfer and others in the Zoology Department of Duke University helped to plan the present building. Research in genetic and evolutionary biology was the original impetus for the colony's existence; behavioural research developed as an adjunct to this primary interest.

Physical description

The Primate Facility building has 12,000 square feet of floor space. The basic floor plan of the building is three towers connected by intersecting corridors that form a T, as shown in Fig. 1. Two of the towers house animals exclusively; each contains five rooms.

The corridor that connects the two animal wings has an electronics

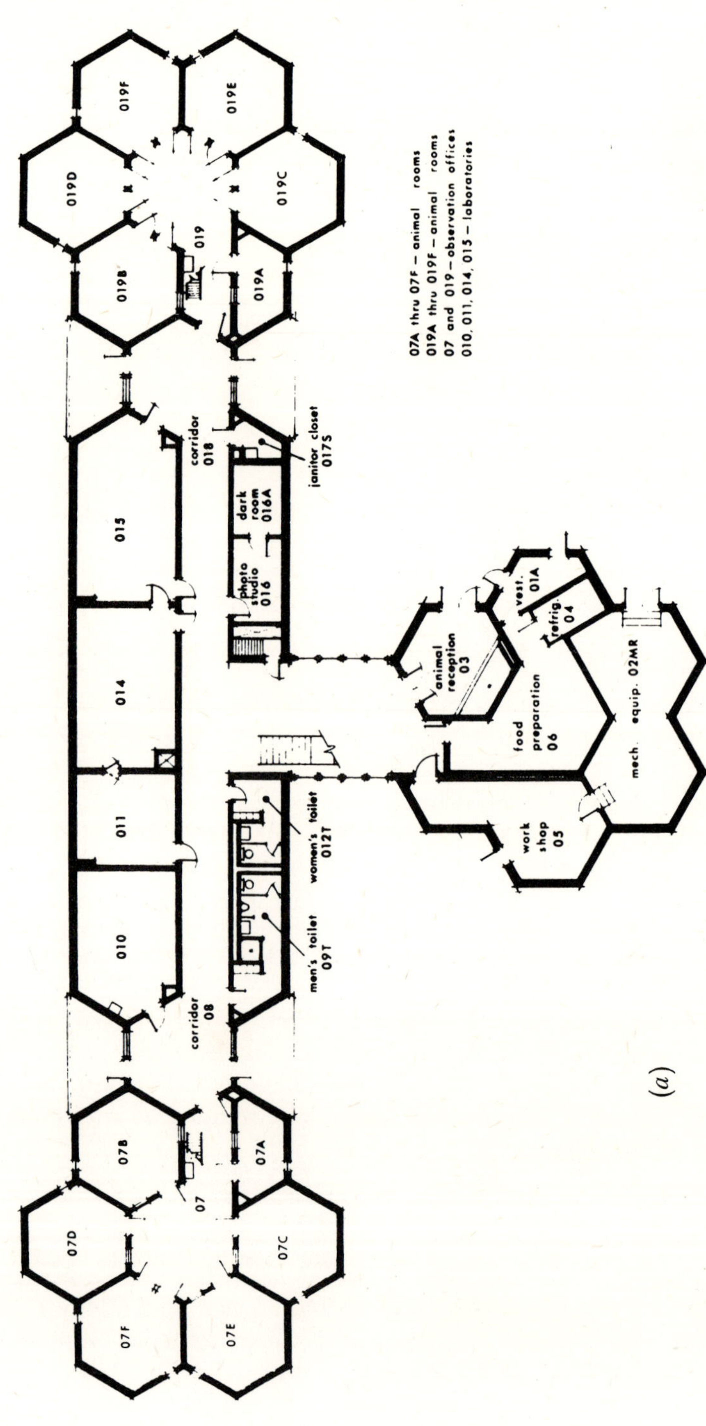
019F
019E
019D
019C
019
019B
019A
corridor 018
janitor closet 017S
015
dark room 016A
photo studio 016
vest. 01A
refrig. 04
animal reception 03
mech. equip. 02MR
food preparation 06
014
011
women's toilet 012T
work shop 05
010
men's toilet 09T
corridor 08
07B
07A
07
07D
07C
07F
07E
07A thru 07F – animal rooms
019A thru 019F – animal rooms
07 and 019 – observation offices
010, 011, 014, 015 – laboratories

(*a*)

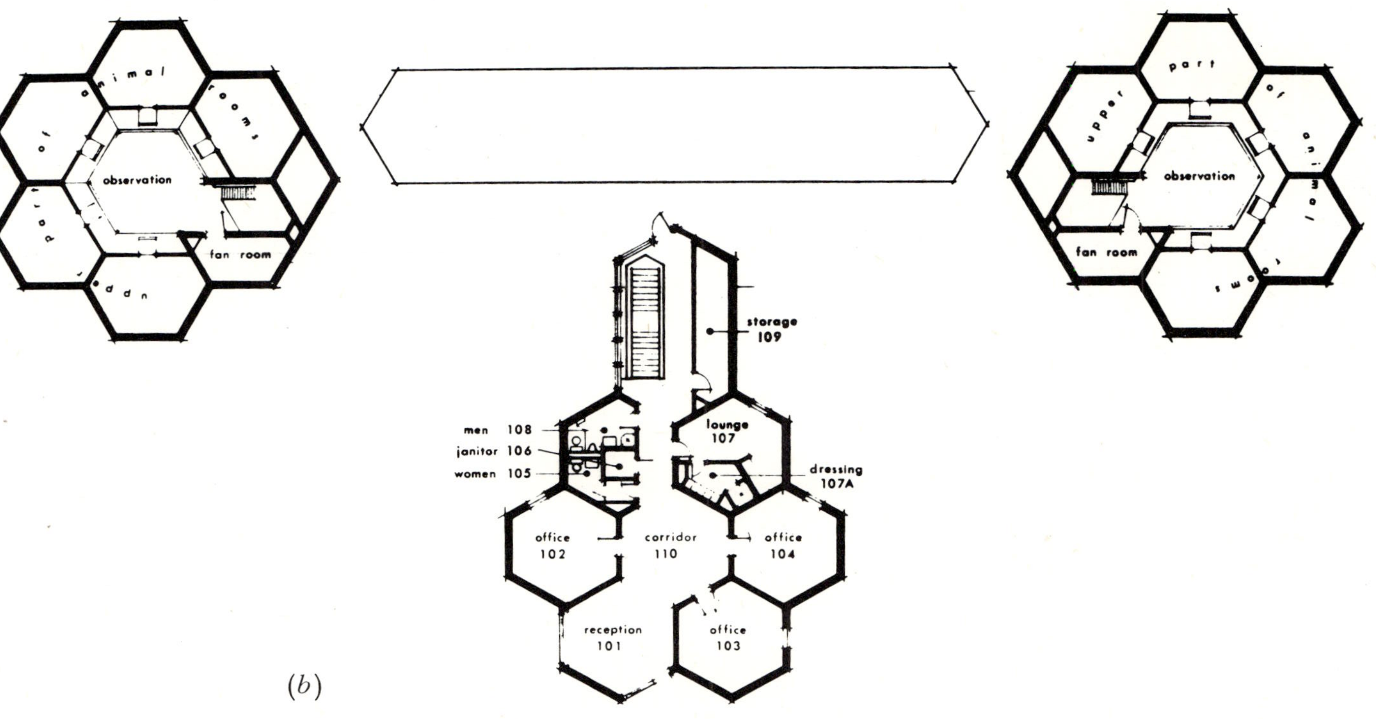

Figure 1. Floor plan of Duke University Primate Facility. (*a*) Lower level. (*b*) Upper level.

shop and individual laboratories for behavioural experiments, cytology, electron microscopy, and anatomy adjoining it. The remainder of the space off the main corridor is devoted to locker and shower rooms, a treatment room, and an isolation room for sick animals. The stem of the T connects the main corridor to the third tower, which contains cage-cleaning equipment, the kitchen, and a workshop on the ground floor, whilst there are administrative offices, a conference room and lounge on the upper floor.

In one of the hexagonal wings (019), two of the animal rooms have been divided into five individual compartments, and two of the rooms have been partitioned in half. The animal rooms are provided with climbing ropes, posts, dowels, and resting shelves (Fig. 2). The floors of the animal rooms are covered with pine shavings. The windows in the walls and ceilings are one-way glass to avoid distracting the animals when they are under observation. In addition to these larger rooms where the animals are housed in conditions resembling a free range, a number of animals are kept in cages of usual size.

There are also three outdoor holding areas. Two of these areas are large enclosures constructed of one inch mesh chain-link fencing. Each encloses about 1200 square feet and is between 8 and 10 feet high. Shelters are provided for the animals within these enclosures. Each shelter has plate glass windows on three sides and wood on the fourth. The animals come and go through a hole cut in the wooden side. The shelter is supported by posts and the floor (wood) is between 6 and 8 feet off the ground. Independent, thermostatically controlled heaters in each shelter maintain adequate temperatures even in coldest weather. Ambient temperature ranges from -18°C to +38°C. The third outdoor shelter, a commercial corn crib made of welded wire panels with a conical metal roof, is 16 feet in diameter and 24 feet high. There are resting shelves and climbing bars at several levels, and there are radiant heaters on the underside of the roof. The two larger outdoor enclosures have been in use since the autumn of 1968. One group of animals has been continuously housed in this enclosure since the summer of 1970. The animals which have lived in these enclosures include *Lemur fulvus* subspecies, *Lemur catta* and *Lemur macaco*.

The animals inside the building are kept under totally artifical light. Most of the animals are kept on a cycle of 12 hours of light and 12 hours of darkness. There are two separate sets of lights; fluorescent tubes for the light part of the cycle and red incandescent bulbs for the dark part of the cycle. In some rooms the light cycle is varied to correspond as closely as possible to the natural light cycle in North Carolina, with approximately nine hours of light in the winter and up to fifteen hours of light during the summer. There is, however, insufficient information to know whether this alteration in light cycle has any effect on breeding. The animals housed under a

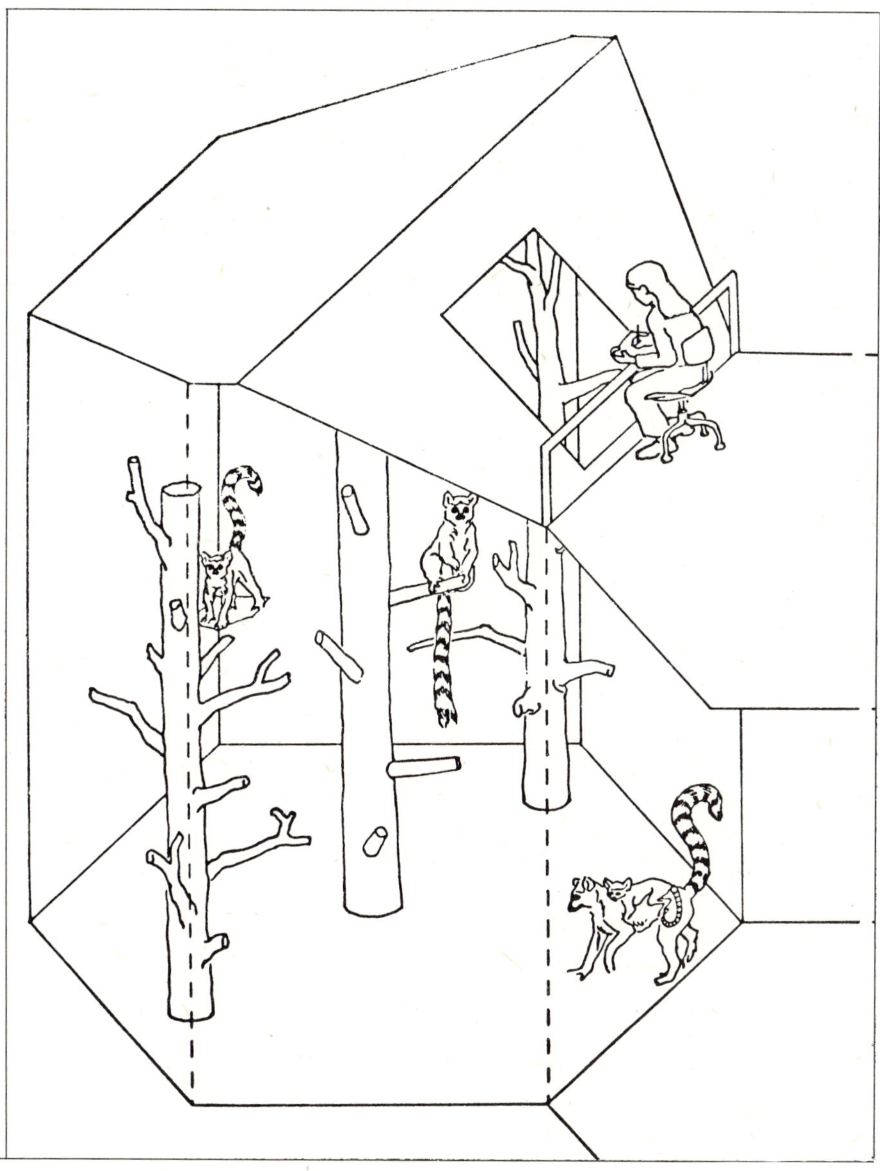

Figure 2. Typical behaviour observation room.

twelve hour light/twelve hour dark cycle breed at about the same season as those animals kept outdoors.

Diet

The basic diet fed to all animals at the Primate Facility is Purina Monkey Chow supplemented with a variety of fresh fruits and vegetables. Twice weekly the animals are fed a mixture of wheat cereal to which is added vitamins. In addition, mineral oil is added to the cereal mixture twice monthly; this helps prevent formation of hair balls in the digestive tract. Certain species require special dietary supplements. *Hapalemur* receives bamboo, which fortunately is obtainable quite readily from the grounds surrounding the building. *Propithecus verreauxi* are fed mango leaves and branches shipped weekly from Florida. These animals apparently need the large quantity of cellulose which comes from these leaves and branches. The Lorisidae and Cheirogaleinae are fed raw liver, canned dog food and hard-boiled eggs weekly as supplements. At times, the diet of these omnivorous animals has been supplemented with insects such as flour beetle larvae (mealworms) and crickets; however, it has not been found necessary to feed insects routinely to maintain their health.

The colony

The colony at present consists of about 180 prosimians: one genus of the infra-order Lorisiformes and six genera of the infraorder Lemuriformes. The galagos are *Galago crassicaudatus crassicaudatus* and *G.c. argentatus.* Sixty-six percent of the galagos in the colony have been born in captivity. Several of these animals are the fifth generation within the colony, and in several instances there are three living generations. The oldest captive-born galagos are now eleven years old. There are ten species of Lemuriformes, including four subspecies of *Lemur fulvus.* The animal population is shown in Table 1. Fifty-seven percent of the lemurs in the colony have been born in captivity, and at present there are four living generations. The oldest lemur born in the colony is eleven years old; another has lived in captivity for twelve years. The extensive pedigrees and the number of generations with known ontogenetic histories make this colony uniquely valuable for genetic and behavioural research (see notes).[1–12] Baseline physiological and haematological data on these animals have been compiled,[1–4,10,11] and information on problems of geriatric disease is beginning to emerge.[5,6]

Table 1 Prosimian population at the Duke University Primate Facility, 1972

Species	*Number of animals*
Galago crassicaudatus crassicaudatus	41
Galago crassicaudatus argentatus	26
Lemur catta	14
Lemur fulvus albifrons	7
Lemur fulvus collaris	3
Lemur fulvus fulvus	9
Lemur fulvus rufus	14
Lemur macaco	8
Lemur mongoz	1
Lemur variegatus	7
Lemur fulvus fulvus x *Lemur fulvus rufus*	19
Lemur fulvus rufus x *Lemur fulvus collaris*	2
Lemur macaco x *Lemur fulvus*	2
Hapalemur griseus	4
Lepilemur mustelinus	1
Cheirogaleus medius	3
Microcebus murinus	10
Propithecus verreauxi coquereli	6

Health and disease

The galagos have been of particular importance in studies of pathology. There is an exceptionally high incidence of chronic renal disease, in most instances taking the form of glomerulonephritis similar to that in man. Evidence of other age- and stress-related diseases similar to those which plague the human population has been found: atherosclerosis, hepatitis, bile stones, peptic ulcers, arthritis and cancer. One case of congestive heart failure and a case of auto-immune haemolytic anaemia in galagos have been seen. The most common health problem in galagos results from fights, in which severe bite wounds may be inflicted, generally on the head and tail. Often these bite wounds, particularly those on the tail, are extensive and cause severe blood loss resulting in shock and death. Bite wounds in galagos also tend to make the animals susceptible to bacterial infection.

Relatively few health problems occur with the lemurs. Several cases of acute haemorrhagic pancreatitis have been seen, but this is the only disease which has been seen on more than one occasion. One case of accidental death in a *Lemur catta* resulted from a fall in which two cervical vertebrae were fractured. Two lemurs died as a result of drug reactions. Complete post mortem examinations are performed, and the eviscerated corpses are frozen and saved for anatomical studies and skeletal preparations.

Special problems have occurred over the use of certain drugs. For antibiotic therapy, penicillin, penicillin-streptomycin combinations, tetracycline and chloramphenicol are used. Problems have occurred with procaine penicillin, particularly when used for galagos, and with a penicillin-streptomycin combination, and therefore this drug is not routinely used. Benzathine penicillin is used with no untoward effects. The antibiotic of choice for most infections is chloramphenicol (Chloromycetin). No adverse effects have been observed with this drug, and no evidence of bone marrow depression, as is reported in man, have been observed. The dose of chloramphenicol usually used is 25 mg. per kg./body weight. On some occasions oral antibiotic therapy has been used and it has been found that pediatric suspensions (generally fruit-flavoured) are quite useful for this purpose. The animals will usually take the drug readily when it is offered through the door of the cage.

For anaesthesia, intramuscular barbiturates have been highly successful. Hexabarbital sodium (Brevital) is used for short-acting anaesthesia and pentobarbital sodium (Nembutal) for longer anaesthesia. Doses of 30 to 35 mg. per kg. body weight are satisfactory. For most minor procedures, when sedation is sufficient, ketamine hydrochloride (Ketalar) is used. Several companies market ketamine in the United States. It is the drug of choice in sedation of cats, the recommended dosage ranging from 2.5 to 10 mg. per kg. For most routine use in prosimians, however, the smaller dose – 2.5 mg. per kg. – is adequate, and it has rarely been necessary to use higher dosages. Two lemurs have been lost following severe reactions to phencyclidine hydrochloride (Sernylan), and variable success has been obtained in using it as a sedative for galagos.

General management

Each of the animals in the colony has a three or four digit identification number. The first or second digit is constant for each genus and among galagos is constant within a breeding lineage. Each animal has a name, since naming the animals leads to better care and individual attention by members of the staff. Various physical characteristics are used to identify the animals visually, and when two animals look very much alike a system of ear notches is used. Ear notching as a means of identification poses a problem among galagos, since fighting often results in bite wounds on the ears and the obliteration of the original notches.

Each individual animal record contains a master identification sheet which contains ontogenetic information, a log sheet for all clinical laboratory data, and a chronological list of all manipulations made on that particular animal. A daily log is used to record manipulations and all animals in the colony are checked visually every day. Any that appear to be wounded, lethargic, or not eating well are examined thoroughly by the veterinarian. Progressive weight records are maintained on all animals, the animals being weighed once every three months; these records are particularly useful as indications of insidious disease.

Lemurs in small cages or enclosures are caught by hand, and those in large rooms or outdoor runs are caught in a nylon net. Heavy leather gloves are always worn when handling the animals. An animal is restrained by holding it around the neck at the base of the skull. Lemurs can also be restrained by a single individual who pulls the animal's arms behind its back and holds the upper arms together at the elbow.

Blood samples are collected from the animals without using sedation. They are physically restrained and blood is drawn from either the saphenous vein or the femoral vein. The saphenous vein is convenient for samples of 1 or 2 ml., and the femoral vein is used for samples of 5 or 10 ml.

Breeding

The management of breeding is centred around the genetic research. Galagos are routinely housed in pairs and occasionally in groups of one male and several females. The lemurs are housed in pairs, in groups of one male and several females, or in groups of several males and several females. The compositions of the groups are arranged so that the transferrin or acid phosphatase phenotype of any offspring can be used to identify unequivocally the male parent. All female lemurs are checked for pregnancy by abdominal palpation every two months. Pregnancy can be diagnosed as early as six weeks. When pregnancy is confirmed, the animal is checked once every four weeks until late in gestation, and then once weekly until parturition.

The large lemurs appear to mate in late autumn and early winter; births occur in late winter and early spring. The months in which most lemur births occur are March and April. The galagos do not appear to breed seasonally. The author does not believe that there is sufficient evidence to postulate seasonal breeding in the large lemurs; he has no data that indicate a breeding cycle correlated with environmental factors.

Cannibalism of newborns has occurred twice in lemurs. It seems likely that this has occurred either because the infant was the object of displaced aggression when adults were fighting or because the infant was not in good health when born. The latter explanation is the more probable. The ill or weak infant may trigger some response in the male or female parent that leads the parent to kill the infant. Cannibalism is much more frequent among galagos than among lemurs. Female galagos must be isolated prior to parturition, or the newborn infant will be killed by any other animal in the cage. Until the infant reaches a certain stage of social development, other adult galagos appear to react toward it as toward a bit of live food, according to experiments conducted by L.R. Rosenson. Rosenson suggested on the basis of these results that in the natural habitat female galagos isolate themselves at the time of parturition.[13]

The highlights of the present breeding programme include the births of several hybrid *Lemur* and three *Propithecus verreauxi coquereli.* Two of the hybrids are offspring of a male *Lemur fulvus collaris* (2N = 50)* and a female *Lemur fulvus rufus* (2N = 60). This pair has produced a male in March, 1970, and a female in July, 1971 (both 2N = 55). These hybrid offspring are thriving and appear to be normal in every way. There is no information about their fertility. Also, a male *Lemur macaco* (2N = 44) has been mated with a female *Lemur fulvus* (2N = 60), and they have had two female offspring (2N = 52), one in April, 1970, and one in May, 1971. The older of the two *Lemur macaco* x *Lemur fulvus* hybrids came into oestrus in the autumn of 1971, but there was no subsequent indication of pregnancy. There are a number of *Lemur fulvus fulvus* (2N = 60) x *Lemur fulvus rufus* (2N = 60) hybrids in the colony, and several of these have been back-crossed to *Lemur fulvus fulvus* or *Lemur fulvus rufus* and have had viable offspring.

The first *Propithecus verreauxi coquereli* was born in the colony in May, 1970. This infant was injured shortly after birth by an adult *Propithecus* in an adjacent cage, and it died within 24 hours of birth. A female offspring was born in December, 1970, from the same parents that produced the animal that died. A male offspring was born to the same parents in February, 1972. This family group, consisting of the parents, the year-old female offspring, and the male infant, are housed together.

The author feels that the success with the breeding programme and with the general management of the colony makes the Duke University Primate Facility a unique institution for breeding of prosimians for genetic, behavioural[7,8,9,12] and other biological research.

Acknowledgments

The author's attendance at the seminar was made possible by an international travel award from the National Science Foundation, No. GB-33212. The Primate Facility is supported by Duke University and U.S. Public Health Service grant RR-00388. Fig. 2 was prepared

* Diploid chromosome numbers.

by Miss Linda Hobbet, and Mrs Jane Hurlburt assisted in the preparation of this manuscript.

NOTES

1 Bergeron, J.A. and Buettner-Janusch, J. (1970). 'Hematology of prosimian primates: *Galago, Lemur* and *Propithecus.*' *Folia primat.*, 13, 155-165.

2 Bergeron, J.A. and Buettner-Janusch, J. (1971). 'Hematology of prosimian primates. II: A comparative study of Lemuriformes in captivity in Madagascar and North Carolina'. *Folia primat.*, 13, 306-313.

3 Buettner-Janusch, J. (1968). 'Man's place in nature'. *J. Invest. Dermatol.*, 51, 309-323.

4 Buettner-Janusch, J. and Wiggins, R.C. (1970). 'Haptoglobins and acid phosphatases of *Galago*'. *Folia primat.*, 13, 166-176.

5 Burkholder, P.M., Bergeron, J.A., Sherwood, B.F. and Hackel, D.B. (1971). 'A histopathologic survey of *Galago* in captivity'. *Virchows Arch. Abt. A Path. Anat.*, 354, 80-98.

6 Burkholder, P.M. and Bergeron, J.A. (1970). 'Spontaneous glomerulonephritis in the prosimian primate *Galago*'. *Am. J. Path.*, 61, 437-450.

7 Klopfer, P.H. (1970). 'Discrimination of young in Galagos'. *Folia primat.*, 13, 137-143.

8 Klopfer, P.H. and Jolly, A. (1970). 'The stability of territorial boundaries in a lemur troop'. *Folia primat.*, 12, 199-208.

9 Klopfer, P.H. and Klopfer, M.S. (1970). 'Patterns of maternal care in Lemurs: I. Normative description'. *Z. Tierpsychol.*, 27, 984-996.

10 Nute, P.E. and Buettner-Janusch, J. (1969). 'Genetics of polymorphic transferrins in the genus *Lemur*'. *Folia primat.*, 10, 181-94.

11 Nute, P.E., Buettner-Janusch, V. and Buettner-Janusch, J. (1969). 'Genetic and biochemical studies of transferrins and hemoglobins of *Galago*'. *Folia primat.*, 10, 276-287.

12 Roberts, P. (1971). 'Social interactions of *Galago crassicaudatus*'. *Folia primat.*, 14, 171-181.

13 Rosenson, L.R. pers. comm.

G. A. DOYLE

The behaviour of the lesser bushbaby

Since 1964 a colony of lesser bushbabies (*Galago senegalensis moholi* – Fig. 1) has been established under semi-natural conditions in our laboratory at the University of the Witwatersrand. The animals have bred freely and to date more than 50 infants have been born, approximately 75% of which have survived into adulthood. It is reliably estimated that this represents a survival rate considerably higher than that in the wild.

Study conditions

The semi-natural conditions in which our bushbabies have been housed and studied have been designed to duplicate as far as is practically possible the natural habitat to which the bushbaby has successfully adapted. While they fall far short of this goal the conditions are, nevertheless, vastly superior to small standard laboratory cages.

The animals under observation are housed in natural groups of one adult male, one or two adult females and their young in four large cages approximately 5′ x 6′ x 7′. The centres of all cages contain branching systems, and ledges and glass-sided nest boxes have been randomly placed on the walls. The floors are covered with a thick pile of wood-shavings. Temperature is allowed to fluctuate between 15°C and 30°C, which is well within the seasonal range characteristic of the natural habitat. The day/night cycle has been changed to suit the observers. The animals are active during 11 hours of red light (between 12.00 and 23.00 hrs.) which alternate automatically with 13 hours of white light when animals retire to the nest boxes to sleep. Daily diet consists of finely chopped-up fruits and vegetables in season, mealworms, milk supplemented with a commercially available vitamin B-complex syrup, and a commercially available food supplement in porridge form containing balanced proportions

Figure 1. Lesser bushbabies (*Galago senagalensis moholi*) photographed at night under natural conditions.

of proteins, vitamins, minerals and salts. Approximately every third day locusts are added to the diet. Animals will eat a variety of other foodstuffs, including nuts, dried fruit, bread and honey, but will not eat dead insects or raw meat. The cages are cleaned and food is placed on the ledges between 09.00 and 12.00 hrs. when the animals are asleep. A detailed description of the Animal Laboratory and methods of care and maintenance have been given elsewhere.[1]

Aim of research

The aim of the present series of studies is to obtain quantifiable data, on a time-sampling basis over long periods of time and under conditions of minimal interference, on various aspects of normal behaviour of the type that cannot be readily and completely obtained in the wild. No *a priori* theoretical assumptions have underlined our studies. Few experimental manipulations have been introduced and the only interferences to which our animals have normally been subjected are the daily removal of infants for weighing, the weighing of adults at regular intervals, the removal of sick animals for treatment and occasional changes in the composition of cage groups. Data in the form of activity counts and time scores are recorded on recording instruments and then entered after each observation session on data sheets which provide for the recording of descriptive data which cannot be quantified. To date approximately 5000 hours of laboratory observations and 500 hours of field observations have been completed.

The ultimate aim then of this series of studies is to provide a detailed and objective ethogram of the lesser bushbaby, and it is based on the assumption that no information of any value on the natural behaviour of a species is yielded by short studies of very few animals under highly artificial conditions. There are, of course, exceptions to this dictum in the form of highly specific studies such as those of Andrew,[2,3] Jolly,[4,5] Bishop,[6,7] and others.

This paper will be devoted to a broad, objective overview of the behaviour and other relevant aspects of the biology of the lesser bushbaby, based largely on quantitative and descriptive data from our own studies. Because of limitations of space, minimal reference is made to those of our own studies which have already been fully reported,[8,9,10] to the studies of others which are implicitly acknowledged, or to theoretical considerations.

General observations

1. The natural habitat

Galago senegalensis has adapted to a rather more generalised habitat than most prosimians, thriving throughout Africa in the drier, wooded bushland and savannah type country from as far north as Senegal, Nigeria, Sudan and Ethiopia, through Central and East Africa and down as far as Rhodesia, South-West Africa, Botswana and the northern Transvaal in the South.[11,12] Like all prosimians it is arboreal, but descends frequently to the ground to cross open spaces, to search for food and possibly to escape from predators.[13] *G. senegalensis* may be found at all heights and is equally at home on the heavier branches and trunks of large trees as it is in the fine branch zone. It thrives in a wide range of temperatures, for instance in South-West Africa where the seasonal variation is from -8°C to +47°C, and in the northern Transvaal where seasonal variation is from -5°C to +45°C.[13,14]

2. Nesting

In the laboratory *G. senegalensis* will use all available nest boxes, but shows a distinct preference for one. Very rarely is nesting material ever carried into the nest box except by the female prior to parturition. No recognisable nest ever results. In the wild *G. senegalensis* sleeps in dense thorny trees, on nests, forks, branches, in the midst of clumps of foliage and occasionally in tree hollows. In general, nests are difficult to locate from the ground. They may be occupied for long periods of time except in the case of females with young, when nesting sites may be changed every 10 to 14 days. Nests are invariably constructed from flat, fresh, soft, broad leaves and leafed branches and placed in thorn trees to form open-topped platforms. Occasionally they are built on a platform of twigs formed by old birds' nests and sometimes an old bird's nest is used without modification. With seasonal changes in leaf cover and temperature there is a gradual change in sleeping places throughout the year as some trees become more or less suitable. In cold weather *G. senegalensis* might expose itself to the direct rays of the sun at the ends of branches, but rarely does so in summer. Variations in type of nest, nest site and location of nest reported for other sub-species in different parts of America probably represent adaptations to localised habitats.[15]

3. Activity rhythms

Like all nocturnal prosimians, *G. senegalensis* begins its activities at dusk and settles down to sleep at about dawn. Both in the laboratory and in the field there is a period of very high activity in the first and second hours after waking, followed by irregular periods of activity and inactivity with a second period of high activity just before sunrise. *G. senegalensis* alternates between periods of rapid movement, sleep, rest, exploration, feeding, toilet activities and social activity independent of any pattern or rhythm. Heavy rain or strong wind make little difference to normal activity.

Regularity and bimodality of waking activity is much more marked in the social prosimians such as *Lemur*, where members of a troop rest, move and feed more or less as a unit,[16,17] than it is in the more solitary species where one animal may be active while another rests, as in *Microcebus* or *G. senegalensis.*[13,18]

4. Ingestive behaviour

Various insects as well as *Acacia* gum are the main food sources, and *G. senegalensis* has never been seen to display any interest in vegetable matter nor in small vertebrates in the wild although a variety of vegetable and animal matter is taken in the laboratory.

Even in the laboratory the most preferred food is insects. When *G. senegalensis* catches live prey with its hands, it shuts its eyes at the moment of impact and puts back its ears as a protection against insects with flapping wings and spurred legs (e.g. locusts). Once the fingers have closed round the prey and the head has been bitten, the eyes open and the ears come forward. Prey is first observed very carefully; the whole prey-catching gesture is very stereotyped and cannot be followed by the naked eye. Bearder confirms this behaviour for *G. senegalensis* in the wild.[13] Food is very rarely taken directly by the mouth and insects never. One hand is almost invariably used both to seize food and to hold it while being eaten. Food too large to be held or picked up with one hand is held in or down by both hands.

G. senegalensis does not use the lower procumbent incisors for eating. In biting or shearing food the canines or premolars are used, as Buettner-Janusch and Andrew[19] and Andrew[20] have shown for both *Galago* and *Lemur.* Premolars and incisors are used in detaching pieces of food only when the food is too large to be tackled directly by the molars.

When feeding in the wild, *G. senegalensis* moves rapidly from branch to branch and from tree to tree. *Acacia* gum is licked or eaten in small pieces and *G. senegalensis* may change its position several times while eating at the same spot. Bark may be chewed to expose the gum.

G. senegalensis drinks by lapping like a cat. It has never been seen to strike the surface of water or milk with the hand and lick the fingers. It will drink both water and milk, the former in small quantities in captivity, though it has never been seen to drink in the wild nor to lick the condensation from the surface of leaves, even in dry weather.

Food-stealing is common in *G. senegalensis* in captivity. Animals spend a great deal of time pursuing each other and all observers agree that food-stealing never leads to aggression.[15] The animal from which food is stolen may try to get it back or simply look for some other food. Food is usually stolen with the hand from the mouth or hand of the other.

Many insects found in the stomach of *G. senegalensis* are ground-living forms,[21] and Bearder[13] reports individuals spending as much as 15 minutes at a time on the ground foraging for insects. *Acacia* gum is found in the forks of trees, on the undersides of branches and near the bases of tree trunks, where it may be licked or eaten for up to 15 minutes at a time. As much as 2¼ hours may be spent in a single tree catching insects when conditions are favourable, while at other times as much as three hours may elapse without feeding. In general, the maximum amount of feeding occurs in the first few hours after dark. However, eating occurs throughout the night on an average of 6 times per hour in the wild. In captivity *G.*

senegalensis also feeds intermittently through the night. Diet and feeding habits vary seasonally. *Acacia* gum is available throughout the year and more time is spent eating gum and looking for insects in the winter than in the summer. During winter there is a noticeable change in the condition of the bushbabies.

Locomotion and posture

Detailed studies of the manipulative behaviour of a number of prosimians, including *G. senegalensis*, in respect of such activities as feeding, grooming, play, agonistic behaviour and general exploration have been reported by Bishop[6] and will not be repeated here.

1. Locomotion

G. senegalensis is a great hopper and leaper. Laboratory studies show that, except for rare occasions when it walks quadrupedally, its preferred mode of locomotion is by hops and leaps and it will jump to and from vertical supports with the same frequency and facility that it will use horizontal supports. Bearder[13] confirms this for *G. senegalensis* in the wild, where the natural habitat does not require exclusive adaptation to a vertical branch zone. In many thorny trees, too dense for jumping, *G. senegalensis* moves carefully and quadrupedally along the smaller branches. In locomotion across large open spaces, any small trees, fence posts or fallen branches may be used as vantage points. In order to progress from one tree to another without going to the ground *G. senegalensis* may make a series of jumps from one fence post to another spaced 10 feet apart. On one occasion *G. senegalensis* was seen walking quadrupedally along 70 yards of fencing wire, the accompanying tail movements appearing to subserve balance. When it has to descend to the ground *G. senegalensis* checks the environment carefully in all directions for as long as 20 minutes before leaving a tree. When on the ground it assumes an upright posture and takes long rapid bipedal jumps. Occasionally, movement along the ground is performed in a broken series of short hops covering distances of up to 60 yards. In the confined space of the laboratory, locomotion on the ground is always by a series of short hops, kangaroo fashion.

Normal standing jumps are performed from a crouched position, propulsion being achieved exclusively by the hind legs, which are brought forward to take the force of the impact on landing. During the jump the hands are usually held against the chest and are then extended to assist in gripping the substrate on landing. A variation of this occurs during very long jumps of 15 feet or more in the wild, when the arms are held forward and above the head to provide extra momentum. In the field as well as in the laboratory *G. senegalensis* carefully assesses the situation before making a long horizontal jump, by moving its head from side to side, checking all directions from a bipedal standing posture and gauging the distance from a number of vantage points. The ears move independently to and fro and the head may rotate owl-like through 180°. However, when a series of quick jumps is made in rapid flight, as in chasing, a succession of long jumps may be made without hesitation and with complete accuracy. *G. senegalensis* may also climb in any position that the substrate requires, upside-down on a horizontal branch, upwards or downwards on vertical surfaces.

2. *Sleeping and resting postures*

Sleeping postures vary greatly in *G. senegalensis*, particularly when they sleep in groups. They may sleep on the side with the head covered with hands and tail, or crouched in a saggital plane with the head bent and resting on the hands, the tail curled round the head. Occasionally animals sleep or rest in a sitting, vertical position or on their backs. During sleep the ears are retracted by means of the transverse folds. In stretching *G. senegalensis* bends the body almost to a lordosis, extending the limbs very much like a domestic cat. It does this frequently, particularly after sleeping, with many variations, one of which consists in hanging by the feet with hands extended downwards.

Maternal behaviour

Where observations of maternal behaviour have been made in the wild they fully support the detailed laboratory data that have already been reported.[10] In respect of the behaviour of the mother in changing infants from one nest to another, particularly in the first two weeks of life, some additional data are provided: Bearder[13] reported that an infant may be carried through several trees to a point some distance from the nest site, where it may be briefly groomed and left to cling, while the mother returns to get the other infant. Infants may be placed together or they may be left as much as 40 yards apart for short periods and sometimes for the greater part of the night. If left alone they may each be visited sporadically for short periods. This phenomenon, known as 'parking', is a common feature of maternal behaviour in the African lorisines.[15] The mother never fails to retrieve both infants back to the same nest site before dawn.

Infant development

G.s. moholi doubles its birth weight in from seven to nine days and a five-fold increase takes place over four weeks. Eyes are open at birth and within an hour infants can crawl on all fours awkwardly and with the belly on the substrate and can cling tightly to the mother's fur with the hands. By the fourth day infants can climb on wire mesh and Bishop notes that the full complement of adult prehensive patterns is present.[7] Before one week of age infants groom themselves and Andersson reports infant twins grooming each other on the second day after birth, in contrast to infant monkeys in which allogrooming does not occur.[22,23] By 10 days of age infants are extremely active in the nest, climb over adults, play and wrestle with each other, take small jumps of up to 6 inches and may stand bipedally with their hands against a vertical support and spend a good deal of time peering out of the nest box opening.

In the laboratory, infants emerge for the first time of their own accord from the nest box at between ten and fourteen days and Bearder notes that it is at about this age that they are first seen on their own in the wild.[13] The age at which the infants first emerge on their own does not seem to be related to the ease with which they are able to get out of the nest box, as Martin noted for *Tupaia.*[24] Infant *G. senegalensis* may get out of the nest box as early as even days, but they immediately get back in. It appears that not until they are at least ten days of age are they ready to face the outside world.

The threat posture has been seen well before two weeks of age. Urine-washing in its complete pattern, following ano-genital smelling and grooming, appears by three weeks and reciprocal ano-genital smelling has been seen at one month. Eating solids and drinking appears before three weeks. Insects are being caught and eaten by four weeks, but Bearder notes that in the wild gum is found and licked before insects are caught.[13] Suckling continues, but by about five weeks solids are being eaten more regularly and hands are employed more frequently in the eating process. By two to three weeks of age, infants run fast quadrupedally on branches and bipedal leaps are becoming longer and more frequent. Although they may frequently lose their balance at this age they seldom lose their grip.

Although at three to four weeks they can actively explore their environment, jump and take solid food, they are still dependent on their mothers for transport from tree to tree in the wild. At four to six weeks they may follow their mother, jump as far as three feet and descend to the ground for the first time, taking 1-foot hops to cross open spaces. After following their mothers from the sleeping trees in the evening, they may spend the rest of the night alone, feeding, playing, resting and moving from one tree to another over short distances. By this age play and locomotor ability have become quite elaborate.

In the wild, infants are probably self-sufficient and independent earlier than in the laboratory, where small cages keep animals together. After six weeks infants leave the sleeping site of their own accord and begin independent movement over an increasing range; but the family still comes together in the morning at the sleeping tree where infants probably still suckle.

Andersson reports that not until a year of age do males display the full adult aggressive pattern, emit the male call for the first time and begin to urine-wash with typical adult frequençy.[22] Females reach maturity earlier than males. The first onset of oestrus, mating and conception at about 200 days appears to be the rule in our laboratory.

Social behaviour

1. Olfactory communication: methods of scent deposition

The highly ritualistic method of depositing urine characteristic of the Lorisidae has been termed urine-washing by Boulenger.[25] In *G. senegalensis* foot and hand on the same side are raised simultaneously from the substratum; the hand is cupped under the uro-genital opening and a few drops of urine discharged into it. The hand then rubs the sole of the foot or grasps it firmly from one to several times; both hand and foot are then replaced on the substrate. The procedure may be repeated on the opposite side. The pattern is identical in both sexes, but it is performed far more frequently by the male than by the female. Under laboratory conditions, where animals are never isolated from each other, the male urine-washes from 3 to 22 times and the female from 1 to 3 times in a 50-minute observation period.

A second form of ritualised urine deposition, employed less frequently by the Galaginae than the Lorisinae, has been called rhythmic micturation.[26] Andersson describes it for *G. senegalensis* as follows:[22] the hindquarters are lowered to the substratum, a few drops of urine are released and at the same time the animal moves forward in a wriggling motion. It is more frequently seen in females than in males. *G. senegalensis* may also on occasion release a copious flow of urine on branches and may then wipe its feet in the urine or faeces thus deposited. This may not be a form of scent deposition, since other animals have not been observed to display any interest in it.

No concentration of cutaneous glands suitable for scent production is reported in *G. senegalensis*,[27] despite what appears to be glandular marking behaviour in this species. The adult male has a conspicuous bare patch on his chest which he rubs against the substratum or against branches. The chest is lowered against the substratum and the animal pushes first with one leg and then the other, sliding the chest over the surface. In chest-rubbing against a branch the arms are clasped tightly around the branch, hugging it closely. The animal then pushes forward with the hind legs, the head turning from side to side as the chest, side of the chin, and inner surfaces of the arms are rubbed along the branch. In the process the lower lip is dragged back and a tail of saliva is deposited. The sides of the mouth and chin may be rubbed against thin branches and ledges. Yasuda et al. report that the lips of *G. senegalensis* are well supplied with sebaceous glands.[27] After chest-rubbing against a branch the animal may bite it several times before repeating the movement. The same process may occasionally be performed by females and by juveniles. Vertical surfaces may also be marked in the same way, and while chest-rubbing against a flat surface the head may be rubbed when the animal comes up against a vertical surface.

Functions of scent-marking

The association between scent-marking, particularly urine-washing, and courtship is particularly obvious. The male will invariably urine-wash after genital-smelling or grooming others, particularly females, or after smelling a spot where a female has just urinated. When the female comes into oestrus, there is an approximate 30% increase in the frequency of male urine-washing and much of the male's urine is deposited directly on to the female. After genital-smelling a female, the male may vigorously chest-rub along her flank and back while adopting a mount-like position and this may be accompanied by urine-washing. In most and probably all prosimians, olfactory cues emananting from the female change and become stronger when the normally sealed vulva opens during oestrus.[11] There is an obvious increase in interest on the part of the male in females in neighbouring cages when they come into oestrus.

Urine-washing is more frequent in the dominant males, and chest-rubbing as well as urine-washing decrease in the male as dominance is lost and increase in the female as she becomes more dominant. The male urinates on strange females and on familiar females returned to the cage after a period of absence. Males generally urine-wash after chasing a subordinate or after staring at a subordinate prior to chasing.

In the nocturnal prosimians, particularly in the absence of sexual dimorphism, individual recognition is more likely to be olfactory than in the diurnal Lemuroidea, as Marler[28] has suggested and as Andersson[22] and Bearder[13] suggest for *G. senegalensis.* In *G. senegalensis* maternal recognition of infants from birth is certainly by olfaction. Repeated observations have shown that mothers are able to recognise their own infants from birth. When infants are removed by the observer for weighing they are returned in the hand for the mother to retrieve. All mothers will approach the hand and smell the infant, but only the real mother will pick it up, irrespective of which gets there first. The same occurs when mothers retrieve infants in distress.

A stranger placed in the cage increases the incidence of ano-genital smelling on the part of those present. Adult males in adjacent cages become aware of each other's presence and try to get at one another even when they cannot see one another. When strange animals avoid one another at first they will smell the places where each has been. A male may attack a female who has been with another male, though juveniles are never attacked under the same circumstances. Bearder suggests that mutual recognition is primarily by scent, which may carry information regarding age as well as sex.[13]

Hediger[29] pointed out that one of the aims of scent-marking in mammals appeared to be to keep the living area smelling of the animal, and Andersson[22] reports that *G. senegalensis*, particularly the male, thoroughly re-marks its cage by urine-washing after it has been cleaned. When placed in a new cage, the male may spend as much as an hour at a stretch chest-rubbing. New objects placed in the area are marked by urine-washing. Scent-marking is not restricted to making novel objects familiar, but also to keeping the familiar smelling strongly of the animal. For instance, a favourite site of scent-marking is the nest box and whenever an animal, but particularly the male, enters the nest box it invariably urine-washes, whether the nest box is empty or not. In the wild, scent-marking is concentrated in the sleeping tree and immediately adjacent area. Most scent-marking occurs after leaving the nest at sunset, which is presumably the time when the environment smells least of the animal. Scent-marking may reinforce territorial attachment as Martin[24] suggests, but it does not appear to act as a deterrent to rival conspecifics.[15]

It has been suggested for a number of prosimians that a function of scent-marking is the establishing of olfactory trails.[15] However, Bearder produces evidence for *G. senegalensis* which shows that scent-marking plays little or no role in establishing olfactory trails in the wild.[13] Most scent-marking occurs before the start of the night's activity in the small area with which the animal is most familiar, and thereafter it occurs only once or twice per hour, during which time *G. senegalensis* may move through as many as 45 to 50 trees over a distance of 200 yards. He reports too that *G. senegalensis* does not always use the same exact route, nor, with the exception of the sleeping area, does it mark at particular places. Finally, he reports that there is an absence of obvious sniffing during its movements in the wild except during social interactions and when investigating certain restricted areas within its home range. He suggests that a thorough knowledge of the home-range or territory is primarily visual.

It has also been suggested that a function of urine-washing is to facilitate the grasping of branches during movement through the environment. However, while male and female are equally active, the male urine-washes a great deal more than the female. Also, jumping or moving through the environment is not as a rule preceded by urine-washing. The evidence noted above that movement over large areas is seldom interrupted by urine-washing is yet another fact inconsistent with this hypothesis.

2. *Vocal communication*

Andersson was able to record a repertoire of 25 different vocalisations in *G. senegalensis* under laboratory conditions, but notes that if only discrete basic sounds are considered, while ignoring intermediate variations and combinations, then the range is reduced to 10 or 11 sounds.[22] Bearder reports that all the loud and many of the soft calls of *G. senegalensis* heard in the laboratory are also heard in the wild.[13] A number of authors have noted that the infant *G. senegalensis* emits tiny high-pitched cries which attract the mother and which cause her to go to the infant, pick it up and return it to the nest box, particularly if it is in difficulty.[15] Andersson reports that the small infant utters intense high-pitched clicks and crackles on loss of contact with the mother and later when it finds itself in difficulty or when it falls.[22] The incidence of clicking decreases as the infants gain confidence and become more independent. Juveniles utter clicks and crackles when introduced to a novel environment and, when frightened or disturbed, they may utter most adult calls except for the high intensity alarm call. In the wild, the female utters a soft maternal call on approaching the infants and the infants reply. These calls may be called contact calls uttered by mother or infant when separated and when they wish to restore contact. When the mother returns to the nest site just before daybreak, she utters a soft coo and three to four week old infants approach making frequent clicks. The infants follow the slowly-moving mother, continuing to make clicks which are not heard in her absence. Infants also utter soft notes while suckling and being groomed and when wriggling into close contact with the mother in search of the nipple.

Faint grunts are given when eating favourite food, a behaviour pattern found in many primates including human infants. Clicks and crackles are elicited by strange objects and by the smell of a strange fellow. A low moan preceded by a steeply rising then falling soft bark is given in warning in *G. senegalensis.* The range of alarm calls is correlated with the degree of intensity of the threatening situation. A typical situation is one in which one or two males start to call in a long-lasting series of increasingly penetrating sounds. Females may.

join in with the males. While calling the animals sit still, but they may periodically orient in different directions.

Some animals may move around and seek food while calling, but the call may change to a more penetrating-sounding chatter if a male discovers an intruder in his territory. The call appears to function to alert other animals. Activity stops, the animals jump higher in the branches and remain absolutely still. The call may continue for an hour or longer. No mobbing call has been reported for *G. senegalensis* similar to that reported for *G. crassicaudatus* by Jolly.[30] The majority of variants in the vocalisations of *G. senegalensis* are emitted during hostile encounters and mostly by the more submissive animal. The sound spectrograph shows that they are shortened versions of discrete calls. It is thought that they occur when the animal under stress is uttering various units of these discrete calls in such a rapidly changing series that it does not have time to give the full unit. The problems encountered in breathing control necessary to cope with strenuous exercise and simultaneous vocalising may account for the variations heard.

3. *Vision and visual communication*

Bearder showed that of 20 *G. senegalensis* none made any jumps over a period of two hours in complete darkness, although considerable movement took place if the environment was familiar.[13] If forced, *G. senegalensis* jumped randomly and landed on the floor. The long accurate jumps in an unfamiliar environment testify to the keenness of its vision. In the natural habitat *G. senegalensis* must be able to see at least over the distance of its maximum jumps (about 15 feet), but probably sees at least 30 to 40 yards. It moves about its habitat as actively and with the same facility in bright light and in almost total darkness. Sight and visual memory coupled with a good kinaesthetic sense enable *G. senegalensis* to move around its natural habitat with minimal reliance on olfactory cues. The status of strangers too is probably determined by visual as well as olfactory cues.

Andersson points out that in *G. senegalensis* the mouth, together with the eyes and various movements of the mobile external ears, give rise to several facial expressions, despite the physical handicap of a simple facial musculature, which, together with body movement and posture, provide the basis for a variety of visual signals.[2,22] Jolly suggests that the highly ritualised scent deposition gestures of *Lemur* almost certainly function as visual signals, both in those species which have scent glands, like *L. catta*, and in those which do not, like *L. macaco.*[16] The same could well be true in respect of the highly ritualised urine-washing displays of the Lorisidae.

4. *Territory and home-range*

In 15 *G. senegalensis* groups Bearder reports a mean home-range of 2.8 hectares (7 acres) varying from 1.25 to 3.95 hectares (3.0 to 9.6 acres).[13] He attributes this relatively large home-range for such a small animal to its great activity and the nature of the vegetation. Natural boundaries are formed by open spaces, roads, etc., and there is considerable overlap of home-ranges. They do not, however, appear to be as stable as in *Lemur*,[16] the size, shape and position of the home-range changing during the year. Like many prosimians and higher primates, *G. senegalensis* advertises its location vocally. It does not appear to occupy a territory in the sense of a core area defended against encroachment by conspecifics as in many other prosimians.[15]

In general, territorial fights in primates are uncommon, belligerence rarely going beyond various types of threat or agonistic display.[31] Conflicts may increase under crowded conditions, as in *Macaca mulatta* in India,[32] and fighting in male *G. senegalensis* is far more common in the laboratory than in the field. In the field, avoidance behaviour and, in particular, the typical male call, ensure separation between potential belligerents. Since the adult male is the dominant figure in each group and since, as Boulière pointed out,[33] territoriality is usually a male dominated activity, rivalry between adult males is probably the most important factor resulting in territorial spacing out.

In *G. senegalensis* the average path length during the night is about 2.0 km. (1.25 miles), representing movement through a mean number of 500 trees. It uses the same general pathways and although its route from one point to another is highly variable it frequently makes the same jumps and crosses the same open spaces. Although it is possible for the animal to traverse its home range in two hours it rarely does so. Field studies indicate that *G. senegalensis* has a thorough knowledge of its environment.

5. *Social groups*

The literature indicates that *G. senegalensis* is frequently seen in the wild singly or in pairs, occasionally in larger groups, and sleeping groups may be as large as seven.[15] Bearder found 238 animals in 119 sleeping groups.[13] Of these 75% were found in groups of two to four, the most common being an adult pair with young, an adult female with young, or a pair alone. Groups or individuals frequently use the same nests or sleeping sites, which form a home base to which they regularly return at the end of the night. The sleeping group does not function as a unit at night, but splits up for individual foraging; 70% of the waking time is spent alone and movement of two or more animals together for any length of time is rare. Before dawn, members of a group meet and move about together before arriving at the sleeping site. Each group lives in a definable area and the composition of the sleeping group varies during the year, probably due to increasing aggressiveness between adults as the young mature. Evidence from both the wild and the laboratory strongly suggests that the natural group in *G. senegalensis* is the family group and that the social organisation is relatively simple. In the laboratory, male and female with young and/or juveniles will live together peacefully. Provided that there is sufficient space, additional females and sub-adult males do not increase friction and females may live together peacefully and produce infants in the same nest.

6. *Contact behaviour*

In *G. senegalensis*, as in other Lorisidae, the need for bodily contact in adults does not appear to be as strong as that reported in some of the Lemuroidea,[16] although it may be equally strong in infancy. Adults display little contact behaviour except in courtship, mating, grooming, aggression, and very occasional play, although they always sleep in close contact even when there is ample space. Social contacts in the wild are brief and follow the same pattern as those observed in the laboratory, enabling the observer to tell the age and sex of the animals involved.

(*a*) *Greeting.* Naso-nasal contact is a common form of greeting in *G. senegalensis.* One animal will approach another slowly with the head stretched forward, ears slightly back and body held low. The individual being approached may remain quiet at first with head held slightly raised and may then respond by sniffing in turn or jumping away. Naso-nasal sniffing may be followed by naso-nasal contact and brief allogrooming of the head and especially chin area. This may be mutual or reciprocal. It may also be followed by high-intensity agonistic behaviour. Naso-nasal greeting is especially common between unacquainted individuals.

(*b*) *Grooming.* After waking and before leaving the nesting area, *G. senegalensis* goes through a sequence of toilet activities, yawning, stretching and intense self-grooming with the toothcomb, during which most accessible areas of the body are covered with much attention to, and much manipulation of, the tail. Evening toilet usually ends with a scratch with the toilet claw. *G. senegalensis* has never been observed to wash its face with its hands after the manner of a cat either in the laboratory or in the field.

G. senegalensis hold one another while grooming with all five digits flexed towards each other but rarely hold with both hands. The groomer usually puts one hand on the back or between the shoulder blades of the fellow. Males may hold the back of the female with both hands and groom it up the back with the toothcomb between the hand-holds. Allogrooming occurs frequently in a wide variety of situations and is not restricted to sexual situations; nor does it serve simply to keep the skin and fur in good condition, since with the aid of the toilet claw *G. senegalensis* can self-groom almost every part of its own body.

(*c*) *Agonistic behaviour.* When the tendency for offensive attack in *G. senegalensis* is very high, the ears are upright, the eyes are rounded, wide open and staring, the mouth is wide open and the tongue and upper canines are partly visible. In defensive threat, on the other hand, the open mouth becomes rounded, the lips are not tense but the teeth remain covered, only the tongue being visible in the mouth, and the head is raised. The eyes are wide open and the ears fully extended at first, becoming flatter as the threatening source approaches. If the threat intensity increases, and particularly if the animal is cornered, a defensive attack posture may be assumed, in which the animal rears up and takes a bipedal stance with the arms held up on either side of the head and fists clenched ready to cuff. This may be followed by a downward pounce accompanied by biting. When threatened by the approach of an aggressive superior, a submissive cringing posture is adopted in which the body is very tense and held low against the substratum, the head retracted against hunched shoulders, eyes narrowed and ears back and flattened against the head. The open-mouth, teeth-covered facial expression in defensive threat is often accompanied by swaying in highly nervous animals in which there is a strong tendency to flee. Andersson notes that this may have been ritualised from an intention movement for jumping, particularly in unfamiliar surroundings.[22] *G. senegalensis* narrows or closes its eyes and flattens its ears when approaching to face a fellow to groom or take food, indicating quite clearly that no aggression is intended.

Overt aggression often takes the form of chasing, particularly if the animals are not equally dominant. Chasing usually occurs between adult females and between an adult male and a maturing male, and it may also occur between an adult male and an adult female where the female is the aggressor. Aggression between females is, however, not as marked as aggression between males, and adult males are not normally aggressive towards sub-adult males. Minor disputes between relatively evenly matched animals may involve sparring matches in which the two animals stand on hind legs, cuff at one another, and grab at each other's hands and limbs. If one gets a hold, the dispute turns into a wrestling match, in which the animals hold and kick one another, rolling over together. This does not involve biting, never leads to serious injury and mainly involves juveniles and opposite sex adults. This type of behaviour is very frequent and may be more in the nature of play.

In the laboratory, relative dominance in a group of *G. senegalensis* is fixed; each animal knows its rank position, behaves accordingly, and the unit is stable. If an animal is removed and, more particularly, if a stranger is introduced, a marked change in the behaviour of the group may occur in which the social structure of the group may be re-organised. In the wild, *G. senegalensis* frequently come into contact with others. The resulting behaviour varies considerably and Bearder investigated this variability in the laboratory through diadic analysis of 130 paired presentations under a variety of conditions.[13] Dominance becomes obvious within the first few minutes of a 40-minute session. Aggression is usually uni-directional, the submissive animal retreating while emitting the full range of submissive vocalisations. Complete submissiveness is usually indicated by the subordinate animal going to the floor. The male is dominant in his own territory and is almost invariably dominant over all females. Sub-adult males are antagonistic towards each other and are usually dominant over females. Dominance in the male is usually benign, whereas when the female is dominant she is highly aggressive. Laboratory and field data suggest that the major factors determining dominance status are sex, age (the greater the age differential, the clearer the dominance status and the less the aggression), and familiarity with the environment (in the laboratory an individual may be sub-dominant in one cage but dominant over the same individual in another cage).

However, little significance can be attached to observations on linear hierarchies under laboratory conditions, since they could well be artefacts of confined conditions in normally solitary animals, as Martin has suggested in respect of laboratory observations on the Tupaioidea.[24]

Serious fighting in which the animals are completely silent only occurs in *G. senegalensis* between more-or-less evenly matched males and very rarely between adult females. Under laboratory conditions, the harmony of the group can only be maintained if it includes no more than one adult male. If adult males are brought together by the removal of a partition between adjacent home cages, vicious fighting will occur, in which severe damage is inflicted. In typical fighting the mouth is open, exposing both upper and lower teeth; ears are folded back and drawn back tightly so that the skin over the forehead and the top of the skull is also drawn back, giving the head a flattened appearance; the skin above the nose between the eyes is deeply wrinkled horizontally and the eyes are nearly closed in a pattern of protective responses similar to those when catching prey. No chasing occurs; animals spring at one another, grapple and fall to the ground, each attempting to gain a foothold on the other's ankles. Locked together, each attempts to throw itself on top of the other, while striking, pulling at fur, and biting at the other's face. Open wounds occur on the feet, hands and face. Loss of fur occurs on the head, neck and shoulders, and fur-plucking may be so violent that the skin is torn in places. Unless separated, the probability of one animal being killed is high. Overt aggression also occurs in the wild, but it is extremely rare for reasons already stated.

(*d*) *Play.* Play wrestling in twin infants appears from the sixth day of age. Like other prosimians, *G. senegalensis* engage in complex locomotor play by themselves at least while still young. They also spend a great deal of time in social play in the laboratory and have been observed play-wrestling in the wild. Adults have only occasionally been observed to play, and then only with juveniles. Social play is frequently associated with mutual grooming; play-wrestling often precedes or is followed by mutual grooming. As with most primates, the behavioural elements of fighting are seen in play-kicking, pushing, pulling at one another in apparent attempts to grasp one another around the ankles or to obtain a hold on one another's ears and hands. The sexual elements seen in the play of monkeys are, however, absent in *G. senegalensis.*[34] Most play contact is ventro-ventral. Wrestling bouts may occur with the animals hanging upside down from a branch suspended by the feet, cuffing and wrestling with the hands or hanging by the hands and kicking with the feet. *G. senegalensis* displays much curiosity towards strange inanimate objects placed in the cage, and, if small enough, will manipulate them playfully with the hands.

Acknowledgments

This research was supported by grants from the University Development Foundation and the University Council, University of the Witwatersrand, as well as by grants from the National Geographic Society, Washington D.C., and the Human Sciences Research Council of the Republic of South Africa.

NOTES

1 Doyle, G.A. and Bekker, T. (1967) 'A facility for naturalistic studies of the Lesser Bushbaby (*Galago senegalensis moholi*), *Folia primat.* 7, 161-8.

2 Andrew, R.J. (1964), 'The displays of primates', in Buettner-Janusch, J. (ed.), *Evolutionary and Genetic Biology of the Primates*, vol. 2, New York.

3 Andrew, R.J. (1962), 'Evolution of intelligence and vocal mimicking', *Science* 585-9.

4 Jolly, A. (1964*a*), 'Prosimians' manipulation of simple object problems', *Anim. Behav.* 12, 560-70.

5 Jolly, A. (1964*b*), 'Choice of cues in prosimian learning', *Anim. Behav.* 12, 571-7.

6 Bishop, A. (1962), 'Control of the hand in lower primates', *Ann. N.Y. Acad. Sci.* 102, 316-37.

7 Bishop, A. (1964), 'Use of the hand in lower primates', in Buettner-Janusch, J. (ed.), *Evolutionary and genetic biology of the primates*, vol. 2, New York.

8 Doyle, G.A., Pelletier, A. and Bekker, T. (1967), 'Courtship, mating and parturition in the lesser bushbaby (*Galago senegalensis moholi*) under semi-

natural conditions', *Folia primat.* 7, 169-97.

9 Doyle, G.A., Andersson, A. and Bearder, S.K. (1969), 'Maternal behaviour in the lesser bushbaby (*Galago senegalensis moholi*) under semi-natural conditions', *Folia primat.* 11, 215-38.

10 Doyle, G.A., Andersson, A. and Bearder, S.K. (1971), 'Reproduction in the lesser bushbaby (*Galago senegalensis moholi*) under semi-natural conditions', *Folia primat.* 14, 15-22.

11 Hill, W.C.O. (1953), *Primates: Comparative Anatomy and Taxonomy*, vol. I: *Strepsirhini*, Edinburgh.

12 Shortridge, G.C. (1934), *The Mammals of South West Africa*, London.

13 Bearder, S.K. (1969), 'Territorial and intergroup behaviour of the lesser bushbaby (*Galago senegalensis moholi*, A. Smith) in semi-natural conditions and in the field', unpublished MSc Thesis, University of the Witwatersrand.

14 Hoesch, W. and Niethammer, G. (1940), 'Die Vögelwelt Deutsch-Sudwest-Afrikas', *J.f. Ornithologie* 88, Sonderheft, 404.

15 Doyle, G.A. (1974), 'The behavior of prosimians', in Schrier, A.M. and Stollnitz, F. (eds.), *Behavior of Non-Human Primates*, vol. 5, New York.

16 Jolly, A. (1966*a*), *Lemur Behavior: a Madagascar Field Study*, Chicago.

17 Petter-Rousseaux, A. (1964), 'Reproductive physiology and behavior of the Lemuroidea', in Buettner-Janusch, J. (ed.), *Evolutionary and Genetic Biology of the Primates*, vol. 2, New York.

18 Martin, R.D. (1972), 'A preliminary study of the Lesser Mouse Lemur (*Microcebus murinus*, J.F. Miller 1777)', *Z.f. Tierpsychol.* Beiheft 9, 43-89.

19 Buettner-Janusch, J. and Andrew, R.J. (1962), 'The use of the incisors by primates in grooming', *Am. J. Phys. Anthrop.* 20, 127-9.

20 Andrew, R.J. (1964), 'The displays of primates', in Buettner-Janusch, J. (ed.), *Evolutionary and Genetic Biology of the Primates*, vol. 2, New York.

21 Haddow, A.J. and Ellice, J.M. (1964), 'Studies on bushbabies (*Galago* spp.) with special reference to the epidemiology of yellow fever', *Trans. roy. Soc. trop. Med. Hyg.* 58, 521-38.

22 Andersson, A. (1969), 'Communication in the lesser bushbaby (*Galago senegalensis moholi*)', Unpublished MSc Thesis, University of the Witwatersrand.

23 Sparks, J. (1967), 'Allogrooming in primates: a review', in Morris, D. (ed.), *Primate Ethology*, London.

24 Martin, R.D. (1968), 'Reproduction and ontogeny in tree-shrews (*Tupaia belangeri*) with reference to their general behaviour and taxonomic relationships', *Z.f. Tierpsychol.* 25, 409-532.

25 Boulenger, E.G. (1936), *Apes and Monkeys*, London.

26 Ilse, D.R. (1955), 'Olfactory marking of territory in two young male loris, *Loris tardigradus lydekkerianus*, kept in captivity in Poona', *Brit. J. Anim. Behav.* 3, 118-20.

27 Yasuda, K., Aoki, T. and Montagna, W. (1961), 'The skin of primates: IV. The skin of the lesser bushbaby (*Galago senegalensis*)', *Am. J. Phys. Anthrop.* 19, 23-33.

28 Marler, P. (1965), 'Communication in monkeys and apes', in DeVore, I. (ed.), *Primate Behavior*. New York.

29 Hediger, H. (1950), *Wild Animals in Captivity*, London.

30 Jolly, A. (1966*b*), 'Observations on *Galago crassicaudatus*', unpublished paper sent to Dr Sauer.

31 Bates, B.C. (1970), 'Territorial behaviour in primates: a review of recent field studies', *Primates* 11, 271-84.

32 Southwick, C.H., Beg, M.A. and Siddiqi, M.R. (1965), 'Rhesus monkeys in North India, in DeVore, I. (ed.), *Primate Behavior*, New York.

33 Bourlière, F. (1964), *The Natural History of Mammals*, New York.

34 Loizos, C. (1967), 'Play behaviour in higher primates: a review', in Morris, D. (ed.), *Primate Ethology*, London.

J. EPPS

Social interactions of Perodicticus potto *kept in captivity in Kampala, Uganda*

Introduction

Although the potto is relatively common within the limits of its distribution, little is known concerning its social behaviour, mainly because of the difficulties attendant on the study of a nocturnal animal that usually inhabits thick forest. It was therefore hoped that a study of pottos kept in outdoor cages in the area of origin, and so subjected to normal fluctuations of rainfall and temperature, might prove useful.

Observations were made of the social interactions of pottos caged at Makerere University, Kampala between February 1970 and May 1971. A total of thirteen pottos, five males and seven females, were kept at Makerere during this time, but one male and one female died of old age, and two females escaped during the first two months after capture. Seven of the pottos, given by Alan Walker, had been in captivity at Makerere Medical School for some time before the beginning of this study. The other six were brought to the Zoology Department by local people, who caught them in small plantations around Kampala. All the pottos used in the study had been caught within 100 miles of Kampala, either in Buganda or Bunyoro districts.

Housing and experimental procedure

The majority of the pottos were kept in a system of home cages, each 4½′ x 10′ x 9′. Each cage was provided with a system of branches and a sleeping box. The cages were well shaded with papyrus matting; the temperature never rose above 27°C, usually cycling between 17.5°C and 26°C. For observations of social interactions, the occupants of the home cages were allowed into a large central observation cage by way of swing doors, which were opened by a pulley system operated from the observation slit at one end of the cage. Light for observation was provided by two 60-watt red bulbs. The home cages were also lighted by red bulbs so that all the pottos could become habituated to red light. They did not appear to be affected by it in any way and were often seen licking insects off the glass around the bulbs. Automatic records of the nightly activity of an animal subjected to red illumination showed an initial decrease, but the animal soon became habituated and activity returned to normal.

For observations of social interactions, any two pottos were allowed into the observation cage together and their actions noted on a check sheet at three-minute intervals. In addition, details of any social behaviour that was seen between check sheet records were recorded on a tape recorder. Interactions between pairs of males, pairs of females and male-female pairs were watched during 2½-hour observation periods. Care was taken to ensure that the animals which met in the observation cage had not previously been in contact with each other. Additional observations were made of one female pair, one male pair and one male-female pair, each of which were transferred to a reversed daylight house for periods of two to three months.

Animals in the reversed daylight house were kept in a cage 6′ x 5½′ x 5′, which contained a sleeping box for each animal, a shelf for the food bowls, and was provided with a system of branches. The room containing the cage was kept dark during the day and was lighted with a 60-watt white bulb between 19.00 hrs. and 07.00 hrs. The hours of day and night were thus almost exactly reversed. The animals kept in this cage took only a week to ten days to become accustomed to the new regime. Light for observation during the artificial 'night' was provided by sunlight filtered through two layers of red cellophane covering a window measuring 4′ x 1′. Using a 12-channel activity recorder connected to a switchboard, continuous records of the behaviour of the animals were made during 1½-hour observation periods. The observer sat behind a curtain three feet from the cage and watched the animals through a peephole. Although the pottos were aware of the presence of the observer, they became habituated to it in two to three weeks.

In addition to direct observations of the behaviour of the pottos, records of nocturnal activity of single pottos were made on an automatic activity recorder. The animal was placed in a cylindrical cage 6′ high and 2½′ in diameter, which was hung on a frame balanced on a knife edge. A thread, attached to the opposite end of the frame from the cage, moved a lever recording on a kymograph drum revolving at 2 mm./sec. (Fig. 1). Any movements of the animal in the cage were thus recorded on the drum. It was found that all the pottos were active throughout the period of darkness, though there was a higher level of activity in the second half of the night when

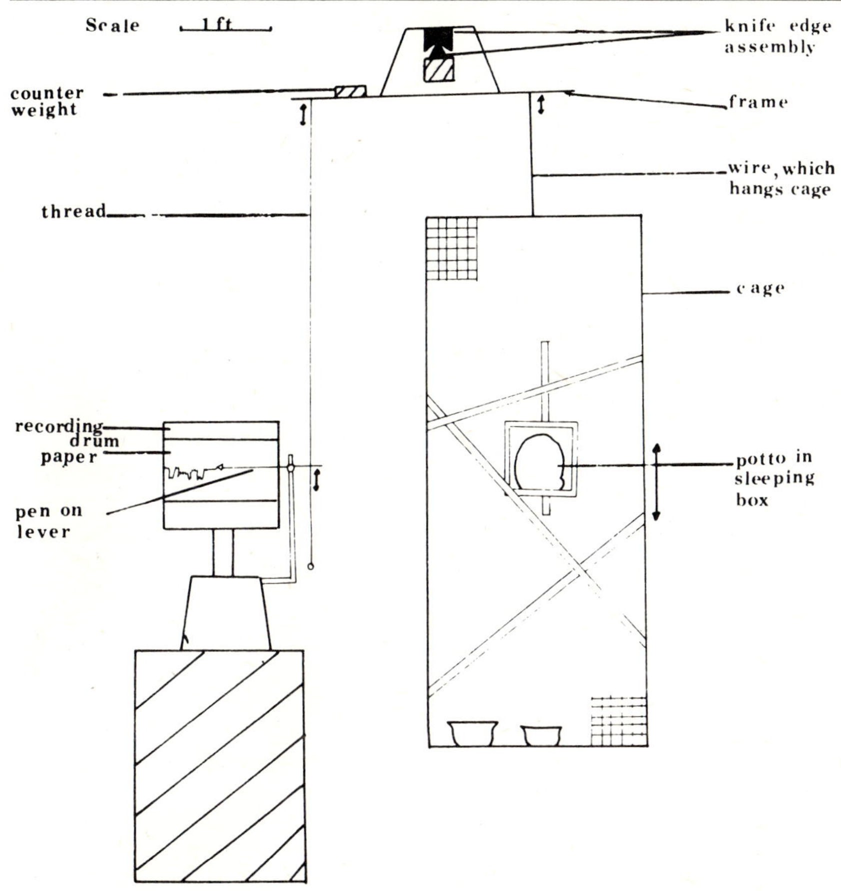

Figure 1. Diagram of the activity recording cage.

temperatures were lower. Records were made of the nightly activity of a single male and of a single female. An animal of the opposite sex to the experimental animal was then placed in a cage close to the activity cage in order to determine whether there was any effect on activity. The activity of a female was unchanged in the presence of a male, but percentage activity time of the male was increased from a mean of 67% to a mean of 90% when a female was caged next to him.

Communication

The pottos had a simple vocal repertoire of six sounds. These were usually heard accompanying aggressive postures, or when one potto was seeking contact with another. Four of these sounds have already been described by Andrew.[1] Grunts and shrieks were heard accompanying fights. Crackles appeared to indicate low intensity threat, and a chattering sound (described by Andrew as a series of screams) usually accompanied defensive threat postures. The spit-click, which sounded similar to the human 'ttt ttt' indicating scolding, may have been what Andrew described as 'loud sharp calls'.[1] The spit-click, together with a metallic clicking sound, appeared to act as a signal of the location of the caller. The metallic click, which was always preceded by a spit-click, was a high-frequency sound (fundamental frequency 15 KHz) with a distinct tonal quality.

Spit-clicks and metallic clicks were frequently heard from a male approaching a female. They were also heard on two occasions when a potto had been moved to a new cage from which the previous occupant had just been removed. The new occupant moved round the cage several times, clicking at intervals, before it entered its sleeping box. These clicks were also heard from a pair of females that were being observed for social interactions in the large observation cage. Before either animal moved out of its home cage into the observation cage, it would hesitate in the doorway and spit-click once or twice. If there was an answering spit-click from an animal already in the observation cage, the caller usually moved back into her home cage; if not, she moved out, still clicking. As she moved out, the spit-clicks were accompanied by metallic clicks, which were answered by the second animal still in her home cage. It could not be

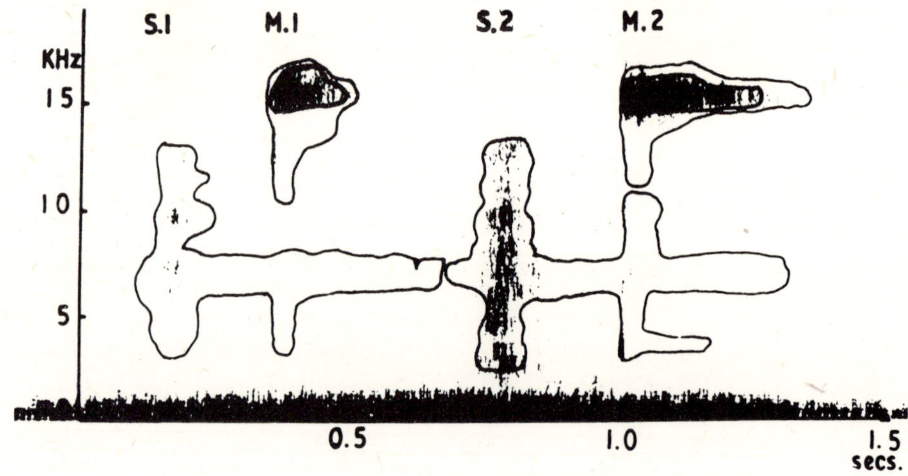

Figure 2. Sonogram of a click call sequence. The thick contour lines outline the fundamental frequency of each click. The thinner lines outline the whole frequency spectrum.
SI = spit-click from the animal furthest from the microphone
MI = metallic click from the same animal
S2 = spit-click from the animal nearest to the microphone
M2 = metallic click from the same animal

determined whether this was a call and answer sequence, but it appeared to be so; one animal would call and put its head on one side in the listening posture; if there was no answering click from the other animal it would click again, and this was usually 'answered' (Fig. 2). Metallic clicks are probably the same as a call, described by Walker, produced by an infant potto which had lost contact with its cloth surrogate mother.[2]

During spit-clicks, metallic clicks, chatters and crackles, the mouth was retracted at the corners. Otherwise no change of expression accompanied vocalisations.

It seems probable that a large amount of potto communication is olfactory. Any potto moving into a new area would urinate on the hardboard along the edge of the cage and on the poles. In addition 'key junctions' (near the entrance to the home cage, or leading from one level in the pole system to another) were marked by perineal rubbing.[3] After they had marked an area, the pottos frequently stopped and sniffed at junctions before moving onto a new pole. Pottos moving into an area that had previously been occupied by unknown animals showed considerable interest in marked areas, sniffing at them vigorously and leaving their own mark on top.

Social interaction

There was relatively little contact amongst any of the pairs of pottos that were observed. The check sheet records revealed that in male-female pairs in the outside observation cage only 12.0% to 19.4% of the activities recorded were of contact behaviour. No contacts at all were recorded for pairs of animals of the same sex meeting in the outside observation cage. In the reversed daylight house, where the animals were caged together continuously, the male-female pairs were in contact for only 15% of the 'night', the two females for only 13%, and no contacts at all were observed between the two males.

The majority of interactions observed between male and female pottos were amicable, though the female usually chattered in threat at the male when they first met. Most of the contacts between males and females were initiated by the male, who would move up to the female, sniff at her, and then (if she did not threaten him) groom her. Grooming was usually reciprocal rather than mutual, each animal in turn standing to groom the other, who sat with head extended towards the groomer. When the male groomed the female, he usually stood on his hind legs at a distance from her (Fig. 3); but when the female groomed the male, though she began by standing away from the male, he would gradually edge forwards until he was sniffing at her genital area (Fig. 4). At this point the female usually

Figure 3. Male grooming female.

Figure 4. Female grooming male.

Figure 5. Urine marking during allogrooming.

chattered in threat and moved off. Allogrooming was frequently accompanied by urine marking (Fig. 5). The groomer would wipe a drop of urine onto his hand and then rub it on the area being groomed.[4] This would result in some mixing of the smells of the two animals.

When a male and female had become familiar with each other, periods of grooming were longer and led to grappling (Fig. 6). This has already been described[5] and appears to be a type of friendly wrestling match; no aggressive vocalisations accompanied grappling.

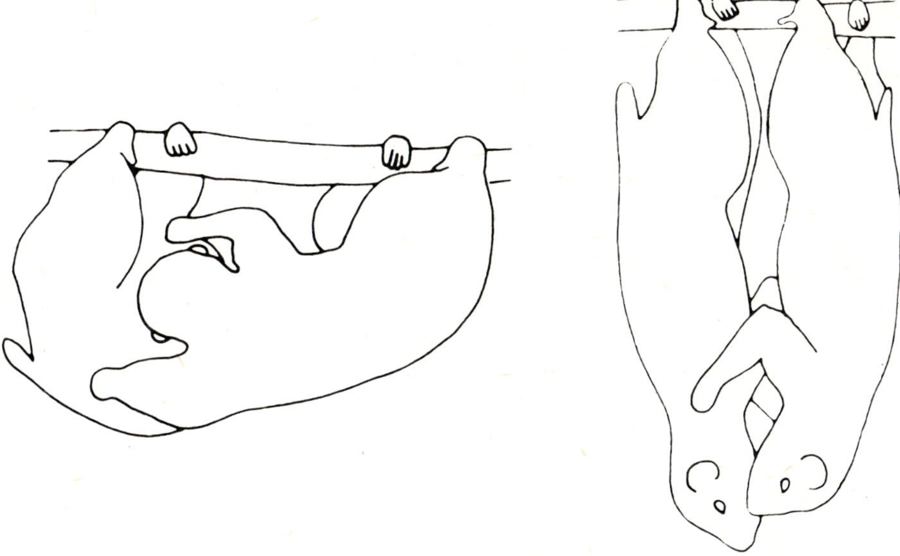

Figure 6. Grappling between two pottos.

Copulation was observed only between one pair; on each of the five occasions that it was seen, it was preceded by grappling, the male finally mounting the female from behind whilst the two animals were suspended from the ceiling (Fig. 7).

Although the male was the prime initiator of contacts, it was the female that determined their number and duration. When a male approached a female, he was frequently repulsed by vocal threats; allogrooming was often terminated in the same fashion, especially if the male sniffed at the female's genital area. The amount of aggression displayed by the female appeared to vary in relation to the oestrous cycle. In the one pair between which copulation was

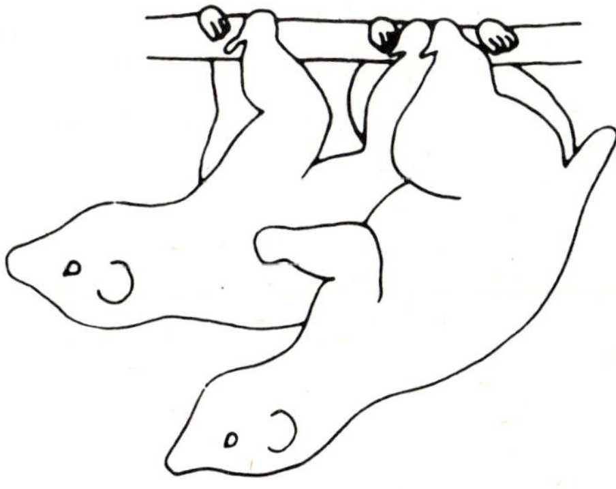

Figure 7. Copulation.

observed, there was a definite reduction in female aggression until just before they copulated, when the female actually sought the male. Two days after copulation had been observed, female aggression rose abruptly, and four days later she would allow very little allogrooming.

Figure 8. Fighting.

Interactions between animals of the same sex that had not met before were always aggressive. If the two animals came into contact, they would fight, wrestling together fiercely and rolling onto the floor (Fig. 8). Fighting was always accompanied by loud squeaks and chatters, and thus could be differentiated from the amicable grappling encounters observed after periods of grooming between a male and a female. A dominant-subordinate relationship was established between all the pairs of the same sex that were observed, usually after the first fight. The subordinate appeared to avoid contacts with the dominant whenever possible; this was frequently achieved by the subordinate remaining motionless under piles of shavings on the floor for a large part of each observation period. Pottos observed by Charles-Dominique[6] in forests in Gabon also avoided contacts with other pottos by remaining motionless, hidden in the vegetation. If a potto became aware of another in the area it was about to enter, it remained completely immobile until the other had moved away. Fights between animals with overlapping territories in the forests were rarely observed. In captivity, however, it is impossible for the animals to avoid contacts and fights do occur. Descriptions of the fights between captive pottos show extreme variety in the form of the fight.[7–10] It is possible that this is due to the fact that fights are rare in the wild state, and therefore no firm pattern has evolved.

Pottos have been observed to hold territories in the forest.[6] Those kept at Makerere exhibited strong territoriality. Even in the relatively small cage in the nocturnal house, it was found that the animals in two out of the three pairs that were kept there established separate areas of movement and rarely trespassed. The male and female met only for allogrooming and at the food bowl. The two males usually remained in their own areas of the cage. If one did move into the other's area, he was immediately threatened by the owner and retreated rapidly. However, the female pair kept in the nocturnal house did not develop separate areas of movement, but their behaviour was probably affected by the fact that they had already been caged together for two years.

An attempt was made to determine whether the olfactory marks left by the pottos served as a deterrent to intruders. It seemed possible that subordinate animals would avoid areas marked by dominant animals; therefore subordinates were allowed into an area that had been marked by a dominant animal, but from which the dominant animal had been removed. Two females that were observed under these conditions showed definite avoidance of the marked area; each moved into the observation cage only three times during a 2½-hour observation period. Each time an animal moved out, it only went as far as a 'key' junction that had been marked frequently by the dominant animal, sniffed this thoroughly and then ran back into its home cage. In a control experiment, in which one of the females used in the previous experiment was allowed access to the observation cage after it had been marked by an unknown female, the intruder spent 70% of the time in the observation cage, sniffing at the marked areas and leaving its mark on top. It therefore seems that pottos are able to recognise the marks of animals with which they are familiar.

Territory experiments with males were less conclusive than those using females. The first subordinate male used had not been in contact with the dominant animal for four weeks. This animal moved out into the observation cage for 54% of the 2½-hour observation period, but he never spent more than three minutes at a time in the observation cage, returning to his home cage at frequent intervals. The second experimental animal moved out into the observation cage only twice after the latter had been marked by the dominant animal. This animal spent 96% of the observation period in its home cage. However, when the same animal was used in the control experiment, it still spent the majority of the observation period in the home cage, only moving into the observation cage for 40% of the observation period after the cage had been marked by an animal with whom it had not previously been in contact. Nevertheless, in the control experiment, the intruder did not exhibit any of the usual signs of stress (holding itself stiffly and moving in jerks) that had been apparent when it was moving around an area marked by the dominant animal. It is possible that the avoidance of areas marked by known animals is less complete amongst males, which, Charles-Dominique has suggested,[6] may have a more rapid turnover of territories in the forest, with juveniles, as they mature, taking over the territories occupied by weak males. Unfortunately, shortage of males prevented the repetition of these experiments.

Discussion

Charles-Dominique's observations of pottos in a forest in Gabon showed them to be solitary territorial animals.[6,11] Encounters between pottos of the same sex were rare, and it seemed that such encounters were actually avoided by the intruder, who remained motionless until the other animal had moved away. In captivity, meetings between animals of the same sex result in fierce fights. It is therefore necessary for the potto to have some mechanism in its behavioural repertoire by which aggression may be overcome when a male and female meet, so that a pair-bond may be built up in order that mating may take place. Allogrooming appears to play a strong part in the formation of the pair-bond. It seems to be pleasurable both for the groomer and for the groomee; pottos that had been groomed were frequently observed soliciting further grooming by holding out their heads towards animals moving off after grooming encounters. Animals that were grooming frequently prevented the groomee from moving off by chattering at it and holding the head down so that grooming could be continued. It is interesting that on the few occasions that a male was heard threatening a female vocally, she immediately responded by holding her head out towards the male, who then ceased threatening and groomed her.

One interesting example, supporting the hypothesis that allogrooming serves as a means by which the pair-bond is built up between pottos, was observed between a male and a female that had been allowed together for seven 2½-hour observation periods. The female was by this time allowing the male to groom her for a short while each time they met. On the eighth night, the female moved up and sniffed at the male whilst he was chattering in threat at a male in an adjacent cage. He immediately turned round and attacked the female; they wrestled together for two minutes, the female shrieking loudly. The female eventually ran away, chattering. After this redirected aggression from the male, the female threatened him whenever he approached and would not allow him to groom her again. The urine marking that frequently accompanied allogrooming must have served a role in the formation of the pair-bond, making the marked animal more familiar and therefore less likely to elicit an aggressive reaction. Further smell sharing took place during the friendly grappling matches that frequently followed allogrooming. Grappling must cement the pair-bond, since all aggressive responses must be suppressed in what is, in essence, a play fight.

Charles-Dominique observed a similar pattern of social interaction between male and female pottos with adjacent territories in the forest.[6] A male potto would visit a female each night, and as she became used to his presence, she allowed him to groom her for gradually increasing periods. Grooming eventually led to grappling. This presumably led finally to copulation, as the female produced infants six months later. However, females in the forest took longer to accept the male. Limitations of space in captivity probably accelerated the pace of courtship and it is possible that, though the cages were kept separate, there had been some olfactory communication between the animals before they actually met.

There has been a great deal of speculation on the role of the nuchal spines of the cervical vertebrae of the potto, which are elongated and project ⅛″ above the body suface.[12,13,14] These spines are covered with a layer of highly innervated skin;[15] stimulation of this area is, therefore, likely to be gratifying. Since allogrooming is concentrated on the area of the spines, it is probable that the hypothesis that they now function as a focus for amicable intraspecific interactions is correct.[14,16,17] It seems likely, however, that the sensory function of the spines is a secondary development and that the primary enlargement was associated with the development of the extensive cervical musculature.[11]

It was encouraging to find that the behaviour of the captive pottos corresponded quite well to that of the pottos observed by Charles-Dominique in a forest in Gabon.[6,11] It seems likely that cage studies of the behaviour of nocturnal prosimians may play a useful role in supplementing field studies in providing information on details of behaviour not easily obtainable in the wild, provided that the captive situation is carefully planned.

Acknowledgments

While conducting the research reported here the author was supported by the O.D.M. 'Study and Serve' scheme. Makerere University Research Grants Committee financed the housing and feeding of the pottos. Lance Tickell supervised the research and supplied help and criticism throughout. Indebtedness to Alan Walker is also gratefully acknowledged, not only for the original pottos, but also for help and criticism throughout the study and especially in the final stages of 'writing up' when the author was living in Nairobi.

NOTES

1 Andrew, R.J. (1963), 'The origin and evolution of the calls and facial expressions of the primates', *Behaviour* 20, 1-109.

2 Walker, A.C. (1968), 'A note on hand-rearing a potto', *Int. Zoo. Yb.* 8, 110-11.

3 Andrew, R.J. (1964), 'The displays of the primates', in J. Buettner-Janusch (ed.), *Evolutionary and Genetic Biology of the Primates*, vol. 2, London.

4 Urine marking of the groomee by the groomer was observed for the first time when the animals were in the full beam of a torch only 1½ feet from the observer. On this occasion, a drop of urine was seen to be deposited on the hand of the groomer each time it was passed back to the urino-genital area. All subsequent observations of marking during allogrooming were from a minimum distance of seven feet, with dim red illumination. It was assumed that this marking behaviour was the same as that observed previously, but in the light of Manley's[18] observations it is possible that at least some of the marking was genital scratch marking.

5 Bishop, A. (1962), 'Control of the hand in lower primates', *Ann. N.Y. Acad. Sci.* 102, 316-37.

6 Charles-Dominique, P. (1971), 'Eco-éthologie et vie sociale des prosimiens du Gabon', PhD thesis, Paris.

7 Bishop, A. (1964), 'Use of the hand by lower primates', in J. Buettner-Janusch (ed.), *Evolutionary and Genetic Biology of the Primates*, vol. 2, London.

8 Blackwell, K.F. and Menzies, J.I. (1968), 'Observations on the biology of the potto (*Perodicticus potto*), *Mammalia* 32, 447-51.

9 Vincent, F. (1969), 'Contributions à l'étude des prosimiens africains; le Galago de demidoff', PhD thesis, Paris.

10 Walker, A.C., personal communication.

11 Charles-Dominique, P. (1971), 'Eco-éthologie des prosimiens du Gabon', *Biol. Gabon.* 7, 121-228.

12 Sanderson, I.T. (1937), *Animal Treasures*, London.

13 Coon, C.S. (1962), *The Origin of Races*, New York.

14 Walker, A.C. (1968), 'A note on the spines on the neck of the potto', *Uganda J.* 32, 221-2.

15 Montagna, A.W. and Yun, J.S. (1962), 'The skin of the primates: 14. Further observations on *Perodicticus potto*', *Amer. J. Phys. Anthrop.* 20, 441-50.

16 Walker, A.C. (1970), 'Nuchal adaptations of *Perodicticus potto*', *Primates* 11, 135-44.

17 Kingdon, J. (1971), *An Atlas of East African Mammals*, 1, London.

18 Manley, G.H., this volume.

J. M. TANDY

Behaviour and social structure of a laboratory colony of Galago crassicaudatus

Introduction

Galagos are probably solitary or live in very small groups – perhaps only families. Sauer and Sauer found groups of two to nine individuals (*Galago senegalensis bradfieldi*) together.[1] Haddow and Ellice found groups of two to four animals (*G. senegalensis*) at a locality in Northern Karamoja, north eastern Uganda, and two to six animals in Tanzania.[2] Vincent found *G. demidovii* in the Congo (Brazzaville) in numbers of one to five.[3] He also found their nests in clusters: 'Adult pairs can be found in the wild outside the breeding season, and adult males and females can be accompanied by subadults of the same sex.'[3] Doyle, Andersson and Bearder, while searching for nests of *G.s. moholi*, found that 'but for the presence of mother and infants on them, they would, in most cases, not be recognized as nests'.[4] This may imply that the males do not stay with females and young. The work of Bearder and Doyle verifies this assumption.[5] Struhsaker found two to four *G. demidovii* in nests during a systematic study of numbers in nests and duration of occupancy.[6] The present author observed two *G. crassicaudatus* together at Athi River, Kenya, and single individuals at several localities in Tanzania during 1970-71.

There have been few behavioural studies of galagos in the wild: (1) *G.s. bradfieldi* in South West Africa by Sauer and Sauer;[1] (2) *G. demidovii* in the Congo (Brazzaville) by Vincent;[3] (3) *G. crassicaudatus garnetti, G.c. lonnbergi, G.c. umbrosus* and *G. senegalensis* in South Africa by Bearder and Doyle;[5] (4) *G. alleni* and *G. demidovii* near Kumba, West Cameroun, by Miller;[7] and (5) *G. alleni* and *G. demidovii* in Gabon by Charles-Dominique.[8] Several publications concern group behaviour of captive animals. Sauer and Sauer studied *G.s. bradfieldi*;[1] Lowther,[9] Doyle, Andersson and Bearder,[4] and Doyle, Pelletier and Bekker[10] studied *G.s. moholi*; Roberts observed

G. crassicaudatus;[11] and various species have been studied by other authors.

The purpose of the present study was to investigate what kind of social structure, if any, would develop among a group of feral *G. crassicaudatus* in captivity. The investigation was intended to be a general one from which could be developed a more intensive study to analyse their communication system. Behavioural data used as indicators of social structure were selected because of their ability to demonstrate positive or negative bonds between individuals.

Materials and methods

This study was conducted at the Physical Anthropology Laboratory of the University of Texas at Austin. The facilities included a thermostatically heated room, 5.4 x 8.3 x 3.1 m., in which were provided seven nest boxes and a variety of tree limbs and other climbing apparatus. The animals were on a natural day/night cycle from September until October 1971, at which time a light timer was installed. Their cycle was then reversed. 'Daylight' began at 18.30 hrs. when white fluorescent tubes switched on, and 'night' began at 06.30. Six 60-watt red bulbs burned continuously and provided enough light during dark periods for observation.

On 1 September 1971, eight feral *G. crassicaudatus* (three females, five males) were released into this room. The animals had probably been in contact with each other only since capture. On 10 September (nine days after importation), B died of pneumonia. Table 1 gives data on localities and dates of capture for the animals. Table 2 gives the animals' weights.

The taxonomy of *G. crassicaudatus* is complicated by a profusion of sub-species. The distribution of the 'sub-species' is continuous in eastern Africa from the Juba River to Natal.[12] It is possible that the

Table 1 **Tag identification, sex, locality data, and date of capture for animals used in this study. Animal names represent colour of tags worn around the neck**

Name	*Sex*	*Locality*	*Capture Date*
Purple (P)	male	Zigi Amani, Tanzania	14-17 August 1971
Red (R)	male	Zigi Kisiwani, Tanzania	2 August 1971
Black (Bl)	male	Fanusi, Tanzania	20 July 1971
Yellow (Y)	male	Gonja, Tanzania	31 March 1971
Blue (B)	male	Zigi Amani, Tanzania	14-17 August 1971
White metal (Wm)	female	Zigi Amani, Tanzania	14-17 August 1971
White plastic (Wp)	female	Zigi Amani, Tanzania	14-17 August 1971
Green (G)	female	Zigi Amani, Tanzania	14-17 August 1971

Table 2 The weights of the animals in grams

Name	*Weight (gm.)*
♂P	1134
♂R	1474
♂Bl	907
♂Y	1304
♀Wm	1134
♀Wp	1077
♀G	1049

subspecific differences represent a cline of continuously varying morphological characteristics. According to Doyle, Pelletier and Bekker,[10] *G. senegalensis* subspecific variations include fairly constant differences in gestation period and number of offspring. The animals used in this study are probably a mixture of what are considered to be *G.c. lasiotis* and *G.c. panganiensis.*[12]

The animals were fed mealworms, fruit, infant formula, peanut butter, Monkey Chow (commercial primate food) and vitamin supplements.

Observation periods varied from five-minute proximity checks to one hour or longer intervals. They were begun with a diary format and later switched to a modified checklist in which sequence and direction of events could be recorded. This paper concerns the period from 1 September 1971 to 23 April 1972. Because tabulated data include only days when all animals were identified, observations for most of the first part of September are excluded. Total data presented represent approximately 70 hours of observation (field hours not included). The galagos exhibited considerable investigative behaviour toward the observer sitting in their cage.

Results

Proximity

Each animal was scored as being in proximity to another if it was sleeping with or sitting by another at the time the observer entered the room. These were random samples which were not taken every time other observations were made. See Table 3 for proximity scores.

♂Y, ♂R and ♀Wm had proximity scores significantly less than their solitary scores. Probability is less than .0001 ($\chi^2 = 30.316$), .0001 ($\chi^2 = 21.053$) and .0012 ($\chi^2 = 11.250$) respectively for their scores representing equal frequencies of being alone versus with other animals. All animals were alone more often than with others. Proximity data indicate that the group as a whole does not associate randomly ($P < .0024$). When each animal's pattern of association with all other members of the colony is considered, proximity scores

Table 3 **Proximity data for about half of the observation sessions. The diagonal row represents the number of times the animal was alone. Side row totals represent the number of times the animal was with other animals**

	♂P	♂R	♂Bl	♂Y	♀Wm	♀Wp	♀G	TOTALS
P	50	5	1	1	8	5	12	32
R	—	62	1	1	3	2	2	14
Bl	—	—	47	5	6	13	7	33
Y	—	—	—	58	2	6	3	18
Wm	—	—	—	—	55	2	4	25
Wp	—	—	—	—	—	49	5	33
G	—	—	—	—	—	—	47	33

between ♂P and ♀G and between ♂B1 and ♀Wp are significantly greater than expected by chance. These pairs do not appear, however, when other interactions are considered. For example, ♂P grooms and is groomed by ♀Wm more often than any other animal.

Agonistic behaviour

Spats, 'spat-calls', cuffs, avoids, displaces, bites, chases, pulls, bipedal stances with or without the mouth open, and open mouth threats were considered agonistic actions. A spat is a combination of the bipedal stance with cuffs and bites (this looks much like a boxing match) and is accompanied by spat-calls. Spat-calls are noise-like sounds somewhat resembling the spitting sounds of young kittens when alarmed. Sounds resembling screams may be produced within such calls. A cuff is a slap at an animal with the hand. When approached by an animal in an agonistic situation, the individual approached may rise from its quadrupedal or sitting position to a bipedal one, the mouth may be held slightly open without emission of sound, and the arms may be raised; or the animal may remain sitting, with the mouth open displaying its teeth.

The following diary extracts illustrate short agonistic episodes:

11 November 1971: 16.15 R muzzles B1 and B1 avoids him. Then R follows B1; B1 spat-calls and partially stands bipedally with his mouth open.

16.20 R is licking himself and P is watching him. P approaches R and muzzles him. R partially rises bipedally, raises his left hand as in a cuff and gives a spat-call. P retreats.

16.25 Wm jumps to the box where P is. He genital-sniffs her. She spat-calls and cuffs him and then jumps away. He foot-rubs.

The highest numbers of agonistic encounters occurred when the animals were first released in the room and during the periods of female oestrus.

Non-agonistic behaviour

Muzzles (touching noses and sniffing), ano-genital muzzles, sniffs (smelling the area around an animal), grooming, 'crack-calls', mounts, follows and copulations were scored as non-agonistic actions. Crack-calls are low-frequency grating sounds given only by males. These were usually directed toward females. Such vocalisation often began when a male was following a female. It also occurred when a male was grooming a female and was not confined to periods of oestrus.

Sexual behaviour, such as crack-calls, copulations and mounts, was lumped with non-agonistic behaviour for convenience in data tabulation. Although a follow or crack-call was initiated sexually, the response from the recipient was usually agonistic. Out of 299 recipients, 293 were females. Since the act was apparently not initiated agonistically, there seems to be justification for including it with non-agonistic behaviour.

The following diary extracts illustrate a sexual episode:

30 October 1971:	16.15	Male P licks female G's genitalia and head. One of them makes a 'crackly' call (consisting of three rasps). P tries to manoeuvre into copulatory position.
	16.20	P has a very large erection and thrusts at G a couple of times, but he cannot get into proper position because the ledge they are on is not wide enough. His hind feet are holding onto the wire below the ledge.
	16.25	P and G have separated. P jumps to the ledge where R is sitting and then climbs back to G. He rapidly licks her head and gives several groups of the 'crack-call'. R urine-washes several times. The calls become more rapid and P has G in copulatory position. R is still urine-washing.

An excellent photograph of this species in the above-mentioned copulatory position can be found in Buettner-Janusch.[13]

Two oestrus periods occurred during this study. One lasted from 21 October to 1 November 1971, and another from 22 March to

27 March 1972 (length based on observed copulations). One infant was born to ♀G on 28 February 1972, but it was killed.

Table 4 summarises total interaction data recorded for each animal. Play encounters as observed by Roberts and Newell were never seen in these animals.[11,14]

Table 4 Total interaction data recorded for each animal

	♂P	♂R	♂Bl	♂Y	♀Wm	♀Wp	♀G
Groom	151	29	49	1	151	17	38
Muzzle	19	14	27	0	24	19	6
Sniff	119	128	184	16	119	167	37
Ano-genital muzzle	62	55	60	0	75	68	6
'Crack-call'	179	116	10	0	126	149	18
Follow	85	75	37	0	113	76	2
Mount	57	30	10	0	70	5	22
Copulate	21	13	0	0	29	0	5
Open mouth threat	2	2	1	0	0	2	0
Bipedal threat	24	23	12	1	8	9	1
Bipedal OM threat	6	4	3	0	3	0	0
Spat	232	141	34	13	93	54	21
'Spat-call'	205	113	30	10	104	96	26
Cuff	61	60	11	3	35	67	7
Bite	10	7	3	0	8	4	1
Avoid	219	106	65	2	254	155	9
Chase	20	16	5	0	7	6	0
Displace	13	5	5	0	4	5	0
Pull	1	2	1	0	0	0	0
Crouch	13	1	0	0	11	2	0
TOTAL	1,499	940	547	46	1,234	901	199
Total Agonistic	806	480	170	29	527	400	65
% Agonistic	54%	51%	31%	63%	43%	44%	33%
Total Non-agonistic	693	460	377	17	707	501	134
% Non-agonistic	46%	49%	69%	37%	57%	56%	67%

Other behaviour

'Loud calls', urine-washing and foot-rubbing are included as 'other behaviour' because their functions are not fully understood by the author. One loud call was heard approximately ten minutes after the animals were released into the enclosure. This call was not heard again until after a month in captivity. These calls occur much more frequently than observations indicate (128 recorded). They are often heard when the author is out of the room working elsewhere. I have never heard this call in the wild. It resembles that of a large bird giving a series of notes first rising rapidly and then falling gradually in frequency and amplitude. It is not possible to describe this call or others precisely until sonograms are available. The animal's body shakes violently during this call and the mouth is held only slightly open. The only animals seen making this call are ♂R and ♂P. It has been recorded eighteen times for the former and only three times for the latter. Occasionally, two occur almost simultaneously, with one male beginning and the other starting to call halfway through the first's call. The barking call frequently given in the wild to intruders has not been heard in our laboratory.

During urine-washing, an animal deposits urine on its hand and then rubs it on the sole of the foot of the same side of the body. Its possible functions are discussed by Sauer and Sauer, Andrew and Klopman, Doyle, and Quick.[1,15,16,17] It occurs in a variety of contexts. Individual animals exhibited significantly different frequencies of this behaviour, as the following table indicates:

♂P	♂R	♂Bl	♂Y	♀Wm	♀Wp	♀G
134	34	68	2	37	53	0

The probability that the distribution of frequencies of this action is the same as the distribution of frequencies of all actions of each individual is less than .0001 ($\chi^2 = 90.481$). Individual frequencies showing positive significant deviations at the .05 level from their frequencies predicted by total actions are underlined; negative significant deviations at the .05 level are circled.

Foot-rubbing consists of rapid rubbing of first one foot and then the other against a branch, box top, floor, etc. This makes a very loud scraping noise. The behaviour may be accompanied by urination or urine-washing. Probability that the following distribution of foot-rubbing is the same as the distribution of frequencies of all actions of each individual is less than .0001 ($\chi^2 = 318.450$):

♂P	♂R	♂Bl	♂Y	♀Wm	♀Wp	♀G
3	82	2	15	149	11	1

♀Wm has been seen foot-rubbing in the same spot where a few seconds previously ♂R had foot-rubbed. Twice, ♂Y moved from his usual place near the air conditioner in the corner onto the top of a nest box. Instead of moving farther, he spent approximately ten minuts foot-rubbing, and then quickly returned to the corner ledge.

Self-grooming was tabulated as a general indicator of the animals' activity levels away from other animals. The higher scores for males may be due to the fact that they are groomed by other animals less than are the females. Probability is less than .0001 ($\chi^2 = 53.135$) that the following distribution of the frequencies of self-grooming are the same as the distribution of frequencies of all actions of each individual:

♂P	♂P	♂Bl	♂Y	♀Wm	♀Wp	♀G
109	80	27	0	44	27	1

Discussion

Proximity

♂Y, ♂R and ♀Wm showed significant differences in solitariness versus association. All were more solitary than social. Y was an isolate or peripheral male characterised by almost total non-interaction with other colony members. ♂R was a subordinate animal which carried its tail in the curled-under submissive posture described by Sauer and Sauer,[1] and his ears were usually carried in a flattened position also indicative of sub-dominant status. ♀Wm was a hyperactive highly aggressive female. She was particularly agonistic toward ♀Wp. Lack of proximity is especially good for indicating these types of negative relationship.

Vincent appears to interpret his Congo data and the data of Haddow and Ellice on numbers of animals captured sleeping together as indicating relatively stable social groups.[2,3] This seems to be supported by Haddow and Ellice's yellow fever immunity tests.[2] In the laboratory, animals often sleep in groups, and these groups show some stability. Proximity scores of ♂P and ♀G indicate that they associate at random with all other individuals, but positively non-randomly with each other. The same type of relationship occurs between ♂Bl and ♀Wp. The former pair consists of a highly dominant male and a very subordinate, almost non-interacting female. The latter pair consists of a young adult male and a female which is subordinate to ♀Wm. The change in ♂Bl's proximity scores to greater solitariness (see below) does not seem to have affected his relationship with ♀Wp.

♂Bl's proximity scores changed markedly from February to April. From September to February he was found alone 24 times versus 31 times with others. From February to April he was seen alone 23 times but only twice with others. This change in proximity scores is statistically significant at the .0002 level. The change might be related to his sexual maturation and resulting difference in social standing in the colony.

Sauer and Sauer have suggested that agonistic dyads will tolerate each other during sleeping periods in order to stay in the same nest.[1] This type of occurrence was very infrequent in our colony.

Agonistic behaviour

During the present study, female antagonisms were high and directed toward the males during oestrus. ♂P, ♂R, ♀Wm and ♀Wp exhibited the highest frequencies of agonistic encounters. The highest scores for ♂R and ♀Wm were with ♂P, and ♀Wp's highest was with ♂R. By contrast, ♂Bl, ♂Y and ♀G had very low scores. ♂Y and ♀G's highest were with ♂P, and ♂Bl's was almost equally divided between ♂P and ♀Wm. Agonistic encounters always took place between two individuals without involvement of others.

Grooming

Grooming sessions were not timed, but were grouped with other actions into a general category of non-agonistic interactions. The increase in frequency of grooming among the animals over time may reflect closer personal bonds developing between animals as fear of the surroundings and of other animals lessened. More than half of the grooming recorded was directed by ♂P to ♀Wm and ♀G, who rarely reciprocated. What little grooming ♂R and ♂B1 did was mainly directed towards the females.

A grooming session often seems to be a stressful situation interspersed with spat-calls from the animal being groomed. But, while calling, an animal may be presenting an arm or bending its head to be groomed. Whether grooming functions as a group cohesive force in galagos, as it does in other primate groups, is uncertain.

Sexual behaviour

When females were not in oestrus, the scrota of the males were inconspicuous and fur-covered. During the oestrus period, the male's scrotum enlarged slightly and a pink-coloured patch of skin with a granular appearance was observed centred in the groove between the two scrotal sacs. This is very similar to that described by Manley for *Perodicticus* and *Arctocebus.*[18] In the females, the clitoris is conspicuous between oestrus periods, but the vulva opens only during oestrus.

During the first oestrus period, which occurred in October, copulations were restricted to ♂P and the females. The number of copulations recorded was low, only one for ♀Wm and five for ♀G. None was recorded for ♀Wp, although she underwent a large amount of investigation by the males and was mounted several times. Usually ♂R was able only to ano-genital muzzle as a female passed. It may be significant that ♂Y exhibited much scrotal swelling and pink colouration, but did not investigate the females. He may have been so intimidated by one or several of the others that his sexual behaviour was suppressed. Little sexual interaction occurred between ♂Bl and the females. That ♂Bl had no scrotal swelling or colouration, combined with his small body size, indicated that he was a subadult. Because of his subadult status, like the subadults of Roberts' group, he may have been more easily tolerated by the adult females than were the adult males.[11] This would explain his high non-aggressive scores and low aggressive ones between September and February.

There were probably no two females at the peak of oestrus at the same time, giving no opportunity to the observer to check for preference on ♂P's part. The males remained sexually excited after the females' vulval opening had closed and they were no longer receptive. This led to many agonistic encounters.

During the night (artificial light 'daytime' for the galagos) of February 28, ♀G gave birth to an infant. It was found dead the next morning, the brain and lower extremities having been eaten. Killing of infants in captive colonies has been thought to be related to over-crowding.[13,19] Females probably separate themselves from other animals during birth of infants.[9,19] Overcrowded conditions were not considered characteristic of the present study; ♀G was, therefore, left to give birth in the enclosure with the other animals. Either the colony was overcrowded (possibly indicated by the high percentage of agonistic interactions), or some other factor was involved. ♀G did not come into a post-partum oestrus as does *G.s. moholi.*[20]

On 22 March, ♀Wm came into oestrus. This precipitated both a sharp increase in fighting between ♂R and ♂P, and what appeared to be a shift in dominance status between them. The formerly submissive ♂R, which had not been allowed to copulate during the October oestrus period, became very aggressive. The aggression took the form of attacks and chases of ♂P and ♂Bl by ♂R, especially when Bl ventured onto the floor. ♂R successfully copulated with ♀Wm several times. ♂Bl shared typical scrotal swelling and pink colouration during ♀Wm's oestrus period. He mounted her several times, but he did not copulate while under observation.

The number of copulations is not considered a reliable means by which to judge dominance or group structure in other social species, because subordinate animals may copulate with females when a dominant animal is elsewhere. This was not observed in my group, possibly because the dominant male could never be any great distance from the other animals. Perhaps females discriminate against subordinates as mates.

Social structure

Field data for *G. senegalensis* and *G. demidovii* indicate that sometimes individuals form small groups. Although these species sleep together in groups, the individuals may separate for nightly foraging, but stay within hearing range of each other. They may regroup at daybreak. Haddow and Ellice used evidence of yellow fever immunity to support such regrouping.[2] They found four out of eleven groups which exhibited immunity. This might suggest little intermixture of groups. Observations in Cameroun by the present author in 1971 agree with such observations for *G. demidovii* and *G. alleni.* When lights were flashed into closely spaced trees, several pairs of red eyes could be seen. Closer inspection sometimes revealed two to four animals in the same tree. Observations by the present author in East Africa seem to indicate a more solitary way of life for *G. crassicaudatus* than is found in the other galagos. Observations by Bearder and Doyle on *G. crassicaudatus* in South Africa have shown one to three to be the most common number found together.[5]

Laboratory data from the present study suggest that 'activity centres' may exist rather than a cohesive group or groups. Flow diagrams are helpful for illustrating these types of relationships for behaviour scored (see Fig. 1). Whether these animals are capable of organising structurally beyond the mother-offspring level is a

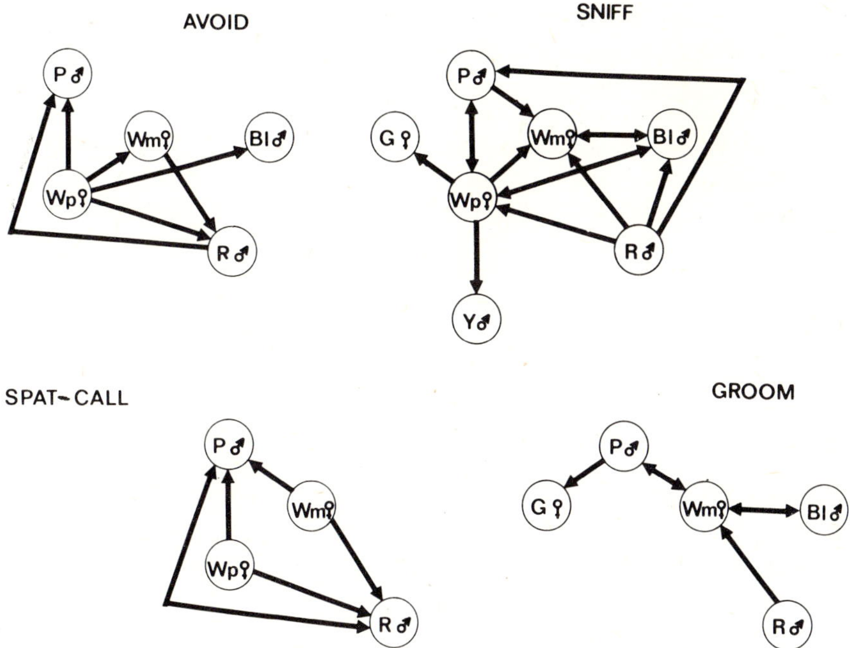

Figure 1. Flow diagrams illustrating four of the behaviour patterns scored. Arrows indicate ten or more interactions, and their directions.

question that needs to be answered. There are strong indications that most interactions between animals are sexual. The frequent sniffs of females by males instead of by other females, grooming of females by males instead of by other females (and vice-versa), and agonistic action of females toward the males are examples of this. A thorough understanding of social structure in *G. crassicaudatus* will require many additional hours of observation.

Urine-washing and foot-rubbing data suggest that more dominant animals are more territorial and that sub-dominant animals have a higher tendency to perform submissive and displacement activities. ♂P and ♂Bl were significantly high ($P < .05$) for urine-washing but significantly low ($P < .05$) for foot-rubbing. Just the opposite is true for ♂R and ♀Wm. ♀G was significantly low ($P < .05$) on both counts, while ♂Y was not significantly high or low for urine-washing but significantly high ($P < .05$) for foot-rubbing. ♀Wp was not significantly high or low for urine-washing but was significantly low for foot-rubbing.

The scent of the animal left with urine-washing might be a good territorial marker (but see Doyle).[16] The loud sound of foot-rubbing, however, might have an immediate effect. The contexts in which the latter has sometimes been seen, such as when a male is rejected by a female or when the isolated male ventures toward the other animals, suggest its role as a displacement activity or as a submissive signal.

Within the group of females, the differences in rate of urine-washing versus foot-rubbing do not seem to indicate dominance relationships as they do among males. For example, ♀Wm is dominant over ♀Wp and ♀G according to other data, but ♀Wm shows high levels of foot-rubbing and low levels of urine-washing. These actions by females may be directed toward males instead of other females. The females may also not be as strongly territorial as the males.

Within the group of males, the differences in rate of urine-washing versus foot-rubbing do seem to agree with other data regarding dominance relationships. ♂R and ♂Y are sub-dominant and ♂P and ♂Bl are dominant. ♂R is dominant over ♂Y and ♂P over ♂Bl. Strong territoriality in males might be adaptive, so that each male has optimum chances of reproducing.

Limitations of present laboratory studies

One must consider the problem of alteration of behaviour in the laboratory compared to that in nature. Overcrowding might cause abnormal levels of agonistic interactions. Possibilities for social stress might be greater in the laboratory than in nature, but physical stress due to parasites and disease may be less in captivity. Some differences of the colony in captivity versus nature are: 1. smaller animal/space ratio than in the wild; 2. opportunities for natural foraging behaviour extremely limited; 3. light cycle altered and animals moved to a different hemisphere; 4. frequent interactions with cage cleaner and observers; 5. probably abnormal age/sex ratio – groups contained neither infants nor young juveniles; and 6. the laboratory provides a more homogeneous environment. Inhibition of animals and alteration of stimuli may cause the absence of behaviour as well as creation of new behaviour (see Kaufmann and Kaufmann regarding this phenomenon in coatis, *Nasua narica*).[21]

The age/sex ratio may have an effect on the type of social structure observed. Such effects should become evident after several years of births and colony growth (which will require the provision of a larger living space).

Communicative behaviour

Few vocalisations have been observed except in a sexual or agonistic context. The spat-call is agonistic and the crack-call has been suggested to have a sexual function because it is usually heard only in male/female interactions. The loud call may function as a spatial indicator much as the morning call of the gibbon.[22] The acoustic qualities of this call are ideal for transmission over long distances and through dense vegetation.

Communication between animals in non-agonistic contexts may be primarily through olfaction. The animals in the present colony continually smell the substrate by lowering the nose to a tree limb or the floor as they walk. This is accompanied by loud sniffing sounds. Initial meeting of another animal usually initiates some sort of sniffing behaviour.

Ear positions are good situation indicators. The ears orient differently during alertness, fights, submissive gestures, etc.[1] Not all possible situations have been categorised in the present study, since this would require more detailed study of this behaviour. Ear positions most probably do function as communication signals. Perhaps in galagos, which seldom change facial expressions, they are analogous to the facial expressions of higher primates.

Terrestrial versus arboreal behaviour

Galagos hop well and are able to cover great distances quickly with this method of locomotion. During periods of extreme stress in the lab, the galagos leave the safety of high ledges and hop about on the floor, even though plenty of limbs are available. This appears to be strange behaviour for an arboreal animal. ♂R, which was very subordinate to ♂P for most of the first part of the study, spent most of his time on the floor. This is a behavioural pattern which developed during several months. ♂R did not do this when the colony was first established. He did not appear to be sick nor injured and could climb and jump well. He is becoming less agile than the others, perhaps because he spends so much time on the floor. Possibly a lack of exercise has resulted in his present 'heavyweight' condition. Considering this behaviour in captivity, it would be interesting to know how much time galagos in the wild spend on the ground. They may be more terrestrial than previously thought. In the farm-bush zone where *G. crassicaudatus* sometimes live, trees are fairly widely separated. This would necessitate locomoting on the ground at least during movement between trees, a situation which would seldom arise in riverine or other forested areas. Bearder and Doyle report on this aspect of behaviour in the field.[5]

Summary

Seven *Galago crassicaudatus* (four males and three females) were observed in early stages of contact and interaction after having been released into a laboratory enclosure. Animals were scored on proximity, agonistic, non-agonistic and other behaviour as possible indicators of social structure. It appears that there are personal preferences among the animals regarding other animals, which are apparent in the interaction data and proximity scores. At present small groupings may be forming, but the tendency seems to be toward solitariness. The animals interact with one another individually rather than as groups. Births of young in the colony may be sufficient stimulus to form family groups. Field data on several galago species indicate solitary and small group habits.

Various communicative methods are discussed. These include an agonistic spat-call, a crack-call probably related to sexual activity and a loud call of unknown function. Olfaction is important in communication as indicated by various types of sniffing and their frequencies, and ear positions may also be important in communication.

The enclosure was constructed to simulate a natural environment, but all stimulus possibilities available in nature were not present and several new ones were introduced. It is not known how such differences affected the animals' behaviour.

Acknowledgments

I would like to thank Y.G. Matola and P. Wegesa of the East African Institute for Malaria and Vector Born Diseases, Amani, Tanzania, for providing the animals used in this study. Thanks also go to Claude Bramblett, without whose help this study could not have been possible, for his aid in obtaining the animals, his suggestions and stimulating discussions, and for critically reading the manuscript. His wife Sharon very generously typed the manuscript. My indebtedness to Larry Quick is also acknowledged. Finally, I would also like to thank my husband Mills, who very patiently read the manuscript and made many constructive criticisms. This study was aided by an Undergraduate Research Grant from the University of Texas.

NOTES

1 Sauer, G.F. and Sauer, E.M. (1963), 'The South West African bushbaby of the *Galago senegalensis* group', *J. S.W. Afr. scient. Soc. Windhoek* 16, 5-36.

2 Haddow, A.J. and Ellice, J.M. (1964), 'Studies on bushbabies (*Galago* sp.) with special reference to the epidemiology of yellow fever', *Trans. Roy. Soc. trop. Med. Hyg.* 58, 521-38.

3 Vincent, F. (1968), 'La sociabilité du Galago de Demidoff', *Terre et Vie* 22, 51-6.

4 Doyle, G.A., Andersson, A. and Bearder, S.K. (1969), 'Maternal behaviour in the lesser bushbaby (*Galago senegalensis moholi*) under semi-natural conditions', *Folia primat.* 11, 215-38.

5 Bearder, S.K. and Doyle, G.A., this volume.

6 Struhsaker, T.T. (1970), 'Notes on *Galagoides demidovii* in Cameroun', *Extrait de Mammalia* 24, 207-11.

7 Miller, C.D. III, personal communication.

8 Charles-Dominique, P. (1971), 'Eco-éthologie des prosimiens du Gabon', *Biol. Gabon.* 7, 121-228.

9 Lowther, F. de (1940), 'A study of the activities of a pair of *Galago senegalensis moholi* in captivity, including the birth and postnatal development of twins', *Zoologica* 25, 433-59.

10 Doyle, G.A., Pelletier, A. and Bekker, T. (1967), 'Courtship, mating and parturition in the lesser bushbaby (*Galago senegalensis moholi*) under semi-natural conditions', *Folia primat.* 7, 169-97.

11 Roberts, P. (1971), 'Social interactions of *Galago crassicaudatus*', *Folia primat.* 14, 171-81.

12 Hill, W.C.O. (1953), *Primates: Comparative Anatomy and Taxonomy*, vol. 1. *Strepsirhini*, Edinburgh.

13 Buettner-Janusch, J. (1964), 'The breeding of galagos in captivity and some notes on their behavior', *Folia primat.* 2, 93-110.

14 Newell, T.G. (1971), 'Social encounters in two prosimian species: *Galago crassicaudatus* and *Nycticebus coucang*', *Psychon. Sci.* 24, 128-30.

15 Andrew, R.J. and Klopman, R.B., this volume.

16 Doyle, G.A., this volume.

17 Quick, L., unpublished data.

18 Manley, G.H., this volume.

19 Sauer, E.G.F. (1967), 'Mother-infant relationship in galagos and the oral child-transport among primates', *Folia primat.* 7, 127-49.

20 Doyle, G.A., Andersson, A. and Bearder, S.K. (1971), 'Reproduction in the lesser bushbaby (*Galago senegalensis moholi*) under semi-natural conditions', *Folia primat.* 14, 15-22.

21 Kaufmann, J.H. and Kaufmann, A. (1963), 'Some comments on the relationship between field and laboratory studies of behaviour, with special reference to coatis', *Anim. Behav.* 11, 464-9.

22 Carpenter, C.R. (1940), 'A field study in Siam of the behavior and social relations of the gibbon (*Hylobates lar*)', *Comp. Psychol. Monog.* 5, 1-212.

U. M. COWGILL

Co-operative behaviour in Perodicticus

Introduction

This study brings together observations concerning a variety of behaviour patterns of four *Perodicticus potto* (2 males, 2 females) gathered over the past eight years. These patterns are co-operative in character and may be divided into two general categories: those that involve the nuclear family only and those that occur between the family and the group. The individual that is most mobile between these two categories is the dominant 'alpha' male potto. His dominance is entirely expressed in either protective types of behaviour or co-operative ones. Despite the fact that some of his behaviour operates to ensure his own survival, it is nevertheless, as far as the family and group of which he is a part are concerned, protective and co-operative.

The death of Genet

The living conditions and type of food supply under which three *P. potto* have been maintained in captivity for the past eight years have been described earlier.[1] On the morning of 9 April 1971, the 'beta' male named M. Genet was clearly not feeling well. He was being groomed by the 'alpha' male and his mate on the shelf. He died sometime in the late afternoon of that day from organic causes of a non-infectious nature. In 1964, when his mate died as a result of pregnancy, it was noticed that the food consumption of the then remaining three animals was rather meagre. As a result of the more recent death (leaving two animals), this observation was again apparent and it appeared necessary to weigh the food before and after consumption to obtain some idea of how much they were eating. It has previously been noted that pottos consume food on a definite bidiurnal cycle.[2] This is confined to the bulk of food they eat, which is primarily bananas. At that time no such cycle was noted in the consumption of other foods. As will be seen later in this paper, foods other than liquids, if offered daily, are consumed on the same cycle. The data given below are all corrected for natural loss.

Initially, the two surviving animals, one 'alpha' male and his mate, were given the same weight of bananas as they had been when three were being maintained. The amount consumed per animal at this time was less than it had been prior to the death of the third. After two weeks, the quantity the pair had been eating was all they were served. It was noticed that they consistently left enough bananas for the absent third. After four weeks, the 200 gm. they were fed daily were reduced to 100 gm. for two weeks, but the pottos continued to leave enough food for M. Genet. The data are set out in Table 1,

Table 1 Average weight (gm.) of bananas eaten by *P. potto* on high and low consumption days. All data have been corrected for natural loss. With the exception of the first line, concerning three animals, the data for the 'length of time studied' category are consecutive

No. of Animals	*Bananas consumed High days*	*Bananas consumed Low days*	χ^2 *periodicity*	*Amount fed per day*	*Length of time studied*	*Expected consumption High days*	*Expected consumption Low days*	χ^2
3	329	237	14.96***	450	7 weeks	—	—	
2	198	127	15.51***	450	2 weeks	220	158	8.28***
2	130	69	18.70***	200	4 weeks	198	127	49.84***
2	71	49	4.03*	100	2 weeks	100	70	14.71***
2	350	288	6.02**	450	3 days	220	158	183.78***
2	200	130	14.84***	450	1 week	220	158	6.77***

*** P = .01, ** P = .02, * P = .05

where the first χ^2 given is based on the hypothesis that there is no difference between every 48 hours, the high days, and the intervening low-consumption 24-hour period. The periodicity appears to be quite stable even when the food supply is low, though it is not as spectacular as it is when the supply is adequate. The same table shows what two pottos would be expected to consume on both days. The χ^2 between the observed and expected values is enormously significant, suggesting the persistence of food saving even when the supply is low. All during this ten-week period fruits, hard-boiled eggs, raw meat and insect larvae were offered to the animals. With the exception of a few larvae and some papaya, all of these foods were rejected. Papaya, a fruit highly valued by these animals, was occasionally offered, but only a half a fruit at a time, which is about 50 per cent less than their normal ration. An average animal portion was always left. To avoid starving further the already depressed and emaciated animals, the original amount of bananas was again supplied. For about three days, they consumed far more than was usual for them, but at the moment they are continuing to consume less per animal than they did when there were three.

Since this behaviour has been noted in two instances following death, though in the first case no quantitative data were collected, it would appear that the tendency to save food for an absent animal is an old and well established one.

The birth of twin pottos

On the evening of 24 September 1971, the two adult pottos left their tree-trunk dwelling and moved to the extreme end of the cage. At 21.00 hrs. they began to copulate, which on the basis of previous data is an unusually early hour for such an activity. All previous births have been preceded by copulatory behaviour initiated in the early evening.[3] On the following morning, dizygotic twin females were born. Data concerning the two babies are shown in Table 2. The

Table 2 Statistics on dizygotic twin pottos

Birth date	*Death date*	*Sex*	*Weight*	*Length*	*Respiration*
25/9/71 07.30 hrs.		♀	34.7 gm.	9 cm.	68
	26/9/71 13.30 hrs.		36.2 gm.	9 cm.	
25/9/71 08.00 hrs.		♀	40.0 gm.	11 cm.	72
	29/10/71 20.00 hrs.		81.7 gm.	13 cm.	

first baby was born outside of the tree trunk on the floor of the sawdust covered cage. The mother used the base of the tree trunk to assist her in her labour. She left the placenta of the first baby at the base of the tree trunk after severing the umbilical cord and moved back inside the trunk. The second baby was born immediately after this move. The mother was quite unconcerned about the second baby. For two hours after her birth, the placenta was partly inside the mother and partly out and the baby was still connected to it. The baby was frantically trying to induce the mother to accept her. By 10.30 that morning the mother had cleaned her off, consumed the placenta, and was attempting to feed two babies. During all this activity, the adult male was moving around at the end of the cage where the births were taking place. During the early evening (17.00 hrs.), the male began to carry bananas and mealworms to the adult female. While the adult female sat up to eat the bananas and thereby left the babies unprotected, the adult male groomed both babies, presumably to keep them warm. The following morning, the lighter of the babies was rejected. It was quite apparent that the adult female could not feed two babies simultaneously. Although all attempts were made to feed the rejected baby, she died one day after birth in convulsions, possibly due to some nutritive deficiency.

Initially, the remaining baby appeared to feed continuously. Twenty-four days after birth she moved back and forth between the two parents and wandered around the cage by herself at night. She was intermittently groomed by one or both parents. When the baby was four-and-a-half weeks old, the mother ceased to lactate. The baby was found still and cold on the floor of the cage. She was fed with a sterile baby doll bottle, containing 15 cc. of cow's milk, 0.5 cc. of a vitamin supplement (ABDEC), 2 gm. of meritine, a concentrated milk supplement, 30 mg. of sucrose and 30 cc. of distilled water. She was fed on demand. Table 3 shows the times of

Table 3 Feeding times and amount consumed by a 4½-week old potto, during hand-rearing

Date	*Amount consumed (cc.)* 03.00 hrs.	07.00 hrs.	10.30 hrs.	13.00 hrs.	16.00 hrs.	20.00 hrs.	22.30 hrs.	*Total*
10/26				15.0	14.6	2.1	2.1	33.8
10/27	4.20	1.1	1.1	1.1	1.1	4.2	2.1	14.90
10/28	9.45	1.5	1.5	1.5	2.1	3.3	3.3	22.65
10/29	3.80	1.1	1.1	0.5	1.1			7.60

her requests as well as the amount consumed. It is interesting to note that, even though the data are scanty, the presence of a bidiurnal cycle is apparent. During her life away from her mother, the peritoneal area had to be massaged to encourage urination. This necessity had been noted previously.[4] Both urination and defaecation were normal. On the evening of 29 October, her temperature began to increase. By 20.00 hrs. it had reached 39°C and she died half an hour later, presumably from a respiratory infection. Attempts were made to revive her, but to no avail.

Malnutrition of adult pottos

After she had given birth to twins, it was noticed that the female adult potto was suffering from a large number of deficiencies, presumably as a result of pregnancy. The male potto was also beginning to show signs of a calcium deficiency but not nearly as acute. The protein deficiency, which was exhibited only in the female, was apparent from cracked and bleeding skin and broken nails. A calcium deficiency was exhibited by muscular tremors. The latter were continuous in the female and intermittent in the male.

There was no doubt that some solution to the problem of finding acceptable food had to be found. Initially, an attempt was made to supplement the diet with calcium by adding calcium carbonate to mealworms and bananas. This caused the pottos respiratory difficulties, and so such treated food was rejected. The problem was finally solved by adding 5 gm. of calcium lactate daily to the drinking water. The pottos refuse to drink Pittsburgh tapwater, partly because of its high sodium concentration (60 ppm most of the time) and partly because of organic compounds such as phenols, which cannot be distilled off. For some time they had been drinking spring-water, which was close to the composition of distilled water.

The protein and vitamin deficiency problem was solved in several ways. The two pottos were hand-fed mealworms. On high consumption days they would eat 15 gm. each, but on the intervening low ones they consumed only 7 gm. each. This helped but did not entirely alleviate the deficiency symptoms. Finally, a mixture known as Nutri-Cal, the composition of which may be found in Table 4, was

Table 4 Composition of dietary supplement (Nutri-Cal)

Substance	*Amount*
Protein	1.0%
Carbohydrates	45.0%
Fats	35.0%
Calories/28.3 gm.	190
28.3 gm. contains	
Vitamin A	5,000 USP units
Vitamin D	250 USP units
Vitamin E	30 IU
Vitamin B_1	10 mg
Vitamin B_2	1 mg
Vitamin B_6	5 mg
Vitamin B_{12}	10 mcg
Nicotinamide	10 mg
Ca pantothenate	10 mg
Folic acid	1 mg
Iron (from Iron peptonised)	2.5 mg
Mn (from $MnSO_4$)	5.0 mg
Mg (from $MgSO_4$)	2.0 mg
I (from KI)	2.5 mg

hand-fed to the pottos daily. It is consumed on a bidiurnal rhythm. The mixture is sold in a tube and has a very thick consistency. It is put on the finger and the animals lick it off. On the average, the female consumed 15 gm. per day and the male about 7 gm.

Despite all efforts to eradicate the malnutrition problem, the female ceased to lactate. By 7 November, eight days after the death of the baby, she began to lactate again. Lactation persisted for about three months, with the male suckling regularly from the female throughout this time (see below).

One interesting observation should be noted here. Namely, the adult 'alpha' male always tastes all the food first. This pattern was apparent when there were a total of four pottos and is still the case with only two. If he decides it is consumable, he transmits that information in an unobvious fashion to the female, who then and only then will try the food in question. For example, Nutri-Cal is offered to the animals at a variety of times during the day. The male wakes up, tastes it, usually eats a portion, wakes up the female by walking around her, who will then also eat. If he decides that he does not want any food, he will still go through the tasting ceremony.

Discussion

The saving of food for an absent individual is such a curious phenomenon that even though it is an example of co-operative behaviour within the group, it is felt that the discussion of this aspect of behaviour should be considered separately. The remaining animals, one 'alpha' male and his mate, were apparently leaving food for a dead, less dominant male. This behaviour persisted in the captive *Perodicticus* even when the food supply was limited. Although there appear to be no comparable data on the subject, it has been noted in chimpanzees that food caught by more dominant individuals is shared with less dominant ones as a result of the latter's 'begging gestures'.[5,6] Of course, less dominant chimpanzees also share their food with more dominant individuals. In the baboon, a highly dominance-oriented species, such sharing has not been noted in the wild.[7] This suggests that the potto may be less dominance-oriented than the baboon.

This type of behaviour might tentatively be considered altruistic. Wynne-Edwards has proposed that altruistic behaviour is that which benefits a group or ensures its fitness and which is relatively independent of the individual exhibiting the behaviour.[8] Klopfer, on the other hand, is critical of this hypothesis as he states that, if the behaviour is disadvantageous to the individual, the genes responsible for it die when the individual fails to survive.[9] However, if the behaviour is advantageous, such action is irrelevant in terms of the group. Klopfer states that any advantage gained by the group is explicable in terms of natural selection alone. Mayr, who is concerned with the question of the widespread existence of altruistic behaviour, suggests that it may operate in populations containing genetically related individuals, i.e. a parent that sacrifices himself for his progeny ensures that safety of his genotype by such action.[10]

The behaviour of the pottos appears to be more similar to that suggested in the hypothesis of Wynne-Edwards where, though the animals go hungry in order to save food for an absent one, they nevertheless by so doing act to ensure the survival of the group. It may be that the ideas of Wynne-Edwards are more applicable to the higher vertebrates. No comment can be offered on the thoughts of Mayr, since the possible existence of kinship between the pottos is unknown. The suggestions of Klopfer, though applicable to other animals, do not appear to apply to pottos. Clearly, sharing of food in captive *Perodicticus* is a social phenomenon that is rather unexpected in a prosimian.

It should be considered that behaviour of this type in primates may be more developed in forest dwellers than in animals that live in a savannah, primarily for the reason that the former do not exhibit such a complicated social structure to ensure protection, since their environment already offers it.

The persistence of the bidiurnal cycle is a rather strange phenomenon. When it was first observed, the pottos had only been in contact with each other for a short period of time.[2] It is interesting to note that even when the food supply is limited, it persists. Similarly, if the cycle becomes disrupted, when the periodicity is restored it becomes so in the rhythm of the already established cycle. The other point to consider here is that this feeding pattern is already apparent in the young, or, at least, by the time the baby is a month old. All these observations tend to suggest that the cycle is not a reflection of prolonged captivity but one that is inherent and probably social in origin and possibly of survival value. Were it a reflection of a variable food supply in the wild, captivity of eleven years with ample available nourishment would tend to eradicate such a periodicity. Though pottos are basically insectivorous,[11] they will consume all stages of the insect life-cycle, so that presumably some stage would always be available in the wild. The other possibility is a low metabolic rate. The body temperature is similar to that of man,[12] but recent data show the metabolic rate to be normal.[13] It would appear that the periodicity in feeding habits may be territorial in origin and reflect some type of social phenomenon, since animals together seem to show the same cycle.

The baby female when fed on demand appeared to consume on a bidiurnal cycle. Possibly therefore this is an inherited phenomenon or perhaps it reflects the distribution of food supplied by the mother. If the latter is the case, then a new-born baby would be taught to feed on the cycle, since that is how food would be available. If this is the case, the bidiurnal cycle could possibly be a built-in population control. Namely, were a baby to be born on a low food supply day, it would have less of a chance of survival than had it been born on a high supply day. By the same token, it is possible to postulate that a baby's birth is arranged in such a fashion so that they are born on high supply days to ensure survival. New-born pottos begin to feed shortly after birth. The second baby that was born was found to urinate about four times a day and, on the second day after birth, produced about 2 cc. of urine at each time. This observation would tend to suggest that, though the baby was suckling, it was also drinking at the same time.

Co-operative behaviour may be divided into that type which is exhibited within the nuclear family and that which extends outside the family but within the group. Since the behaviour within the family appears to be somewhat different from that within the group, co-operative behaviour might well be separated into two general categories.

Within the nuclear family, much of the co-operative behaviour, though not all, is confined to the general activity of infant-rearing. It is also apparent that parturition is initiated by copulatory activity. During labour and the actual process of birth, the 'alpha' male rushes around 'trying to be useful'. Once birth has occurred, he grooms the baby and examines it extensively. Any time the mother found it necessary to raise herself on her perch, the male would groom the baby. During the first few days after birth, the 'alpha' male would carry food to the mother. After the baby was two weeks old, the mother would sometimes leave the tree trunk to drink fluid and often would leave the baby with the male potto. Other times she took the baby with her. Many of these observations have also been noted by Anderson.[14]

After the baby died and lactation was resumed, as a result of improved nutrition, the male suckled from the female. Initially, the suckling occurred six times a day. By mid-January the frequency had declined to four times daily. A similar observation has been noted by Anderson.[15] In the latter case the male was occasionally noted to suckle from the female in the presence of a live baby.

An explanation for the nursing adult male may be found in an examination of the nutrition of *P. potto.* An analysis of food eaten either by the pottos or the insects they eventually eat is given in Table 5. Comparative elemental data on the composition of insects may be found in Spector and Bowen.[17,18] In the wild, pottos live in a low-calcium region.[19] The types of food they are known to consume are all high in potassium. Although the varieties of insects they eat in the wild are not known, those they prefer in captivity are all high in potassium. Since zinc is known to accumulate in the mammalian prostate gland,[20] and in its absence spermatogenesis is adversely affected, it was thought wise to check that aspect of potto nutrition. Interestingly, zinc is accumulated by all the insects examined, and the concentration of zinc in the insects is higher than that of copper. The analyses clearly show that a diet consisting primarily of bananas will not be adequate. It is also possible that insects living in a low-calcium region will contain less of that element than the analyses show. It would appear that the types of insects *P.*

Table 5 The quantity of some elements in insects and food consumed either by *P. potto* or by the insects *

Food	*Ca*	*Mg*	*P*	*K*	*Na*	*Fe*	*Cu*	*Zn*
	Dry Weight mg/100 gm.							
Bran	120.0	62.5	1187.5	980.0	15.0	8.5	3.1	1.20
Bananas	36.8	87.5	9.0	1201.0	5.0	3.0	2.0	0.53
Pellets	955.7	237.5	1350.0	762.0	12.5	20.6	3.0	1.08
Insects								
Periplaneta americana L.	116.6	100.0	822.5	590.0	7.5	4.8	3.8	11.0
Blaberus giganteus L.	138.8	87.5	615.0	501.8	10.0	9.7	3.4	13.3
Blatta orientalis L.	214.2	125.0	1230.0	1055.6	12.5	9.8	3.4	14.2
Blattella germanica L. (male only)	359.8	112.5	1265.0	997.9	7.5	17.8	3.0	15.7
Tenebrio molitor L. (adults)	71.3	225.0	1632.5	879.2	1.5	7.6	3.9	12.8
T. molitor (mealworms)	62.7	200.0	1167.5	672.6	2.5	6.3	4.0	14.3

* Bran (crude) is consumed by *T. molitor*. Pellets are eaten by roaches. Insect data are averages obtained by drying (110°C, 1 week) 100 individuals with the exception of *B. giganteus* where only 15 were available. The chemical procedure employed is that described by Cowgill.[16]

potto is most likely to value are those high in potassium and low in sodium. This is rather an interesting point, since recently it has been suggested that the set-point for body temperature of a primate is maintained by the physiological ratio of sodium to calcium within the brain stem.[21] An excess of sodium in the absence of calcium perfused into the brain stem causes a rise in temperature, while an excess of calcium in the absence of sodium results in a fall in temperature. Pottos will not eat food high in sodium. They have, however, recently developed an appetite for milk, thus partly alleviating the problem of supplying them with adequate calcium. However, since the availability of calcium is low in the region where they live, such suckling behaviour would supply a higher calcium diet to the male.

Moreover, continuous suckling of this order may ensure the survival of future progeny in the sense of developing 'maternal behaviour' in the absence of the baby or ensuring its continuation in the presence of an infant by occasional suckling.[22] In the event of the death of a progeny, such behaviour may also ensure the survival of the female by preventing the difficulties that may in time develop when the results of lactation are not utilised. Some zoo animals which are permitted to lactate in the absence of their young may die of mammary tumours at a later time.[23,24] It has also been suggested that long-term nursing of infants in man contributes to a lower incidence of breast cancer.[25] More recent data tend to dispute this suggestion.[26] However, there does not seem to be any available information in man on the relationship between non-utilised lactation and the subsequent incidence of mammary tumours. It is therefore possible that suckling of an adult may have some survival value.

Co-operative behaviour within the group really is exhibited in quite a different form from that observed within the family. Initially, when the pottos first came, the 'beta' male lost his mate and became depressed and so isolated himself from the rest of the group. The 'alpha' male and occasionally his mate visited him each evening for a period of seven years.[1] When the 'beta' male became ill, he was found in that state with the other two pottos on the shelf, his head in the lap of the 'alpha' male and the rest of him in the lap of the female. When he finally died and his body was removed from the cage, the remaining two spent at least an hour each evening searching for him. This searching behaviour persisted until the birth of the twins. At the same time they saved food for him even to their own detriment.

Apart from the above patterns, which appear to act to make the captive group (whatever its size) cohesive, there are other forms of behaviour that are exhibited entirely by the 'alpha' male which serve to ensure the survival of the group. For an example, when the pottos first arrived, each was in a separate compartment in a carrying case. When the cases were placed inside the cage and the individual doors to the compartments were opened, the 'alpha' male came out by himself and investigated the entire cage. Then he visited each of the other animals, first his female, then the 'beta' male and then the 'beta' male's mate, at which point they all left their compartments. Any time the pottos have been moved over the past years, the same procedure has been repeated. Each evening since they arrived, the 'alpha' male potto has left the tree trunk first, examined the food, tasted it, and then returned to the trunk. This procedure, if the food supply were limited, would ensure the survival of the 'alpha' male. If, on the other hand, the food was in some way unhealthy to consume, this behaviour would ensure the survival of the group and not necessarily that of the 'alpha' male. When any of the animals have been ill, the 'alpha' male has tasted the food and then made all attempts to induce the sick animal to partake of the nourishment. Although these observations tend to indicate the behaviour of the 'alpha' male towards the rest of the group, they are nevertheless co-operative in character.

At this point in the discussion it might be well to consider the relevance of behavioural data obtained from captive animals in relation to that in the wild. Visiting in *Perodicticus*, for example, has been noted in captivity as well as in the wild,[1,27] though in the latter case this type of social activity has been confined to male-female contacts exclusively, while in captivity it has involved male-male as well as male-female visiting. It is not impossible to imagine that most of the types of behaviour mentioned here could be observed in the wild. The point that needs to be made is that this kind of behaviour is possible in prosimians that are this low in the evolutionary scale. How food saving would be noticeable to the observer in the field remains to be seen.

Acknowledgments

I gratefully acknowledge the support of the National Science Foundation which as a result of support of other work made the present study possible. I am also indebted to the Bio-Medical Sciences Support Grant awarded to the University of Pittsburgh by the National Institute of Health for travel aid to attend the Research Seminar in Prosimian Biology. I wish to thank Gulf Research and Development Company of Pittsburgh for supplying the cockroaches. J. Iben gave freely of her time and knowledge in trying to solve the problem of malnutrition among the pottos.

Editor's note

The results presented in this study and the conclusions drawn by the author, with respect to altruistic behaviour in P. potto, *if accepted, would place* P. potto *in a position quite unique amongst the primates and particularly amongst prosimians. At the conference, no-one was able to offer even a tentatively acceptable alternative explanation to the one offered by the author and, in fact, very little discussion took place. On reflection it is felt that results and conclusions of this nature are so controversial that the conventions of scientific research would require that, before they are accepted, the study should be repeated at least twice, carefully, using larger samples and extended over a longer period of time. It should also be duplicated exactly in a closely related species. Only when these requirements have been met can any definitive statement on the subject be added to the literature. This comment in no way reflects on the integrity of the author nor on the value of her research.*

NOTES

1 Cowgill, U.M. (1964), 'Visiting in *Perodicticus*', *Science* 146, 183-4.
2 Cowgill, U.M. (1965), 'A bidiurnal cycle in the feeding habit of *Perodicticus potto*', *Proc. nat. Acad. Sci. Wash.* 54, 420-1.

3 Cowgill, U.M. (1969), 'Some observations on the prosimian *Perodicticus potto*', *Folia primat.*' 11, 144-50.

4 Walker, A. (1968), 'A note on hand-rearing a potto', *Int. Zool. Yb.* 8, 110-11.

5 Van Lawick-Goodall, J. (1968), 'A preliminary report on expressive movements and communications in the Gombe Stream chimpanzees', in P. Jay (ed.), *Primates*, New York, 313-74.

6 Van Lawick-Goodall, J. (1971), *In the Shadow of Man*, Boston, 200-7.

7 Hall, K.R.L. and DeVore, I. (1965), 'Baboon social behaviour', in I. DeVore (ed.), *Primate Behavior*, New York, 53-110.

8 Wynne-Edwards, V.C. (1962), *Animal Dispersion in Relation to Social Behavior*, New York, 1-23.

9 Klopfer, P.H. (1969), *Habitats and Territories*, New York, 96-102.

10 Mayr, E. (1970), *Populations, Species and Evolution*, Cambridge, Mass., 114-7.

11 Jewell, P.A. and Oates, J.F. (1969), 'Ecological observations on the lorisoid primates of African lowland forests', *Zool. Afr.* 4, 231-48.

12 Wislocki, G.B. (1933), 'Location of the testes and body temperature in mammals', *Quart. Rev. Biol.* 8, 385-96.

13 Sucklin, J. (1969), 'The retia mirabilia of *Perodicticus potto*: a functional study', Unpublished PhD thesis, University of East Africa.

14 Anderson, M. (1971), 'The potto family', *Yale Alumni Mag.* 35, 24-7.

15 Anderson, M. (1971), 'A watched potto never grows: a chronicle of the prenatal and first months of *Perodicticus potto*', *Discovery* 6, 89-98.

16 Cowgill, U.M. (1966), 'Use of X-ray emission spectroscopy in the chemical analyses of lake sediments determining 41 elements', *Developments in Applied Spectroscopy*, 5, 2-23.

17 Spector, W.S. (ed.) (1965), *Handbook of Biological Data,* London and Philadelphia.

18 Bowen, H.J.M. (1966), *Trace Elements in Biochemistry*, New York, 70-2.

19 Papadakis, J. (1969), *Soils of the World*, New York, 146-50.

20 Mann, T. (1964), *The Biochemistry of Semen and of the Male Reproductive Tract*, London, 44-8.

21 Myers, R.D., Veale, W.L. and Yaksh, T.L. (1971), 'Changes in body temperature of the unanaesthetized monkey produced by sodium and calcium ions perfused through the cerebral ventricles', *J. Physiol.* 217, 381-92.

22 Cowie, A.T. and Tindal, J.S. (1971), 127 and 272. *The Physiology of Lactation*, Physiological Society Monograph 22, London, 127, 272.

23 Personal communication from J. Iben, a Pittsburgh veterinarian, who stated that a female lion in the local zoo died of mammary tumours some time after giving birth to young she was not allowed to nurse.

24 Heidrich, H.J. and Renk, W. (1967), *Diseases of the Mammary Glands of Domestic Animals*, Philadelphia, 312. Neoplasms may occur in female dogs who have never given birth or who have not done so for some time or who have repeatedly produced stillborn infants or who have undergone pseudo-pregnancy or false lactation.

25 Wynder, E.L., Bross, I.J. and Hirayama, T. (1960), 'A study of the epidemiology of cancer of the breast', *Cancer* 13, 559-601.

26 MacMahon, B. et al. (1970), 'Lactation and cancer of the breast', *Bull. Wld. Hlth. Org.* 42, 185-94.

27 Charles-Dominique, P. (1971), 'Eco-éthologie et vie sociale des prosimiens du Gabon', *Biol. Gabon.* 7, 121-228.

P. H. KLOPFER

Mother - young relations in lemurs

Introduction

Lemurs are lovely creatures, but this, of course, is rarely considered a sufficient reason for their study. As is the case with other primates, the justification for study is believed to be in the light they shed upon human behaviour, and particularly those kinds of human behaviour which some men would manipulate or control. However, there are countless problems in extrapolating from monkey to man. Most of these have been adumbrated often enough. Yet there remains a persistent inclination in each of us to indulge in extrapolations which would have us first identify an animal's activity as akin to ours, and then interpret the evolutionary origins of our acts in terms of the animal's performance. A robin indulges in a display that *we* (not the robin) label 'territorial defence'. The emotional state which we associate therewith is imputed to the robin; the robin's behaviour then becomes an analogue or even homologue of our own.[1,2] Hinde points to a related aspect of this problem when he underscores the difference between the statements, 'the rat ran to the goal box' and 'the rat ran to box X'.[3] More recently, Lehrman has traced the history of the belief of many U.S. Congressmen in the biological justification for defeating bills to establish child care centres.[4] Motherless rhesus monkeys display a depressed countenance that to psychiatrists is reminiscent of a condition in human foundlings, and is labelled anaclitic depression. The emotional concomitants of separation are assumed to be similar in monkey and man, as are the long term effects. *Ergo*, mothers being vital to the well-being of rhesus monkey infants in the same way as they are to human infants, surrogate mothering (or child care centres) are contrary to sound biological principles. Lehrman asks how the Congress would have responded had studies of bonnet macaques been presented to them rather than those of rhesus. In contrast to rhesus, bonnet monkeys accept surrogate mothers fairly readily.

The point, surely, is that we study monkeys not to understand man but in order to generate 'the rules of the game'. The game is the ecological game, staged in the evolutionary theatre.[5] What factors determine the origin, spread and preservation (or loss) of particular patterns of behaviour? Can one predict which factors will be significant for any particular species?

Consider, for example, the question of the exclusiveness of the mother-young relationship. Among goats, a newly parturient doe will generally accept all of her own kids, licking them and permitting them to nurse. Normally, she will violently repel, by butting and biting, any kids not her own, even though the alien kids were born simultaneously with her own, and at a distance of but two metres.[6,7] On the other hand, domestic milch cows can be relatively easily induced to accept alien calves. Lest it be thought this is but an artefact of domestication, the roe deer (*Capreolus capreolus*) has been reported to accept alien fawns, and moose cows (*Alces alces*) even attempt to lure alien calves to themselves.[6,7,8] Why these differences? One possibility lies in the fact that if the species in question is organised into a herd with high variability in the nearest-neighbour distances, young animals that wander randomly through the herd will be unlikely to encounter all the adults. If the young form small clusters within the herd, feeding at intervals throughout the course of their random wanderings, some mothers may not be nursed at all in the course of a day or two. The long-term effect of this on the gross milk production of the herd, to say nothing of the effect on the individual's udder, is surely detrimental. That would be a poor play, from the standpoint of every member of the species. By maintaining an exclusive nursing relationship, a constraint would be applied to the randomness of the youngsters' walk. Periodically the kids would have to take 'time out', and return to their own mothers, each of whom would then be assured of being nursed at sufficiently regular intervals. The sole disadvantage to this scheme lies in the inevitable fatal end of a youngster whose mother dies. If it cannot yet feed itself, it will starve. This apparently happens with insufficient frequency to counterbalance the benefits of the exclusive relationship.

If this kind of cost-benefit consideration is indeed one of the rules of the game, we should be able to predict the circumstances under which exclusive mothering would not occur. Surely this must be so in the most extremely different case of a stationary herd, with low variability in nearest-neighbour distances, and a higher rate of maternal mortality. Such a situation is posed by Northern elephant seals (*Mirounga angustirostris*), which form dense aggregations ('pods') of a few dozens of animals on certain Pacific beaches.[9,10] Baby seals do in fact nurse from mothers other than their own, and lactating seals are as tolerant of older aliens as of their own newborn.[9,10,11] I would expect that computer simulation would allow us to predict more precisely the break-even point, where the cost of exclusivity no longer exceeds the benefit. Other variables might also be important, but the process of identifying these is part of the rule-generation game.

Lemurs have particular advantages as players in this game. Their confinement to a rather restricted part of the globe, similar history, physical and ecological similarities, all combine to provide an array whose dimensions are probably fewer than would be the case for a comparable number of cercopithecine monkey species. As stated in an earlier report, 'the purposes of our studies are to uncover the factors that determine whether a mother will care for her young and whether that care is restricted exclusively to her own or extends to alien youngsters'. From the standpoint of the infant, we are interested in knowing how a separation from the mother affects its development under circumstances where a surrogate mother may or may not be available. Finally, we want to examine the long-term results of different rearing conditions. All this is to be studied against the different social backgrounds provided by three particular species of *Lemur; L. catta, L. fulvus* and *L. variegatus.* These three, while similar in their morphology, and, we assume, their physiology, are neither entirely sympatric nor identical in their patterns of maternal care. In *L. catta*, animals other than the mother may hold the infant before it is a week old and independent exploration begins before it has aged three weeks. In *L. fulvus*, however, other animals are generally prevented by the mother from taking over the infant until four weeks after its birth, a time which coincides with the commencement of independent explorations by the infant. The *L. variegatus* infant, finally, is denied early contact of any kind, except with the mother, until it is fully mobile, at about seven or eight weeks. Before this, it receives only intermittent attention even from its mother.

What is the significance of these different patterns of care? Are the more communally reared *L. catta* infants better buffered from the trauma of parental separation? Are the 'independent' *L. variegatus* better able to invade and colonise new areas? I do consider knowledge of the answers to be interesting *per se*, but the agencies that support this work can take comfort in the fact that by learning the answers we can bring the rules of the game to light. Those rules, surely, apply to all productions appearing in the evolutionary theatre, including the production we call man.

Summary of previous findings

Our first report merely described the pattern of events in captive groups of *Lemur fulvus* and *L. catta*, and in one group of *L. variegatus.* Those findings provided an indication of the character and degree of difference between the three species. In general, they allow the *L. fulvus* infant to be characterised as dependent for its care on a single individual, its mother, and relatively shielded from direct contact with other animals. *L. catta* infants, on the other hand, experience considerably more handling by other animals from their earliest days, while our single *L. variegatus* infant lived a solitary life until it was able to perform effective locomotion.

We subsequently sought assurance that our conditions of captivity were not unduly distorting the behaviour we were studying. To this end, several hundreds of hours of *Lemur catta* activity were logged under natural conditions, on the same protocol charts as those used for the captive groups.[12] The comparison of data from wild *L. catta* at Berenty, Madagascar, and captive *L. catta* at the Duke University Primate Facility showed the following: the start of a consistent increase in the proportion of time the infant spent off the mother (when in captivity) occurred at about 30 days of age. Specifically, the percentage of the observation time when either of two infants was not in contact with its mother rose from less than 10% by day 28, to over 15% by day 32 and over 30% by day 48 (one of the two reached the 30% mark on day 40). Wakeful contacts with other adults, simultaneously with maternal contact, occurred from the start. Contacts with others when separated from the mother were not noted until day 8, and then only infrequently. This corresponds closely with the situation at Berenty, where a significant number of wakeful contacts with animals other than the mother was also noted only during the September-October period. The first occasion on which a captive infant moved from the ventral surface to the back of its mother was day 3; this corresponds to the earliest age noted both by Jolly and by us at Berenty.[17] In the colony, the first independent exploration was noted as early as day 18, though this did not occur regularly until about day 30 (day 27 and 37, respectively). At Berenty, too, only two occasions of total separation were noted before day 30 (approximately), and several hundred thereafter. Thus, with respect to the specific indices of development which we used, the captive and wild animals appeared strikingly similar. While subjective reports must be treated sceptically, let me nonetheless record that the observers did indeed agree that this was the case, even with regard to other, less carefully measured attributes.

In order to assess the degree of buffering afforded infants of the different species, we examined in captivity the effects of mother-young separation in different social constellations (mother and infant alone; mother, father and infant; mother, father, infant, plus other adults and juveniles). At the time each infant attained 150 days of age, and again at 200 days, its mother was removed for seven days. A full series of observations has until now only been reported for *L. fulvus.* Where the infants were reared alone with their mothers, the infants remained close to their mothers for their first 80 days. The first separation of 1 m. distance did not occur until day 50; the maximum separation was permitted in their chamber on about the 110th day. The removal of their mothers at the 150th day led the infants to alternate between stupor and frenzy, but within a week following the reunion the animals had resumed their previous patterns of activity. The infants took the initiative in the re-establishment of relations, their mothers seeming relatively uninterested. For *L. fulvus* infants reared with both parents it appeared that there was a slight advance in the age by which the infant departed from its mother's back or lap. In other respects there were no differences from the mother-infant pairs. On removal of the mother, the infant increased the number of its grooming contacts with the male, but otherwise gave no indication of trauma. The infants shifted their attention back to their mothers on the latter's return. The infant kept with its parents and other animals was generally similar in behaviour. Again, following a week-long separation from the mother at 150 days, the infant seemed more affected by the reunion than the mother.

A final methodological consideration involved the effects of varying the number of animals in a group on the space available to any one of them. (In separate studies, at the Duke University Primate Facility, Miss Laura Vick reported that removal of one animal from a group immediately changed the status of the most subservient female from that of a peripheral to central animal, and similar phenomena have been noted by C. Chandler, personal communication). Perhaps even more importantly, one alters the character of the available space. Wherever any degree of dominance or attraction exists between two or more animals, the area surrounding them must be viewed as charged, analagous to the surround of atomic nuclei. The shape of the space available to other animals will then depend on their valence and will be altered as they move about. Complexities of the substratum, partitions, overhanging ledges, holes, may function as insulators, distorting the fields of charge and permitting distributions that would otherwise be impossible. Hence, the experiments on the infant separation trauma and its amelioration require an assessment of the role of the substratum on social spacing.[13]

Hence, the activities and distributions of four juvenile *L. fulvus* in a chamber with many artificial trees, and in the same chamber with fewer artificial trees, were compared. The number of active (and resting) periods did not differ significantly between the two environments (53 and 54, for the 'impoverished' and 'enriched' rooms, respectively). There were small differences between rooms in the amount of clumping, however. All the animals were alone 13 times in the impoverished room, compared with 28 in the enriched room (confidence intervals of these proportions are separate with confidence coefficients between .90 and .95). Clusters of 2 animals were seen 72 (impoverished) versus 53 (enriched) times (confidence intervals of these proportions are separate with confidence coefficients between .90 and .95). Large clusters were infrequent, and did not differ significantly, though during resting the enriched room seemed to promote more clumping (groups of 3 and 4: 12 *vs.* 5; difference too small to show significance, if any). The total number of animals involved in any size clump in all observations was 192 (impoverished) *vs.* 169 (enriched), which is not significant. Since the frequency of activity was unchanged between environments, differences in cluster frequency are not likely to be an artefact of the animals' speed or frequency of movement. The impoverished room did have fewer perches and rest areas, of course, though there were a sufficient number to allow total separation (3 shelves, a chair, a feeding ledge). Separation did not occur as often; under impoverished conditions, the animals were more often paired (whether resting or active) than under enriched conditions. The large clumps, on the other hand, seemed to occur more often among resting, 'enriched' animals. Hence, the effect of the artificial forest is not merely to provide a greater array of potential perches. Were that the case, one would expect fewer, rather than an equal or greater number of the 3 and 4 animal clumps. Nor did the 'forest' alter activity frequency, and thus, indirectly, the groupings. The observations themselves reveal no qualitative differences in the nature of the activities in which the animals engaged in the two environments. In a subtle, as yet undefined manner, the 'forest' did cause small differences in the groupings to emerge. Less dispersion of individuals in the impoverished room, and a predominance of pairs over other clumps of another size.[13]

In summary, observations of *Lemur catta* mother-infant pairs in a forest preserve in Madagascar provided assurance that the conditions of captivity in the Duke University colony were not seriously distorting the behavioural relations between mothers and their infants. Intensive observations of captive, individual mother-infant pairs of *L. fulvus* suggested that the frequency and distance of the infant's excursions from its mother were increased with the presence of other lemurs. The temporary absence of the mother upon the infant's attaining an age of 150 days was readily tolerated if another familiar animal was present, but otherwise proved traumatic. However, no subsequent effect of a separation of one week's duration was noted. On reunion, the infants seemed more eager to re-establish contact than their mothers.[13]

Answers to a number of other more specific questions are being sought. These include queries as to the proximate factors determining whether a mother will recognise neonatal young; whether she will accept them; whether she will discriminate between her own and alien young; how differences in maternal responses influence the infant's response to separation from its mother; and the basis for the success of surrogate mothers.

Subsidiary questions include the roles of other adults and peers in the development of young, and the importance of sex and individual differences in development. Obviously, the data collected thus far and presented below do not yet suffice to provide other than speculative answers.

Method

The observation chambers of the Primate Facility, within which the animals lived, were hexagonal rooms with a volume of about 82 cubic metres. The ceiling tapered to a height of 5 m.; the maximum distance from corner to corner across the floor was 5.3 m. Corner shelves and a single central post with several horizontal branches provided perches. Observations were made through one-way glass either from an overhead mezzanine or at ground level.[13] The rooms are not fully soundproof, but only especially loud sounds distract the animals.

The animals were maintained on a fixed schedule of 12 hours of light, 12 hours of dark (dim red light). While nocturnal observations were made to establish activity patterns, the maternal data were collected entirely during daylight hours. Observations were evenly spaced throughout those periods of the day when the animals were most likely to be active. (Nocturnal activities were observed with a light amplification device). These observations were supplemented by videotape records, which served to provide confirmation for periods of presumed inactivity. Each observation period, scheduled for a specified and different time each day, was of 60 minutes duration (in contrast to the 30-minute periods in the field). Data were compiled by noting the position of the infant at half-minute intervals and recording whether and with whom it was in contact, distance from the mother, and manner of behaviour, particularly whether grooming or being groomed. These data themselves were grouped into 30-minute data periods that facilitated the various comparisons.

The animals were generally introduced into their chambers several weeks prior to the birth of an infant. Formal recording of data commenced upon the discovery of a newly-born infant, usually within hours, always within 24 hours after birth. Upon attaining 150 days of age, the infant was subjected to the First Separation, which involved removal of its mother to separate quarters for a seven day period. At 200 days, there was a Second Separation, also of seven days' duration. Observations were made just before and just after the removal and return of the mother. At 215 days the observations were terminated.

Data

The data presented here are a supplement to those summarised above. They are based on observations of the following two groups:

(*a*) *L. catta*
♀ 534F Bilbo, infant ♂ 584M Cato
♀ 546F Balin, infant ♀ 585F Calpurnia
♂ 527M Socrates, and ♂ 548M Fredegar

(*b*) *L. fulvus*
♀ 554F Bathsheba, infant ♀ 572F Miriam
♀ 552F Shulamite, ♂ 553M David and
♂ 551M Ishmael

Mother, father, infant and others in L. catta

A group of four adult lemurs, two of each sex, was placed in the chamber on 10 March 1971. Bilbo's infant, a male, Cato, was born on 15 March; Balin's infant, the female Calpurnia, on 16 March. The graphs, Figs. 1*a*, *b* and 2*a*, *b* reveal the changes in the maximum distance between mother and infant at various ages and the proportion of the observation times the infant was not in contact with its mother.*

The two infants did differ in the age at which they first began to separate themselves from their mothers. Calpurnia was over 40 days old before she regularly moved more than an arm's length away, while Cato's explorations began fully 10 days earlier. However, both animals were showing maximal utilisation of the space at their disposal by the same age (approx 100 days). No change in this measure was evident subsequent to the separation and return of their mothers.

The graphs showing the % of time spent off their mothers reflect parallel behaviour. Calpurnia's rate of becoming 'independent' did not begin a rapid rise until around the 50th day; Cato preceded her by about 15 days. However, both achieved a plateau at about the

* In preparing graphs from the data sheets certain conventions were adopted for convenience' sake; observation hours with less than five minutes of activity were discarded; each graph point was made to represent a 4-day average, with dashed lines connecting points for which there were 2 or less observations in 4 days; during the 7-day separations, the averages were based on 4 and 3 days, respectively; distances or times during separations are based on next nearest animal to infant.

same time, around the 100th day.

Other major events are recorded on Table 1. In general, this captive, confined group proved similar to both those confined groups examined previously[11] and to wild groups at Berenty. It is clear, however, that there is considerably more variation than originally expected between infants (and mothers) in the ages at which certain activities first appear. Some of the differences may be more apparent than real, the result of sampling errors for instance. These are of particular significance in the case of infrequently occurring behaviour, especially where a group is being observed for only two to four hours out of 24. As the sample size grows, however, it will be possible to provide estimates of the intra-specific variations. Some other miscellaneous observations that were noted include the following (arranged chronologically).

25 March to 1 April 1971: Each mother frequently attempts to seize the other infant, attempting to keep both simultaneously. Even when there are no overt attempts at kidnapping, the mothers do much grooming of one another's infants.

13 May 1971: Kidnap attempts are no longer always resisted. Mothers occasionally nip their own infants when these seek their lap and make no protest when the other mother carries their own infant away. There is much *reciprocal nursing*, Bilbo even nursing Calpurnia while the latter is still in her mother's – Balin's – lap! Bilbo was the more aggressive kidnapper of the two mothers, but Balin actually had possession of both infants more often. She was also the more protective mother, inhibiting her infant's attempts at exploration. In fact, her infant, Calpurnia, was much later than Cato to move away from her mother.

8 May 1971: First noted wrist-marking from the axillary gland (by Cato).

8 June 1971: First noted mounting attempts by Cato (on Calpurnia) with pelvic thrusting.

10 June 1971: Calpurnia, the female infant, appears to dominate Cato. She is the one that initiates play and is always more active.

10 July 1971: First noted overhead tail flicking by Cato towards Calpurnia.

Effects of separation: there was an initial reduction in the infants' activity upon the removal of their mothers, but by day 3 their normal patterns had reappeared. Upon the mothers' return the

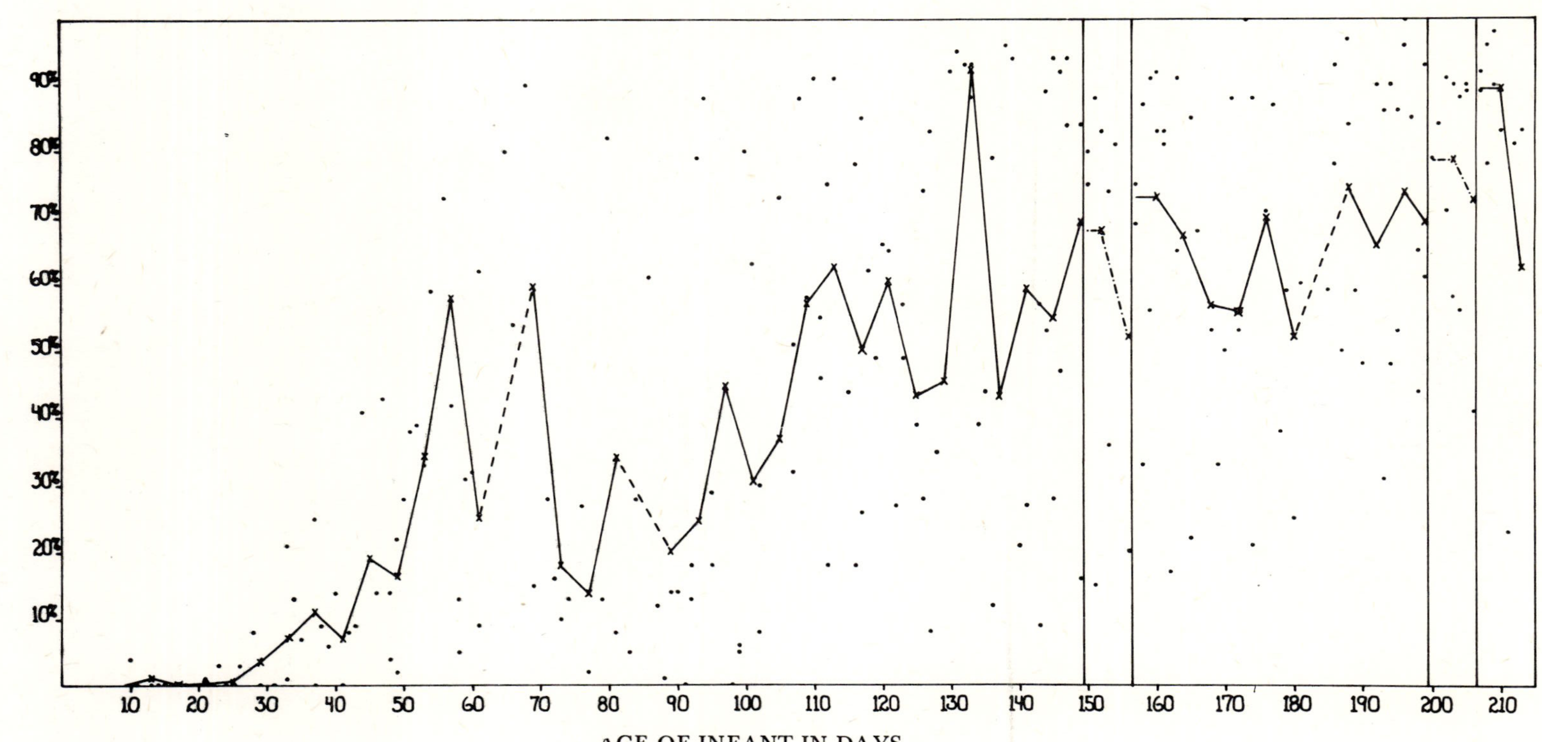

Figure 1. *L. catta*; Bilbo and ♂ infant Cato. (*a*) % time infant not in contact with mother.

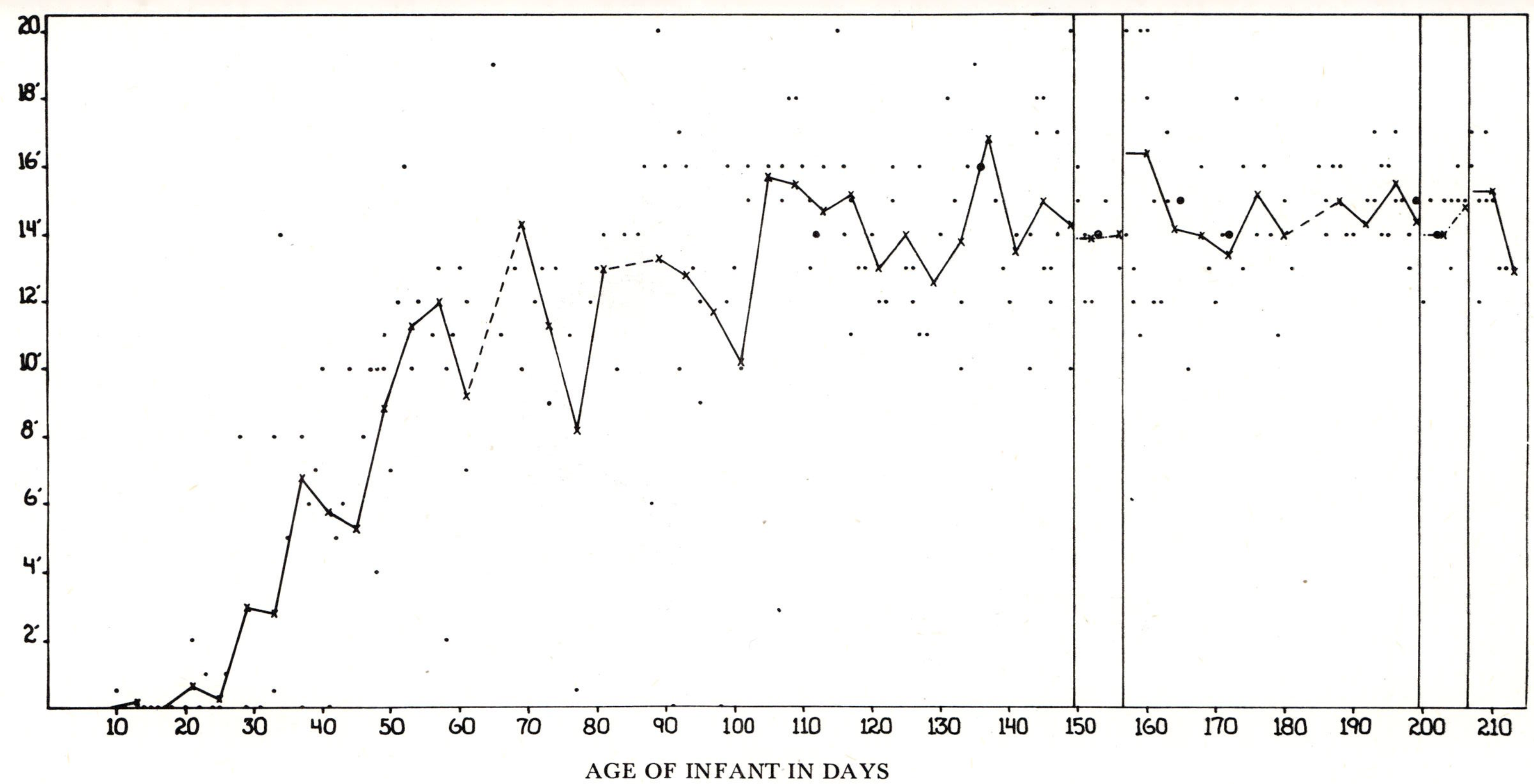

Figure 1 (*b*). Distance in feet of infant from mother.

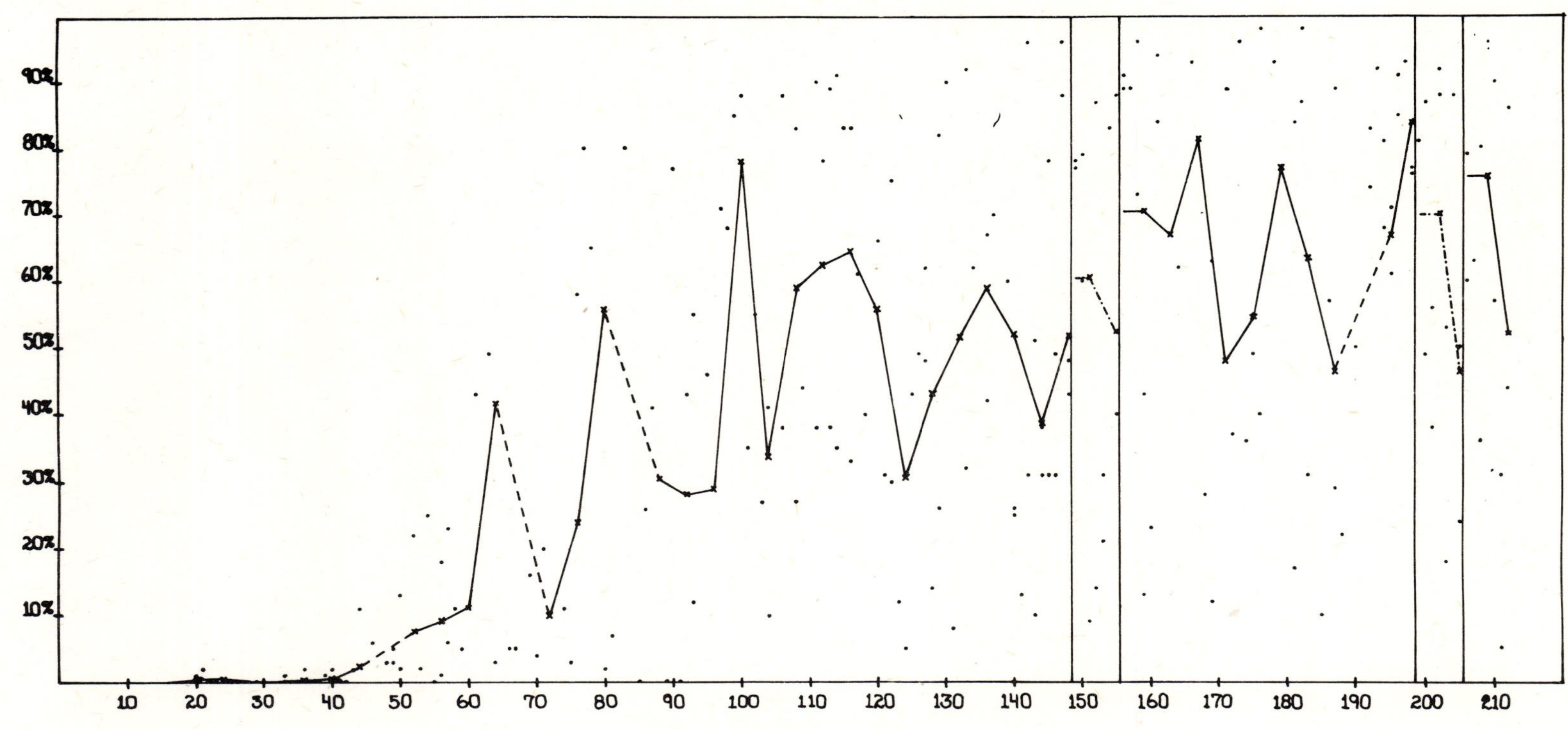

Figure 2. *L. catta*; Balin and ♀ infant Calpurnia. (*a*) % time infant not in contact with mother.

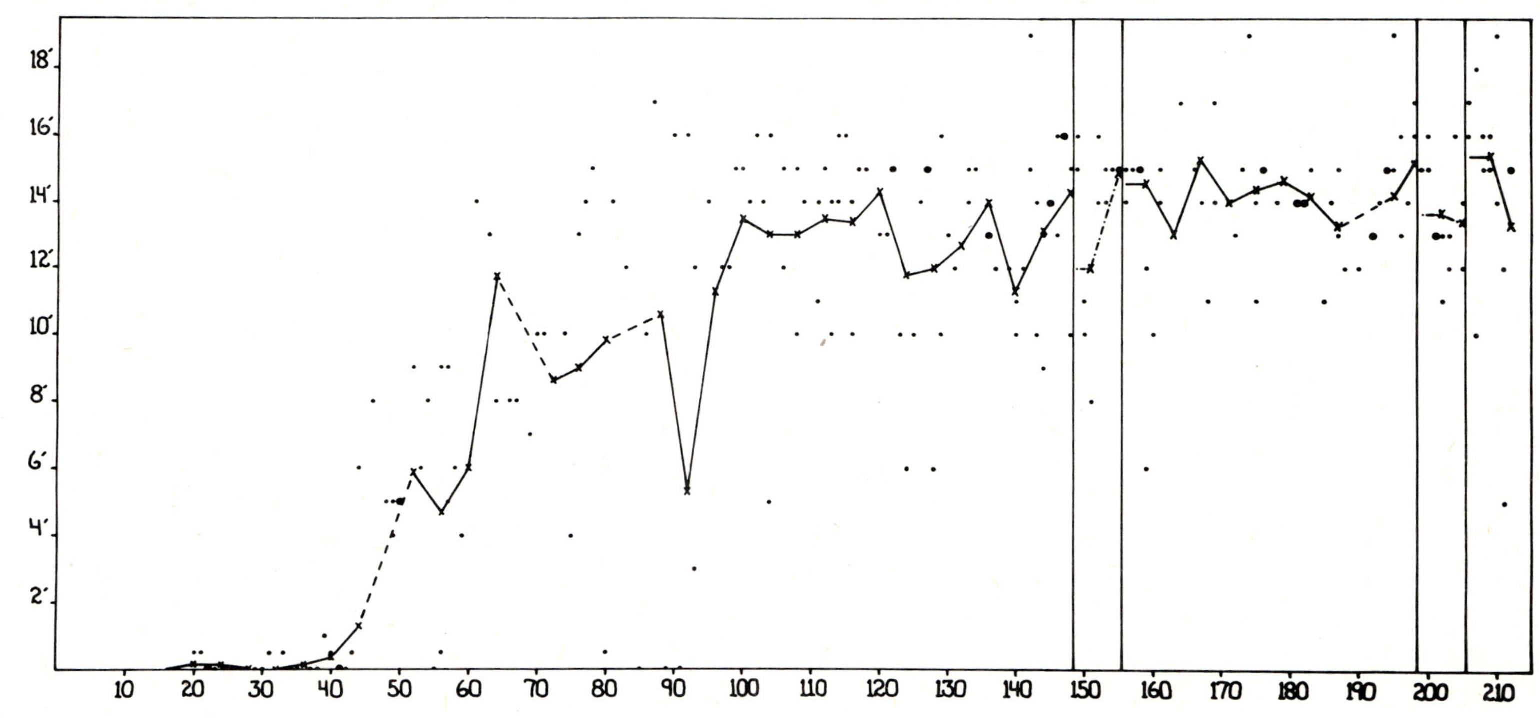

Figure 2 (*b*). Distance in feet of infant from mother.

Table 1 Major events in infant development

		in L. catta *(group)*		*in* L. fulvus *(group)*		*in* L. fulvus *(solitary)*		*in* L. fulvus *(pairs)*	
Sex of infant:		♂	♀	♂	♀	♀	♀	♂	♀
Age of infant when 1st groomed or handled by another animal (not its mother)		3 days;	2 days	7 days;	3 days	8;	2 days	?	?
Age of infant when 1st held by another animal (not its mother)		10 days;	20 days	46 days;	28 days	44;	30 days	?	?
Age of infant when 1st without contact with any other animal		22 days	33 days	49 days;	28 days	44;	30 days	49 days;	49 days
Age of infant when 1st off its mother for at least 30% of an observation hour		44 days;	61 days	? ;	82 days	77;	97 days	87 days;	72 days
Infant-mother and mother-infant grooming frequencies for 1 week before and 1 week after 1st separation	I-M	1.%*-2.1%	1.5%-1.5%	? ;	.8%-4.9%				
	M-I	4.8%-7.8%;	6.5%-3.2%	? ;	10. %-6.9%				
Infant-mother and mother-infant grooming frequencies for 1 week before and 1 week after 2nd separation.	I-M	2.5%-2. %;	4. %-3.7%	? ;	1.2%-5.1%				
	M-I	6.1%-3.3%;	3. %-7.1%	? ;	2.4%-4.9%				

*Mean % of 60-second intervals during observation hour when grooming occurred.

infants first fled from them, but after several minutes sought to climb onto their backs. In this, they were actively repulsed by their mothers. By the end of the first day following the return of the mothers, the pre-separation patterns were fully re-established. On 5 September both mothers were seen to be in full oestrus.

Relations to males: the infants repeatedly teased the males, bouncing on them, batting and cuffing them, or even biting. The males were very gentle in their responses, usually backing away or feinting in response. Their occasional bites never seemed to injure the infants, even to the extent of eliciting a scream.

Mother, father, infant and others in L. fulvus

A group of 4 adults, two of each sex, was moved into the chamber on the day of the infant's birth, 21/11/71. The graphs, Figs. 3 *a, b*, reveal changes in the distance between mother and infant as a function of age, and her growing independence. They indicate a slower initial rate of development than in *L. catta.*

The second adult female of the group, Shulamite, sat pressed close to Bathsheba, the mother, from the time of the infant's birth. She very probably groomed the infant from the day of its birth, though this was first observed only when the infant was 3 days old. At 11 days of age, the infant briefly climbed into the female Shulamite's lap, this occurring when she and Bathsheba were grooming one another. After a few minutes, Bathsheba took her back. Shulamite made repeated efforts to groom or seize the infant, but usually these were repulsed by Bathsheba. Once, when the infant was in contact with Shulamite, she was seen trying to nurse from her. Bathsheba seems to have been more tolerant of Ishmael's than Shulamite's attentions.

The age at which the infant was first alone, that is, without physical contact with others, was at 28 days, but it was over 80 days before this degree of independence was manifested for as much as one third of an observation hour.

The two separations (at 150 and 200 days) did lead to a subsequent slight increase in mother-infant grooming interactions, but this effect was transitory and no other changes were noted.

When the infant was 50 days old, Shulamite gave birth to twins. After 24 hours, one of the twins was found to have been badly bitten. Shulamite and the surviving twin were thereupon removed from the group. No change in the infant's behaviour was detected as a consequence of Shulamite's removal.

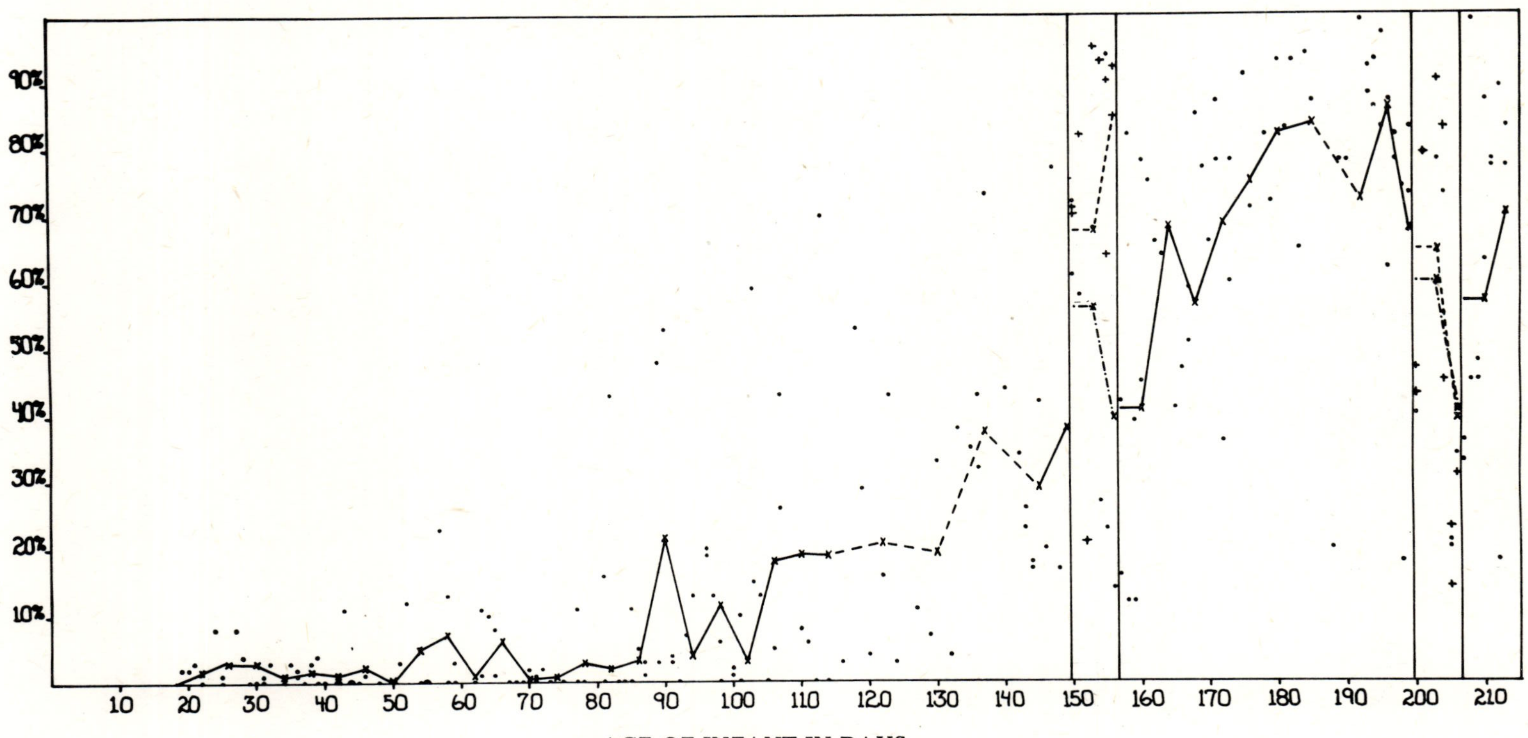

Figure 3. *L. fulvus*; Bathsheka and ♀ infant Miriam. (*a*)% time infant not in contact with mother.

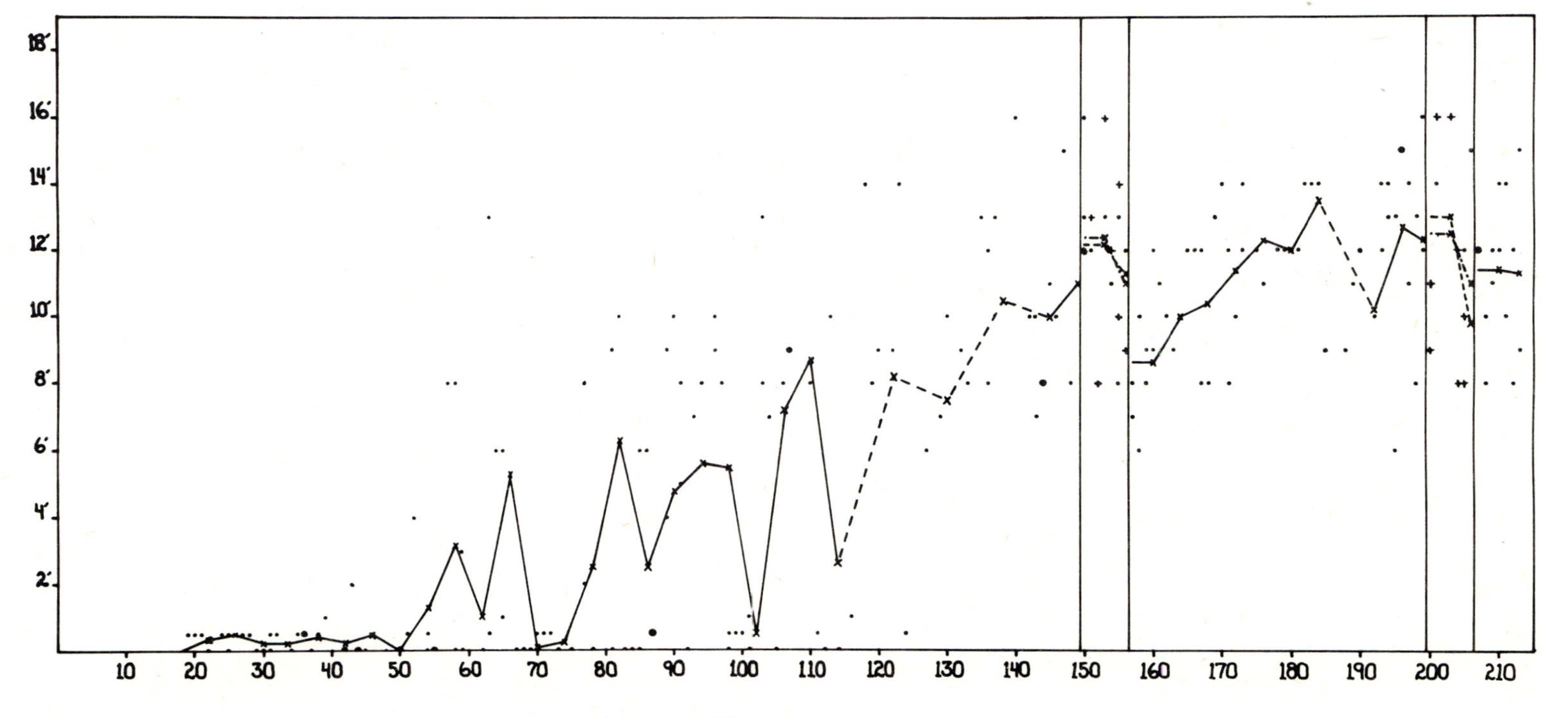

Figure 3 (*b*). Distance in feet of infant from mother.

Comparisons

With sample sizes as small as are these, and individual differences looming ever larger, generalisations must necessarily be tentative. The following relations appear to stand out: The age at which particular events first occur does not seem to differ between *L. catta* and *L. fulvus*. What does differ is the rate of increase in the recurrence of the behaviour in question. Thus, the first time an infant sits alone seems not to vary specifically, at least within pairs or groups (22 and 33 days for *L. catta*, 49, 28, 44 and 30 for *L. fulvus*), but infant *L. fulvus* consistently take longer before they are independent of the mother for 30% of their activity time.

Taking both species together, differences related to sex, as reported for some other primates,[14] have not yet been detected, nor is there any striking change in developmental rates as a consequence of rearing the *L. fulvus* infant in the company of its mother alone. (No data are yet available for solitary *L. catta.*) Solitary rearing produces, as its principal result, a reduction in activity during the separation period, a result that is hardly cause for surprise.

General conclusions

What can then be said about the functional significance of the differences that are seen? What in fact are the 'rules of the game'?

The ecology of *L. fulvus* and *L. catta* has been investigated by Petter and Sussman,[15,16] with particular attention to differences in their use of the habitat in areas where both species occur. Not unlike certain woodland warblers of the north-eastern United States, the two species have subdivided a common area. *L. catta* spends much time on the forest floor (the exact proportions of the active day apparently varying with season and weather).[17] *L. fulvus* is exclusively arboreal, slower moving, and with a more restricted vertical range within the trees as well. What the ultimate and proximate reasons for these, and other, differences may be remains a subject of investigation. Sussman's data do indicate that these differences are not a case of character displacement and hence susceptible to rapid change. That is, in areas where *L. fulvus* or *L. catta* occur alone, their respective behaviour *vis-a-vis* the habitat is apparently indistinguishable from their respective behaviour where the two overlap. We thus must suspect that whatever constraints limit their behavioural repertoires, these are relatively inflexible. It is, therefore, reasonable to expect that there would be concomitant effects upon maternal care patterns.

A leaping, vertically wide-ranging, arboreal-terrestrial animal might be far more likely to lose an infant, or, if on the ground, have greater need to spurt into the treetops than a more phlegmatic branch-walker. Would there then not be an advantage to promoting motor precocity or the acceptance of substitute mothers, or both? An actively-moving mother, as of the species *L. catta*, must profit more than a slow-ranging *L. fulvus* from not having continually to carry her infant. At the same time, the sudden appearance of a predator could pose an unacceptable risk if the more independent infant were not readily snatched into the bosom of the nearest adult, whether its mother or not. High rate of activity and widely ranging movements, in short, promote infant precocity; precocious infants have greater requirements for surrogate mothers for protection than do less precocious babes. That, in short, is the suggestion, tentatively put, as to the rules of this particular game.

Acknowledgments

This work was done at the Primate Facility of Duke University's Field Station for Animal Behaviour Studies, supported by NSF GB4000 and NIH FR 00388 and HD02319, as well as a Research Scientist Award. I also record my appreciation to Linda Hobbet, Martha Klopfer, Lee McGeorge, Charles Chandler, and June Bronfenbrenner for their many hours of observation and compilation. The good health of our colony is largely due to the ministrations of Dr Jan Bergeron and his staff.

Editor's note

Due to the author's unavoidable inability to attend the conference, this paper did not have the benefit of being discussed and, excepting minor editorial changes, is included in the form in which it was originally submitted.

NOTES

1 Leach, E. (1966), 'Don't say "boo" to a goose', *N.Y. Review of Books*, 7, no.10.

2 Klopfer, P.H. (1969), *Habitats and Territories: A Study of the Use of Space by Animals*, New York.

3 Hinde, R.A. (1966), *Animal Behaviour*, New Jersey.

4 Lehrman, D.S. (1971), Unpublished Lecture, Duke University.

5 Hutchinson, G.E. (1965), *The Ecological Theatre and the Evolutionary Play*, New Haven.

6 Klopfer, P.H. (1971), 'Mother love: what turns it on?', *Amer. Sci.* 59, 404-7.

7 Klopfer, P.H. and Klopfer, M.S. (1968), 'Maternal "imprinting" in goats: fostering of alien young', *Z.f. Tierpsychol.* 25, 862-6.

8 Altman, M. (1958), 'Social integration of the moose calf', *Anim. Behav.* 6, 155-9.

9 Bartholomew, G.A. and Collias, N.E. (1962), 'The role of vocalizations in the social behaviour of the northern elephant seal', *Anim. Behav.* 10, 7-14.

10 Klopfer, P.H. and Gilbert, B.K. (1967), 'A note on retrieval and recognition of young in the elephant seal, *Mirounga angustirostris*', *Z.f. Tierpsychol.* 23, 755-760.

11 Lluch-Belda, D., Irving, L. and Pilson, M. (1964), 'Algunas observaciones sobre mamíferos acuáticos', *Instituto Nacional de Investigaciones Biológico-Pesqueras, Comisión Nacional Consultiva de Pesca y Industrias.* 10, 1-23. Mexico, Conexas.

12 Klopfer, P.H. (1972), 'Patterns of maternal care in three species of *Lemur*: 2. Effects of early separation', *Z.f. Tierpsychol.* 30, 277-96.

13 Klopfer, P.H. and Klopfer, M.S. (1970), 'Patterns of maternal care in three species of *Lemur*: 1. Normative description,' *Z.f. Tierpsychol.* 27, 984-96.

14 Hinde, R.A. (1971), 'Development of social behavior', in Schrier, A.M. and Stollnitz, F. (eds.), *Behavior of Nonhuman Primates*, vol. 3, New York.

15 Petter, J.J. (1962), 'Recherches sur l'écologie et l'éthologie des lémuriens malgaches', *Mém. Mus. nat. Hist. nat.* 27, 1-146.

16 Sussman, R. (1972), Unpublished PhD Thesis, Duke University. (See also this volume.)

17 Jolly, A. (1966), *Lemur Behavior*, Chicago.

H. COOPER

Learning set in Lemur macaco

Although prosimians have been described as curious and manipulative,[1,2] little work has been done on assessing their learning skills in a carefully controlled situation, as has been done with the higher primates.[3,4] Part of this is due to the difficulty involved in obtaining and working with these reputedly nervous animals.

Lemurs can, however, be fruitful subjects for laboratory experiments in learning and problem-solving if they are carefully adapted to the experimenter and the experimental situation. The present study is a preliminary note on learning by one species of Malagasy lemur, *Lemur macaco.*

The learning set procedure, first introduced by Harlow,[5] has been shown to be a convenient inter- and intraspecific measure of learning ability. Learning set refers to progressive interproblem improvement between problems of the same type. Typically, the subject is given a large number of problems in which the stimulus elements change from problem to problem, but which are of equal difficulty and which all involve the same principle, such as object discrimination, discrimination reversal, or oddity. Performance is measured in terms of percentage correct responses, on individual trials or a series of trials, and the acquisition of learning set is indicated by progressive reduction in errors in successive problems.

This method has been shown to be fairly sensitive to interspecies differences within the primate order.[3,4,6] The ability to form a learning set is not restricted to the primates, however, as cats,[7] rats[8] and squirrels[9] can also develop this type of learning, though only to a limited extent. Learning set is also a function of age, as in macaques where the ability does not appear until about 200 days of age.[10]

Studies in discrimination ability of lemurs have been few in number.[11,12] A more recent study comparing learning through many discrimination reversals on a single problem[13] showed *Lemur* to be inferior to *Cercopithecus* in this particular problem. Similar results were found when these two species were compared for extinction.[14] The actual learning set procedure, however, allows great possibilities for analysis of response tendencies.

Methods

1. Subjects

The animals used in this study consisted of 2 *Lemur macaco albifrons* and 1 *Lemur macaco albifrons x fulvus* hybrid. All these animals were tame, and were each given 2-3 weeks of habituation to the transport and test procedure before the first problem was begun. Animals were housed in individual cages and were trained to enter a transport cage which had been adapted to the test apparatus. All subjects were experimentally naive before testing.

2. Apparatus

The test apparatus consisted of a Wisconsin General Test Apparatus,[5] suitably modified for use with lemurs (see Figs. 1 and 2). The animal compartment measured 50 cm. x 40 cm. x 40 cm., just large enough to accept the transport cage. Bars were mounted horizontally on the transport cage, 4 cm. apart. Two sliding screens, one opaque and one transparent, separated the animal from the test compartment, which measured 40 cm. x 40 cm. x 30 cm. A third screen containing a one-way mirror was located at the end of the test compartment to allow observation. A sliding plexiglass tray, 40 cm. x 30 cm., was used to present the stimulus objects. Two food wells, 4 cm. in diameter, were located on the tray 20 cm. apart, 1 cm. from the edge of the tray. The food wells were 1 cm. deep watch-glass shaped depressions, allowing easy retrieval of the small food rewards. In addition, the test tray protruded into the animal cage by 3 cm., allowing the animal to respond with the snout or the hand. Lighting was provided by a 100-watt bulb overhead, and a ventilation system was installed at the rear of the apparatus. All surfaces were painted grey.

The stimulus objects consisted of 500 randomly collected pieces of junk (beads, toys, bottle tops, etc.), mounted on plastic plaques measuring 6cm. x 6cm. and painted to match the test tray. The plaques were attached to strings to prevent the subjects from taking the stimulus objects into the cage. The response was a simple displacement of the object in any direction.

Objects were randomly paired, and then ordered in sequence randomly for each animal.

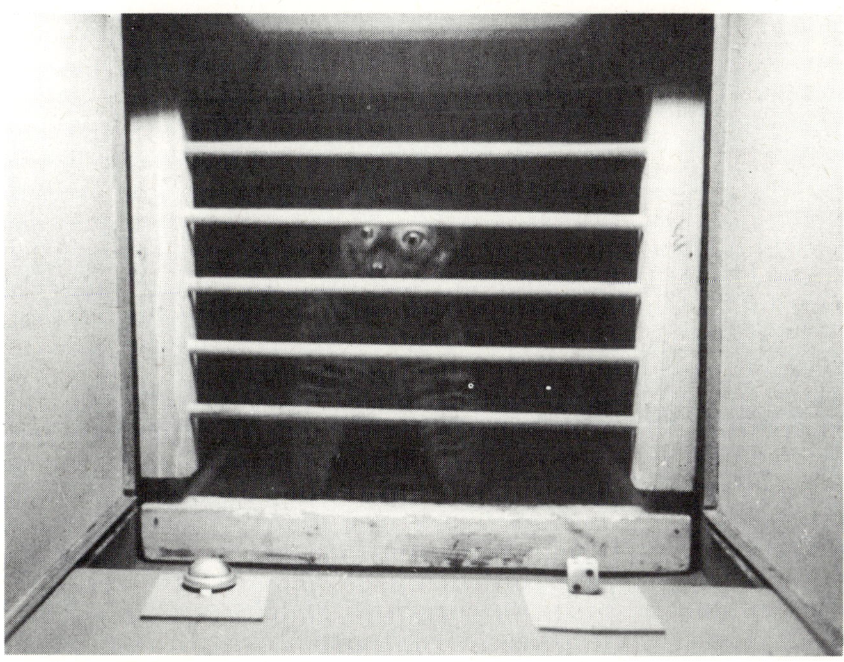

Figure 1. A female *Lemur macaco albifrons* attentively watching the stimulus objects before the test tray is pushed forward.

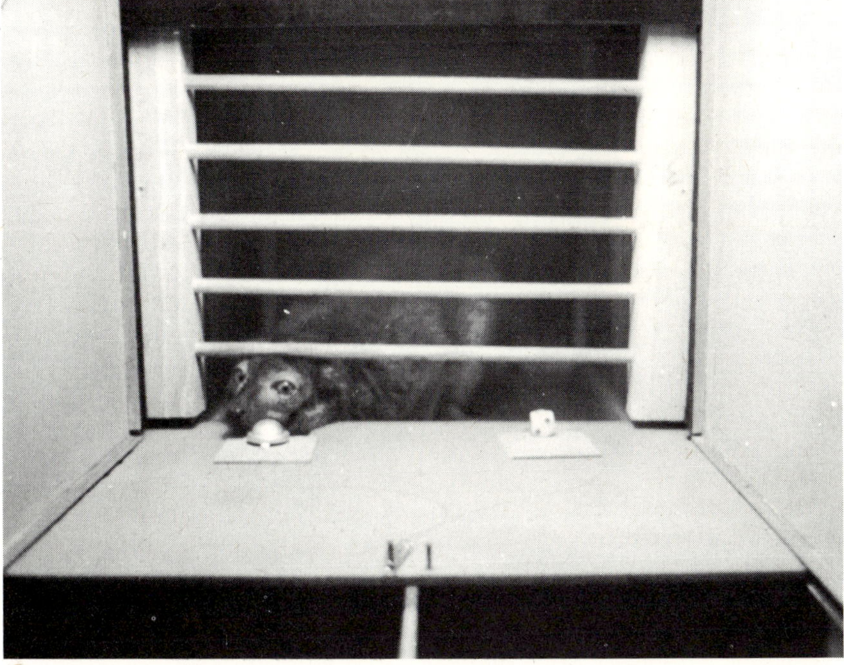

Figure 2. *Lemur macaco albifrons* displacing the stimulus object with the snout.

Procedure

Subjects were deprived of food for 16-20 hours before testing. Immediately following the test session, animals were allowed to eat their usual diet, consisting of bananas, apples, oranges etc., for 2-3 hours. Water was available *ad libitum.*

The reward consisted of dried raisins, a favourite food of the subjects. Adaptation to the apparatus was accomplished by first allowing the animals to find the food in the food wells, and then gradually covering the wells with neutral objects until the animals displaced the object within 20 seconds, 30 times in succession.

The form of response varied with each subject, some animals preferring to displace the object mainly with the hand, others mainly with the snout. Retrieval of the raisin was almost always accomplished with the mouth, only occasionally being transferred to the mouth by hand.

The subjects gave up to 100 responses a day, sometimes divided into two test sessions per day.

Testing was divided into three phases. The first 10 problems were presented for 100 trials each, for a total of 1000 trials. Between the 5th and 6th problems, each subject was given up to 200 trials in which the raisin was placed under one of two identical objects, to determine if the animal was utilising olfactory cues. Response was at chance level throughout.

The second phase consisted of 64 problems, each learned to a criterion of 17 correct responses in 25 or less consecutive trials. The third phase consisted of 192 additional problems presented for a fixed number of 6 trials each. Throughout testing, left-right position of the rewarded object was balanced both for trials within each problem, and for equivalent trials between problems.

The actual procedure in presenting the stimulus objects involved lowering of the opaque screen so that the experimenter was hidden from the animal's view. The tray was baited with a raisin associated with the positive object, and the stimulus objects were placed over the food wells. The screen containing the one-way mirror was then lowered, the opaque screen raised, and after a delay of two seconds the tray was advanced towards the animal to allow a response. A non-correction procedure was used throughout, only one response per trial being permitted. Time between trials was 15 seconds, and between problems 40 seconds. In the inter-problem interval, a 'free' raisin was given to the subject.

Results

Intraproblem learning curves are illustrated in Fig. 3. It indicates a gradual progressive improvement between successive blocks of problems. In the first problem block which consists of 10 problems run for 100 trials each, performance remains about chance level. This

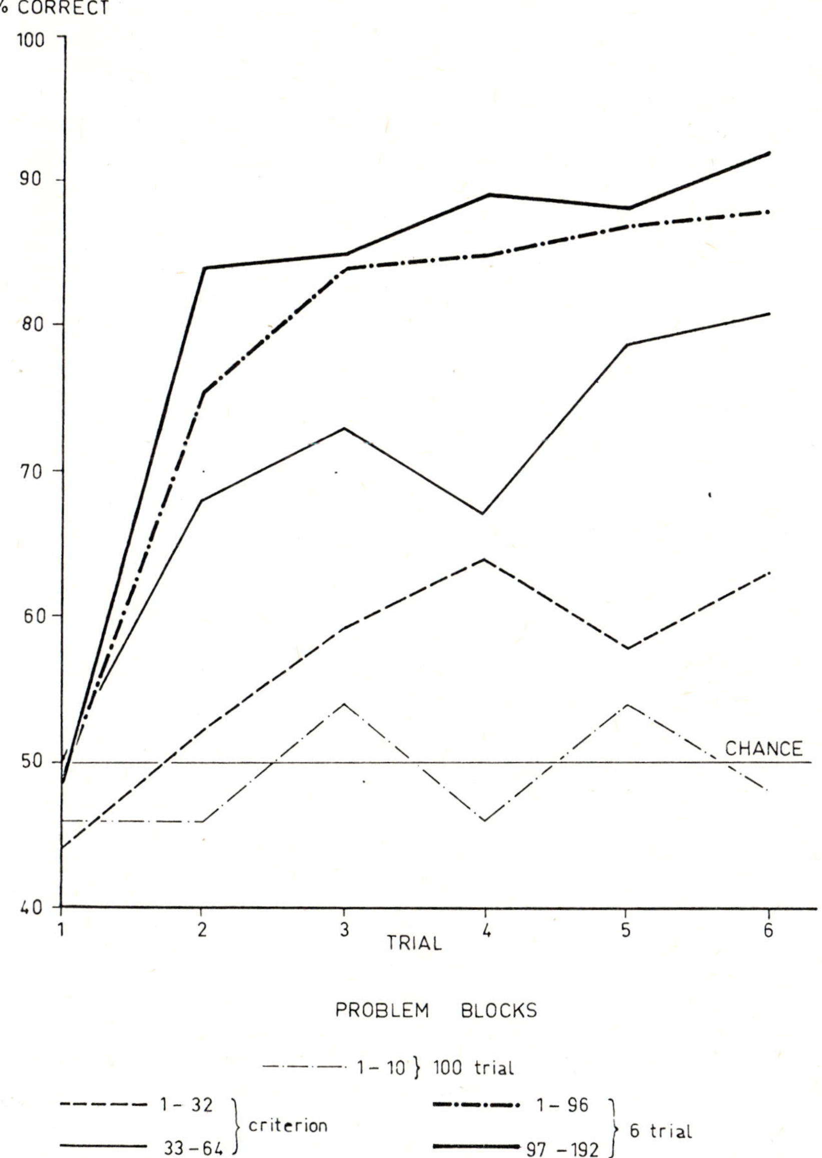

Figure 3. Intraproblem learning curves for *Lemur macaco*.

block of problems was designed thoroughly to adapt the animals to the experimental procedure, and to reduce frustration by ensuring a minimum of errors once a given problem was solved.

The second series of problems consists of 64 problems run to a criterion of 17 correct responses in 25 or less consecutive trials. This rather light criterion assured a minimum of learning on each problem (68%), while allowing problems which were solved rapidly to be completed without unnecessary overtraining. Problems were limited to a maximum of 50 trials. Dividing into two blocks of 32 problems each, improvement occurs over the initial problem blocks, and within the two blocks of criterion problems. Trial 2 performance for the first 32 problems is still at chance level, 52% correct, and for the second 32 problems has progressed to 68% correct. The average number of trials each animal required to solve the criterion problems was 23.5, 27.0 and 23.2 for the first 32 problems, and 19.3, 20.8, 19.6 for the second 32 problems, respectively.

The third phase consisted of 192 problems presented for 6 trials each. This gradual reduction in problem length, between each phase, from initially long problems to problems of shorter length, is considered to be the most effective for learning,[6] although learning has also been shown to be independent of problem length.[15,16] Improvement continues, especially at the trial 2 level, where 76% correct responses were made on the first block of 96, 6 trial problems, and 84% correct responses were made on the second block of 96 problems, of this phase.

Discussion

These curves illustrate that *Lemur* has a fairly efficient ability to form a learning set. Initial problems, where performance remains at chance level for the first six trials, are in contrast to later problem blocks, where 84% correct responses are being made on the second trial. On the early problem blocks, the learning curves are continuous S-shaped curves, and can be described by a single function. On later problems, however, learning curves are discontinuous, and can best be described as two functions; a rapid gain in performance between trials 1 and 2, followed by the gradual improvement between trials 2 to 6. The ability to solve problems in one trial, as measured by trial 2 performance, is typical of primates, but not non-primates,[4] and apparently also exists in *Lemur.*

Although the intraproblem curves are discontinuous, the inter-problem learning curves (Fig. 4) are continuous. Performance on any individual trial improved in a gradual and continuous manner, through each series of problem blocks. Learning set develops in an orderly fashion with experience, even though this learning is manifested as a discontinuous curve within individual problems.

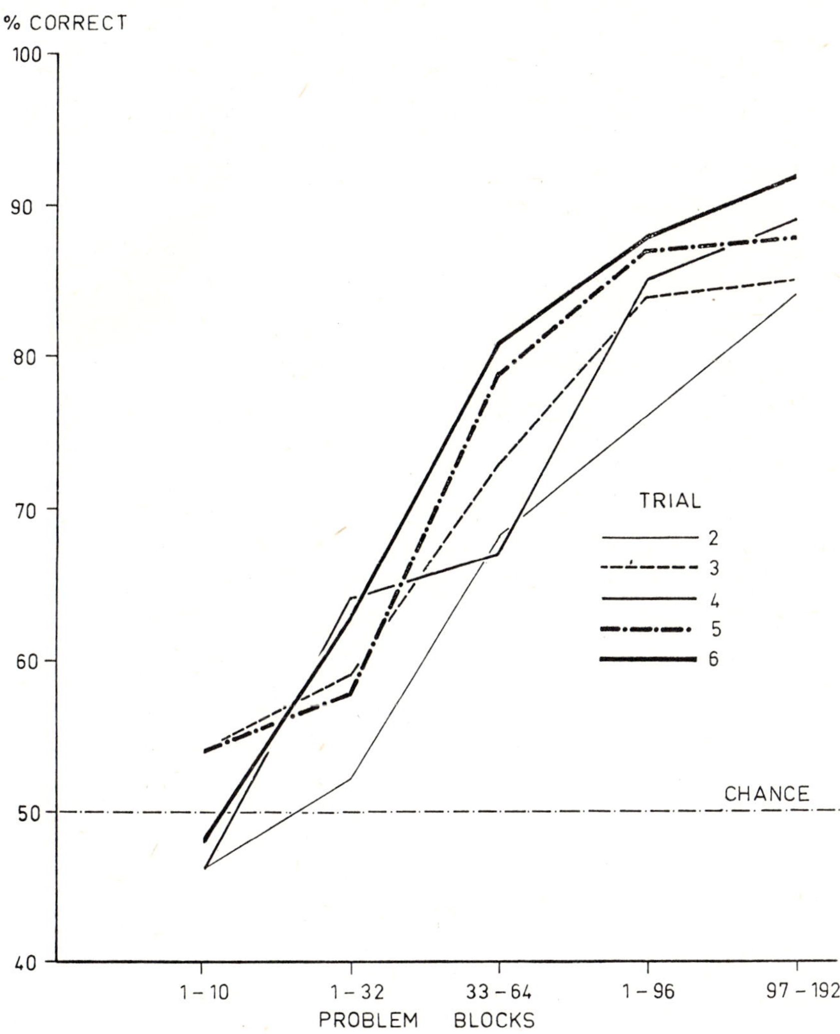

Figure 4. Interproblem learning curves for *Lemur macaco.*

Although one of the objectives of comparative psychology is assessing differences in learning ability between different species, this study of learning in *Lemur* is not directly comparable to similar studies with other primates in other laboratories, and therefore under different conditions. For the moment it is important simply to point out that *Lemur* shows the ability to form a learning set that follows the typical primate pattern.

NOTES

1 Jolly, A. (1964), Prosimians' manipulation of simple object problems', *Anim. Behav.*, 12, 560-570.

2 Ehrlich, A. (1970), 'Response to novel objects in three lower primates: greater galago, slow loris and owl monkey,' *Behaviour*, 37, 55-63.

3 Warren, J.M. (1965), 'The comparative psychology of learning', *Ann. Rev. Psychol.*, 16, 95-118.

4 Miles, R.C. (1965), 'Discrimination learning sets', pp. 51-95. In: A.M. Schrier, H.F. Harlow and F. Stollnitz (eds.), *Behaviour of non-human primates.* Vol. 1. New York: Academic Press, 51-95.

5 Harlow, H.F. (1949), 'The formation of learning sets', *Psychol. Rev.*, 56, 51-65.

6 Harlow, H.F. (1959), 'Learning set and error factor theory', pp. 492-537. In: S. Hoch (ed.), *Psychology: a study of a science.* Vol. 2. *General systematic formulations, learning, and special processes.* New York.

7 Warren, J.M. and Baron, A. (1956), 'The formation of learning sets by cats', *J. comp. physiol. Psychol.*, 49, 227-229.

8 Koronakos, C. and Arnold, W.J. (1957), 'The formation of learning sets in rats', *J. comp. physiol. Psychol.*, 50, 11-16.

9 Rolling, A.R. (1963), 'Successive object discrimination reversals by squirrels', Paper read at Eastern Psychol. Ass., New York.

10 Harlow, H.F., Harlow, M.K., Rueping, R.R. and Mason, W.A. (1960), 'Performance of infant rhesus monkeys on discrimination learning, delayed response, and discrimination learning sets', *J. comp. physiol. Psychol.*, 53, 113-118.

11 Jolly, A. (1964), 'Choice of cue in prosimian learning', *Anim. Behav.*, 12, 571-577.

12 Glickman, S.E., Clayton, K., Schiff, B., Guritz, D. and Messe, L. (1965), 'Discrimination learning in some primitive mammals', *J. gen. Psychol.*, 106, 325-335.

13 Rumbaugh, D.M. and Arnold, R.C. (1971), 'Learning: a comparative study of *Lemur* and *Cercopithecus*', *Folia primat.*, 14, 154-160.

14 Arnold, R.C. and Rumbaugh, D.M. (1971), 'Extinction: a comparative study of *Lemur* and *Cercopithecus*', *Folia primat.*, 14, 161-170.

15 Sterritt, G.M., Goodenough, E. and Harlow, H.F. (1963), 'Learning set development: trial to criterion vs. six trials per problem', *Psychol. Rep.*, 13, 267-71.

16 Rumbaugh, D.M. and Prim, M.M. (1963), 'A comparison of learning set training, methods and discrimination reversal training methods with the squirrel monkey', *Amer. Psychol.*, 18, 408.

PART I SECTION C
Olfactory Communication in Prosimians

R. J. ANDREW and R. B. KLOPMAN

Urine-washing: comparative notes

Urine-washing is unusual amongst mammalian fixed action patterns in that its complexity and peculiarity allow it to be recognised with near certainty as homologous in a restricted number of primate genera. The odd distribution of the behaviour and its possible systematic implications were first pointed out by Hill.[1] It is the purpose of this communication to consider to what extent urine-washing has the same form and causation in the various genera which show it, and to examine some broader questions. The first author has seen the behaviour in all of the species mentioned, which may make cross-species comparisons more reliable; in the case of *Galago demidovii, Loris* and *Perodicticus* fuller observations were made by the second author.

Data relate to captive animals in all cases, and are further restricted in *Loris, Aotus, Saimiri* and *Cebus* by the small numbers available for study (see Appendix). This is most serious in *Cebus*, where only one male was watched intensively. However, published sources independently confirm the existence of urine-washing in all these genera.

The movements of urine-washing may be most conveniently described in *Galago senegalensis* and *G. crassicaudatus*, where hand-reared individuals were watched extensively and at very close quarters, and the movements are relatively leisurely. The foot and hand of the same side are invariably involved, except in very incomplete intention movements, when one hand alone may be raised a short way from the substrate, sometimes repeatedly. The original description by Hill of both hands being rubbed together is almost certainly an error.[1] Hand and foot are raised almost or quite simultaneously. The foot passes under the genital region, with the sole turned inward and the big toe upwards; usually it appears not quite to touch the genital region. The hand is approximated to the foot. It may not quite reach the foot, but more usually the foot is grasped one or more times, with the palmar surface of the hand pressed against it. Urination begins as the hand and foot arrive under the belly, and probably ceases from the time when they are lowered to the time when the hand and foot of the other side have been raised in their turn.

The main variations in the form of the movements are in their degree of completeness; the addition of foot-rubbing (below) may perhaps be considered as the highest intensity form of the pattern. Raising the hand alone has already been noted, and urine-washing may also stop after the hand and foot of only one side have been raised. However, variation is also caused by interaction with other patterns. Thus urine-washing is often preceded by side-to-side swaying of the forebody as a part of parallax movement. As urine-washing develops, the two hands are alternately and repeatedly raised from the ground as part of the swaying, without any movement of the feet.

Foot-rubbing occasionally follows urine-washing. However, it also may follow rhythmic micturition (below) or ordinary micturition without urine-washing, so that it is probably best considered to be a separate, although closely related pattern. In its fullest form, several rubbing strokes of one hind foot over the substrate in the sagittal plane of the body are followed by a similar series of strokes from the other foot.

Urine-washing is basically similar in all the Lorisoidea studied, except that foot-rubbing has so far been seen only in *G. crassicaudatus. G. demidovii* urine-washes very frequently, and performs it rapidly, usually with at least one alteration between the sides. Finger flexing movements are unusually vigorous, as the hand grasps at the foot. Urine-washing in *Loris tardigradus* has been described by Ilse,[2] and no deviation from the pattern as already described was observed in the present study, except that incomplete movements might involve the foot alone. *Perodicticus potto* was not seen to urine-wash. This negative finding is confirmed by more extensive observations on another colony (G. Manley, unpublished data).

In the Pithecoidea, *Aotus trivirgatus* urine-washes in a manner almost identical to *Galago*, including the characteristic movement of rubbing and pressing the hand against the raised foot of the same side. *Saimiri sciureus* also rubs the hand against the ipsilateral foot in urine-washing, with the usual alternation between the limbs of each side.

The form of the pattern in *Cebus* is of particular interest. Even *Saimiri* has a limited repertoire of hand movements: the same two-handed grasp, in which the hands tend to rotate outwards, opposing each other, is used to hold food steady, to break it and to turn it in the hand. *Cebus* is thus the only genus, of those in which urine-washing has been examined, capable of varied and highly dexterous hand and arm movements. The male observed intensively was tentatively assigned to *C. albifrons*, but the behaviour has also been seen in *C. apella.* Urine-washing involved the usual basic movements with hand and foot raising, first on one side and then the other, but was considerably more variable than in any other species studied. Either the hand or the foot might be the first to be raised. At low intensities, it was common for the foot alone to be raised and placed with the sole under the penis and the thumb directed up the flank. Occasionally the same movement might be followed by scratching rather than urine-washing. More rarely, the hand alone was thrust under the genital region, the animal at the same time lying down. The foot might be rubbed repeatedly over the penis area, or be held relatively or quite still whilst the hand grasped at it with repeated flexings of the fingers. Both foot and hand sometimes appeared to move together. A very common variant was to bring the forearm, elbow or even the upper arm below the foot, so that one of these regions, rather than the hand, was wetted. The movement was a clasping one and was sometimes accomplished lying down; in general legs and arms were flexed at the beginning of urine-washing. Variation in foot movements also resulted in the spreading of urine widely over the body. Lateral rubbing movements of the foot might be replaced by forward strokes of the palm up over the flank, sometimes as high as the shoulder. Bouts of each type of stroke might alternate, and the hand was sometimes rubbed on the foot between bouts. Arm position was also adjusted to spread the urine widely: thus the arm might be progressively drawn out so that first the elbow, then the forearm, and finally the hand was wetted. Nolte has suggested that the behaviour of rubbing the body with pungent or acid substances shown by a *Cebus apella* might be related to urine-washing.[3] The present animal showed great interest in the areas damp with urine, often sniffing or licking at them. Functionally, the changes in form result in the presentation of scent on the body surface of the animal rather than a particular part of the substrate, which is likely to be more appropriate in a species where social interactions are of great importance. However, the most interesting aspect is the way in which an originally stereotyped movement has acquired variability during evolution, apparently as part of a general increase in the range of possible limb movements.

A number of other components were also commonly added to the basic pattern. After the initial crouch (if it occurred), and the commencement of urination, the tail was often coiled more tightly (sometimes twice), and shaken laterally; sometimes small amplitude lateral shaking of the body or even violent shuddering followed. Similar shuddering in association with urination in both horse and man has been observed by the first author, but its causation remains unknown.

The second topic on which comparative data may be of interest is that of *causation.* The main types of situation in which urine-washing occurs will be examined, but no attempt will be made to discuss motivational models. In the Lorisoidea, the situations may be classified as follows:

1. *Mobbing*

In *G. crassicaudatus*, *G. senegalensis* and *G. demidovii* urine-washing is common whilst gazing at a strange distant object; (its evocation by the introduction of a strange object into the home cage may be regarded as a variant of this situation). The gaze is usually intent and quite uninterrupted by the urine-washing. It is often accompanied by lateral swinging of head and forebody such as would yield parallax information. The animal may be silent or (particularly in the case of *G. senegalensis*) may give mobbing calls.

2. *Exploration of a strange area*

Rhythmic micturition (below) is more frequent than in 1, probably because it can be combined with forward locomotion in a way in which urine-washing cannot. Urine-washing is also common in all lorisoids observed in this situation.

3. *Defence and attack*

Urine-washing when approached by a superior of which the animal is nervous has been seen in *G. crassicaudatus* and *Loris*. Urine-washing may also immediately precede a leap to attack another male in *G. crassicaudatus* (although foot-rubbing is more usual in this situation; L. Rosenson, personal communication), or it may follow defence or a fight.

4. *Social interactions*

A hand-reared male *G. senegalensis* commonly urine-washed in the pauses between unusually vigorous grooming of the first author's hand. The association was very clear: two or three seconds of scraping with the lower incisors on first arrival might be immediately followed by urine-washing, after which scraping immediately resumed. No sexual responses were directed to the first author by the male at any time, but he sought body contact freely. No comparable association was seen in other species. This association has now been much more extensively documented by Doyle,[4] who has shown that male *G. senegalensis* commonly urine-wash after genital-smelling or grooming a conspecific, in particular a female in oestrus.

5. *Evocation by stimuli associated with urine-washing*

The female *Loris* often urine-washed immediately after the male had done so. The scent of urine may be involved, as it may also be in the concentration of general urination at a particular point in the cage (a folded towel). Ilse observed that captive *Loris* usually urine-washed on the same horizontal pole,[2] so that performance of the behaviour at particular scent posts is perhaps especially characteristic of this species.

6. *As part of normal urination*

It is possible in *G. demidovii*, where urine-washing was very frequent indeed, and in *G. senegalensis*, where urine-washing was sometimes seen in a very tame individual in the absence of any obvious cause except a full bladder, that the need to urinate may sometimes induce urine-washing.

In *Cebus* urine-washing was an almost invariable response to the appearance of the first author after some hours absence. It was never closely associated with threat (frown and *orbicularis oris* contraction) or grins, but no general conclusions as to causation are possible. In *Saimiri* the behaviour was typical when exploring a strange area, or when abandoned and giving contact calls. Latta et al. found urine-washing to be associated with external disturbances (cf. mobbing, above) and excitement in a colony of *Saimiri.*[5] It rose in frequency in both male and female during periods of copulation, but there was no direct association with copulation, and Latta et al. attribute the increase to general excitement.

Two other behaviour patterns, which have sometimes been classed as 'scent-marking' along with urine-washing, are widespread and apparently ancient in the mammals. The situations in which they occur provide useful material for comparison with urine-washing. The first pattern involves movements of rubbing the sides of the face and flanks on the ground or on fellows. It seems to have been primitively associated with copulation. Roberts has described the appearance of such movements in a generalised marsupial (*Didelphis virginiana*) as an invariable immediate precursor of copulation attempts induced by stimulation of the anterior hypothalamus.[6] The movements are specifically directed to furry surfaces, so a particular tactile sensation may be sought. Very similar behaviour occurs in equally generalised Eutherian mammals, *Solenodon* and *Tupaia.*[7] In *Tupaia glis* a male will often rub flanks and/or belly on the ground as it approaches an oestrous female to copulate, or, even more commonly, after smelling or licking her anogenital region. At full intensity, these rubbing movements pass into rolling on the ground. When similar movements are directed at the female rather than at the ground during copulation periods, they begin with wiping first the chin and throat, and then the side of the face on her, and end (much as in the opossum) with flank-rubbing. The indication that the behaviour is in part motivated by scent signals is confirmed by the fact that a hand-reared female would perform such rubbing after sniffing at a pool of urine (not necessarily from a *Tupaia*). Andrew[8] suggested that rubbing chin and throat should be considered quite a separate movement, used chiefly in marking scent posts; it is also as common as the first phase of flank rubbing.[8] It should be noted that flank rubbing may be inconspicuous, or even absent, in some other species of *Tupaia.* Sorenson[9] describes what appears to be flank-rubbing only in *T. longipes.*

Flank-rubbing motivated by scents is familiar in the rolling of the domestic cat at the scent of catnip and of the dog at scents associated with carrion and dung. In the female cat, the association of rolling and flank-rubbing with copulation remains clear. An increased tendency to seek rubbing contact over the body surface also occurs in greeting behaviour in these two species, and very generally through the mammals, both in social greeting and in association with copulation.

A second pattern of importance is the movement of drawing the anogenital region forward over the substrate. When rubbing on a vertical object, the movement may be first up and then down. *Tupaia glis* shows such 'anal rubbing' very clearly. The movement may follow defaecation, and may well have evolved from a purely defaecatory movement: a number of insectivores deposit faeces by such movements on vertical objects.[8] However, in *T. glis* it is likely that glandular secretion is being deposited. Anal rubbing may be directed to branches which appear to serve as scent posts, or (at least in a hand-reared female) to social companions. The association with copulation appears to be absent, although anal rubbing in the hand-reared animal was associated with grooming the hand or foot of the human rearer. A new *G. senegalensis* was very persistently and vigorously rubbed at first introduction. Rubbing could also be reliably elicited, along with the vertical leaps and body shakes given by many mammals in great excitement (e.g. rabbit, horse), by tickling her belly; novel stimuli may act in part, then, through their arousing properties. Neither flank nor anal rubbing occurs in the conspicuous mobbing displays of *Tupaia.*

Within the prosimians, *Lemur* shows characteristic anal rubbing remarkably like that of *Tupaia*, both in form and in the way in which it is directed to strange conspecifics (*L. fulvus*).[8] Causally it overlaps with urine-washing in that it is also given at the approach of a frightening stimulus object.[8] *L. fulvus* also shows what is probably the remains of flank-rubbing, in that males may rub the sides of the face (and top of the head) in female urine. Other than the above example, this pattern does not appear amongst the prosimians.

When compared with the above two patterns, urine-washing differs most obviously in its evocation (at least in most species in which it appears) by distant disturbing visual stimuli. It also differs in that it is not usually given in association with bodily contact with a conspecific. *G. senegalensis* is an exception here: not only is urine-washing common following genital smelling or grooming but it may be associated with scent marking by chest rubbing-against the female.[4]

Piloerection along back and tail in cat and dog (cf. also Von Holst, for *Tupaia*)[10] provides interesting resemblances in causation to urine-washing. It is also evoked by distant strange objects, and in particular by 'strange' animals. It also occurs in the preliminaries of defensive threat, but disappears when threat proper develops (Prescott, personal communication). In dogs, back hair erection is particularly prominent when a distant source of movement or sound is difficult to locate. Prescott has pointed out that such piloerection along the tail of the cat probably serves to release scent, as well as providing a visual cue.[11]

The origin and evolution of urine-washing remains mysterious. Simple urination (and defaecation) in a strange environment or in response to a strange object is probably widespread in mammals. In the Lorisoidea a slight elaboration of this has led to 'rhythmic micturition',[2] in which the anogenital region is repeatedly touched to the substrate, often as the animal moves slowly forward in exploration of a strange area. This has been seen in the present study in *Galago crassicaudatus, G. senegalensis, Loris* and *Perodicticus*, where it is slightly more elaborate (back legs are spread, for the pelvic lowering, and the short tail is flicked during micturition). However, comparable movements of deposition appear to occur outside Lorisoidea in species which are not known to urine-wash. A tarsier, for example, was several times seen to touch its genital area to the vertical post to which it was clinging and deposit drops of urine, whilst staring at a distant disturbing object. Sprankel[12] has suggested that urine-washing may derive from stepping movements following urination such as he saw in *Tupaia.* The first author has not seen behaviour definitely of this type in *T. glis*, and it appears that neither has Sorenson,[9] in all species of *Tupaia* watched by him. (A few drops of urine were often produced during anal marking but no obvious stepping occurred). One difficulty for such a derivation is that the simultaneous raising of foot and hand on the same side of the body is unusual in locomotion (although it does occur in the 'amble'). A movement such as the repeated rubbing of first one hind foot and then the other on the substrate, which has been described in *G. crassicaudatus*, would provide an alternative origin for the foot movement. Hand raising may be compared with the raising of one forepaw, held flexed backwards at the wrist, which can be seen in animals as widely separated as dog ('pointing') and *Tupaia* (where it is sometimes associated with urination), when trying to localise a particular somewhat distant stimulus. Finally, the side-to-side swaying of the body in parallax may have helped to constrain foot and hand movements to one side of the body at a time.

The final question to be considered is the systematic distribution of species which urine-wash. In discussion at the conference it proved possible to extend the number of species for which there is good evidence of the presence or absence of urine-washing, and this information has been summarised in what follows. The first point of importance is that in the Lorisoidea it is likely that urine-washing has been lost in certain lorisine species such as *Perodicticus*, and probably *Nycticebus* and *Arctocebus* (Manley, discussion). The pattern is present not only in the three species of Galaginae already cited (*G. crassicaudatus, G. demidovii* and *G. senegalensis*), but also in *G. alleni.*[13] Its presence in *Loris* shows that it is characteristic of Lorisinae as well as Galaginae. The possibility that urine-washing may have originally been present in the ancestors of *Perodicticus*, which is suggested by the wide distribution which has just been described, is strengthened by the description of genital scratching[13] in male pottos. This movement is very like the first phase of urine-washing and could easily have been derived from it. Urine-washing by *G. senegalensis*[4] after genital-smelling or grooming other animals, particularly oestrous females, provides an example of the elicitation of urine-washing in the same situation as that in which genital scratching occurs. Charles-Dominique (discussion) suggests that lorisoids which move chiefly on broad branches (e.g. *Perodicticus*) may not need to use palm and sole to spread urine on the substrate in the way which is necessary in animals which move amongst slender twigs, and adds *Euoticus* as a possible example of another lorisid in which urine-washing is absent for this reason.

Clearly, if urine-washing has been lost in some members of the Lorisoidea, similar loss might explain the patchy distribution in other groups. However, the possibility that the presence or absence of the pattern may have systematic implications deserves examination. In the Lemuroidea, urine-washing is probably present in *Microcebus murinus.* The pattern is certainly much rarer in this species than in comparable Lorisoids: the movement was never seen in the present colony, nor in a hand-reared male who was observed intensively by the second author. However, Charles-Dominique and Martin record urine-washing for *Microcebus*,[14] and Manley has shown that the characteristic movements of rubbing palm and sole together, first on one side of the body and then on the other, are involved.[15] It is possible that some components of the pattern are present in *Cheirogaleus*, in which the belly may be rubbed in urine (Perret, discussion). It is not clear why the pattern should be absent in other Lemuroidea, if it occurs in the Cheirogaleinae: *Cebus* (Pithecoidea) shows that it can be retained even by species with a monkey-like *facies.* Such evidence should be borne in mind in any consideration of the possibility of a polyphyletic origin of the Lemuroidea. Polyphylety is suggested much more strongly in the case of the

Ceboidea. Urine-washing is present in those of the Cebidae whose behaviour is well known (*Saimiri, Cebus* and *Aotes*) and could therefore be general in the family. Its apparent absence in *Lagothrix* and *Ateles* might be associated with their advanced type of locomotion, but this would not apply to marmosets (Hapalidae), in which it has not been seen by the first author, nor are there any published reports of it in the literature. Early independence of Hapalidae and Cebidae would be quite consistent with their marked differences in structure.

Appendix : species studied

Aotus trivirgatus: two wild-caught, female; *Cebus albifrons:* one male and one female, wild-caught; *Galago crassicaudatus:* 30-40, including both wild-caught and laboratory bred – three hand-reared animals in particular; *G. demidovii:* 8, wild-caught, male and female; *G. senegalensis:* 12, wild-caught and several laboratory bred – one hand-reared male; *Loris tardigradus*, 4, wild-caught, male and female; *Microcebus murinus:* 6, wild-caught, male and female; *Perodicticus potto:* 5, wild-caught, male and female; *Saimiri sciureus:* 2, wild-caught, female; *Tarsius:* 1, wild-caught, male; *Tupaia glis:* 8, wild-caught, male and female – one hand-reared female.

Acknowledgments

The authors wish to express their thanks to Professor D. Poulson and to the Department of Biology at Yale University for provision of the animal quarters which made these observations possible.

NOTES

1 Hill, W.C.O. (1938), 'A curious habit common to lorisoids and platyrrhine monkeys', *Ceylon J. Sci. B* 21, 66.

2 Ilse, D.R. (1955), 'Olfactory marking of territory in two young male loris, *Loris tardigradus lydekkerianus*, kept in captivity at Poona', *Brit. J. Anim. Behav.* 3, 118-20.

3 Nolte, A. (1958), 'Beobachtungen über das Instinktverhalten von Kapuzineraffen (*Cebus apella L.*) in der Gefangenschaft', *Behaviour* 12, 182-207.

4 Doyle, G.A., this volume.

5 Latta, J., Hopf, S. and Ploog, D. (1967), 'Mating behaviour and sexual play in the squirrel monkey', *Primates* 8, 229-46.

6 Roberts, W.W., Steinberg, M.L. and Means, L.W. (1967), 'Hypothalamic mechanisms for sexual, aggressive and other motivational behaviors in the opossum, *Didelphis virginiana*', *J. comp. Physiol. Psychol.* 64, 1-15.

7 Herter, K. (1957), 'Das Verhalten der Insektivoren', *Handb. Zool.* 8, 10.

8 Andrew, R.J. (1964), 'The displays of the primates', in J. Buettner-Janusch (ed.), *Evolutionary and Genetic Biology of Primates*, vol. 2, New York.

9 Sorenson, M.W. (1970), 'Behavior of tree shrews', in L.A. Rosenblum (ed.), *Primate Behavior*, vol. 1, New York.

10 Von Holst, D., this volume.

11 Prescott, R.G.W., personal communication.

12 Sprankel, H. (1961), 'Uber Verhaltensweise und Zucht von *Tupaia glis* (Diard 1820) in Gefangenschaft', *Z. wiss. Zool.* 165, 186-220.

13 Manley, G.H., this volume.

14 Charles-Dominique, P. and Martin, R.D. (1970), 'Evolution of lorises and lemurs', *Nature* 227, 257-60.

15 Manley, G.H. (1966), 'Prosimians as laboratory animals', *Symp. Zool. Soc. Lond.* 17, 11-39.

G.H. MANLEY

Functions of the external genital glands of Perodicticus *and* Arctocebus

Introduction

The existence of highly specialised concentrations of glands on discrete areas of the body-surface of lorisid prosimians has been recognised for some time. In the lorises of South-East Asia and India, the slow loris *Nycticebus coucang* and slender loris *Loris tardigradus*, the principal glandular fields in both sexes are located on the medial side of the upper arm (the flexor surface) and have been termed the brachial organs. In the two African members of the group, the potto *Perodicticus potto* and angwantibo *Arctocebus calabarensis*, they are found in the genital region, forming part of the scrotal surface of the male and lying closely adjacent to the vulva of the female.

In the case of the potto and angwantibo, the subjects of the present communication, some information is available concerning the morphology, histology and cytochemistry of the genital glands, particularly for the first of these two species, the potto. On the matter of function, however, our knowledge is far from adequate. We know little or nothing of the roles these specialised structures play in the lives of the animals bearing them. Speculation has been couched in general terms or, in a few instances, specific functions have been suggested; but, as far as the author can ascertain, no one has yet described patterns of behaviour directly and consistently associated with these organs. The topic was very briefly touched upon in an early work-in-progress report.[1]

The material presented here comes from a comparative study of the behaviour of the Lorisidae, carried out over a number of years on animals housed in relatively large cages in a nocturnal room, i.e. under a reversed-lighting regime. Details of housing conditions, cages, diet, etc. will be found in earlier papers.[2,3] The number of individuals of each species available for observation and manipulation had to be small, and this drawback, and certain others attendant upon captive conditions, is clearly limiting. It follows that conclusions drawn in this paper may be neither definitive nor exhaustive and should be viewed in that light.

Perodicticus potto

Observations on pottos covered a period of 19 months between April 1962 and November 1963. All the animals were of the western subspecies *potto* (see Hill).[4] The principal study animals comprised an adult male, and an adult female and her female offspring, which became full-grown in the course of the study. Initially the male, and the female with her infant, were housed separately; after four months the young female was isolated and the two adults housed together; six months later the young female joined them, being introduced to the male for the first time. Additional limited information came from two more adult females acquired late in the study period.

1. The external genital glands

Descriptions and illustrations of the external genitalia of the potto may be found in a number of publications[2,4–10] and histological and cytochemical studies of the glands have been conducted by Montagna and his colleagues.[7–9,11] A summary description may be appropriate here.

The major glandular areas lie on the scrotum of the male and the pseudoscrotum of the female. In the male, the scrotum has an extensive central area of closely applied tessellae, largely naked, the intersecting fissures being somewhat pigmented and, under normal conditions, narrow. Laterally and to some extent posteriorly the scrotal surface loses its sculpturing and becomes haired. In the female of this species a pseudoscrotum is present, i.e. the rima pudendum bisects a substantial mound with the clitoris situated at its anterior end, and the appearance of the glandular organ, although smaller, is extremely similar to that of the male. It has been shown that the tessellated areas in both sexes have rich fields of large apocrine glands surrounded by nerves containing abundant acetylcholinesterase.

In the female alone, an additional pair of discrete flask-shaped glands is found, flanking the vulva and opening into it[8] or close to the base of the clitoris.[10,12] These glands produce a thick, strongly-smelling secretion of semi-fluid keratinous composition. The observations recorded in the present paper almost certainly refer to the superficial glandular organs of both sexes and not to the female structures just mentioned.

Certain short-term changes in the conformation of the genitalia which are relevant here have been observed. These concern the male;

equivalent changes in the female genitalia were not noticed (compare changes accompanying oestrous cycles).[2] Under certain conditions and then only briefly, the male's testes project ventrally, the valley between them deepens and the fissures between the tessellae widen appreciably to as much as 2 mm. or more. Such changes have been noted (i) immediately after rousing from sleep, when the phenomenon is probably an effect of sustained warmth, (ii) during sexual behaviour, and (iii) during genital-scratching grooming (GSG) (see below).

2. *Genital-scratching grooming (GSG)*

Mutual grooming is extremely common in pottos living amicably together in captivity. Typically, the grooming individual grips the fur of the passive groomee and, using the characteristic licking and raking (dental comb use) actions of prosimians, attends to the fur lying between the clenched hands. Whereas most allogrooming is of this relatively uncomplicated type, the potto may incorporate an additional element of a quite unexpected and specialised nature to perform a behaviour pattern called by the author 'genital-scratching grooming' (GSG). In this, as ordinary grooming proceeds, the groomer is seen to move alternate hands caudad towards the genitalia, specifically the scrotum or pseudoscrotum. With the tips of the fingers applied to the glandular surface, the animal then scratches it in a typically human and completely unprosimian manner, i.e. the fingers are clenched and unclenched so that their tips are repeatedly drawn across the scratched surface (Fig. 1). The hand is then returned forward to regrasp the fur of the groomee, the other hand moves back, and so on. Regrasping appears to be simple and without special movements, but, even in the absence of these, one must suppose that the secretions of the scrotum or pseudoscrotum are transferred to the groomee's fur. The grooming animal's attention, in terms of nose and head orientation and gaze, is concentrated on the area being groomed and never diverted towards the genitalia during these scratching operations. Immediately after a GSG bout, however, the groomer often bends the head down to lick the scrotal or pseudoscrotal surface.

It should be emphasised that scratching actions of the kind described here occur in no other situation in the potto, nor have they been observed under any circumstances in the angwantibo, the lorises, or other prosimians such as *Galago senegalensis*. Normal scratching, of course, employs the 'toilet claw' of the foot.

The number of backward hand movements in a GSG bout varies from one to very many. Regular left-right hand alternation occurs in the majority of cases (88%; $n = 48$); otherwise one hand only may be used, or the movements of each hand grouped into blocks. With the fingertips touching the glandular area, 1-5 clenching-unclenching

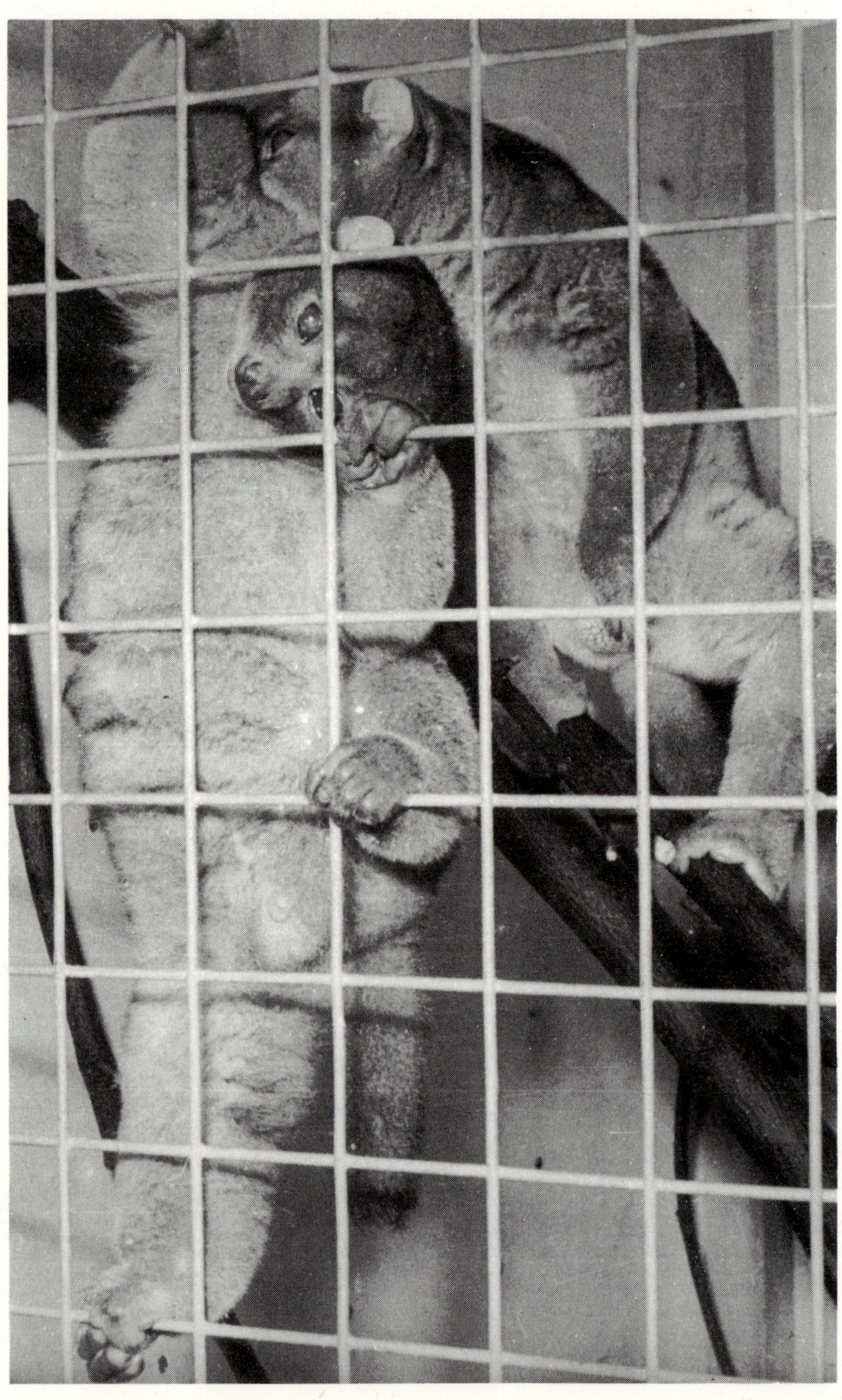

Figure 1. *Perodicticus potto potto.* Adult male (upper right) performing 'genital-scratching grooming'.

scratching actions are performed. In the majority of bouts observed the genitalia are scratched at least once and often many times (80%; $n = 61$), but sometimes no scratching at all is effected. This is because any hand movement may be incomplete, varying from no more than lifting the hand in an intention movement of genital-scratching to making scratching actions in the air immediately above the genitalia; not uncommonly, incompleteness takes the form of scratching a leg which may block the hand's path to the genitalia.

Licking of the scrotum or pseudoscrotum after GSG occurs more often than not (62%; $n = 50$) and, at this point, the groomer has been seen to lick its own *fingers* as well. After a GSG bout the groomee may lick or closely smell the groomer's genitalia. Attention to the genitalia of the groomer is occasionally also observed following apparently ordinary grooming bouts.

Both protagonists in a reciprocal grooming session may be seen performing the genital-scratching pattern, but only the individual actively grooming at the time does so. Like ordinary grooming, GSG takes place with the animals in virtually any position and the groomer may make caudad scratching movements even when fully suspended by feet alone.

With the somewhat limited data available, the parts of the body groomed during GSG are not clearly different from those covered by ordinary allogrooming (Table 1). Although the male seemed to concentrate upon grooming the other's back, little or no such focussing could be discerned in the female.

Table 1 Body areas groomed by two adult pottos during ordinary allogrooming (OG) and genital-scratching grooming (GSG). Percentages shown are of the totals indicated in brackets at the feet of the columns

Region groomed	*OG ♂*	*GSG ♂*	*OG ♀*	*GSG ♀*
Head and face	19.2	10.3	10.4	0.0
Neck	25.6	7.7	21.7	39.4
Back	16.8	58.9	24.5	9.1
Ventrum and flanks	12.0	7.7	9.4	27.3
Armpits	16.8	12.8	19.8	24.2
Limbs	4.0	2.6	12.3	0.0
Genitalia	5.6	0.0	1.9	0.0
	(125)	(39)	(106)	(33)

It will be evident from what has gone before that the GSG pattern is not confined to one sex but is performed by both males and females. However, the impression was gained that the male in the present study showed this behaviour more often than did the female and that this was not related to the frequency distribution between the sexes of allogrooming as such (about equal). In the captive conditions of the present study, at least, GSG was by no means rare and could be observed daily for many days on end.

GSG is not a component of sexual behaviour in this species, nor has any obvious relationship between its occurrence and the regularly recurring oestrous periods of the adult female been detected:[2] no exact counts were made but male and female certainly performed the scratching pattern both during the female's oestrus and throughout the dioestrous interval.

Table 2 summarises the GSG relationships developed among the three principal study animals in the different phases of the study. It reveals that mother and growing infant lacked such behaviour and that, later, the young but full-grown female was not seen to mark her mother; otherwise all other GSG relationships were established.

Table 2 GSG relationships among three pottos in different study phases
1. Adult female with growing infant (adult male isolated)
2. Adult male with adult female (young female isolated)
3. Young female, now full-grown, joins adult pair

		Groomee (recipient)		
		Adult ♂	*Adult ♀*	*Young ♀*
Groomer showing GSG				
Phase 1	Adult ♀	absent		–
	Young ♀		–	
Phase 2	Adult ♂		+	absent
	Adult ♀	+		
Phase 3	Adult ♂		+	+
	Adult ♀	+		+
	Young ♀	+	–	

The initial establishment of GSG relationships is also of significance here and could be traced in the behaviour of the animals following introductions. Although few introductions were possible in the present study, their results are highly suggestive.

In the case of the adult male and adult female, the full, permanent introduction was effected after a series of seven brief 'trial sessions' during the preceding three months. In these, aggressive interactions predominated but declined over time to a low level on the occasion of the full introduction. Then, a reciprocal grooming relationship (but without GSG) was established on the day following introduction, a full play relationship was established on the sixth day, and only after this, on the sixth day, did the male start marking the female with the GSG pattern. The female was first observed marking the male in this way several days later. In short, it would seem that the GSG relationship became established once the initial agonistic barriers had been fully overcome by mutual grooming and play.

When the full-grown young female was first introduced to the adult male and rejoined her mother, the process was accelerated apparently by the lack of aggression and hesistancy on the part of the young female. In this case the latter initiated grooming immediately and play occurred very soon afterwards, so that these relationships between the animals were established within the first half hour of the introduction. The male then GSG-marked the new female 35 minutes after the time of introduction and repeated the pattern at least 12 times in the following two hours. The young female first marked the male on the following day and was herself marked by her mother from the sixth day. The unusually high frequency of GSG marking shown by the male towards the new female should be noted.

3. Angstgeruch

Besides the GSG and ordinary allogrooming contexts, in which there is no evidence of any motivational conflict in the groomer, certain conflict situations in the potto seem to activate the genital glandular fields. In most instances this has been inferred from the attention given to the scrotum or pseudoscrotum by the animal itself (licking) or its cage-mates (licking or closely smelling); in some, secretion from the glands has been clearly visible. Unlike GSG, however, special associated behaviour is lacking.

No single interpretation is sufficient to account for the various occurrences but they tend to fall into a number of groups. The intention here is to concentrate on one group only, after dealing summarily with the others. Thus, on the criteria given above, stimulation of the scrotum or pseudoscrotum was observed with:

1. Physical thwarting of normally available locomotor patterns, e.g. preventing the performance of a cage stereotype in the adult female by simply blocking off a hand-hold; temporary leg malfunction in the same female making locomotion difficult and apparently painful. Neither overt fear nor social factors are evident here.
2. Thwarting of an aroused social tendency by inappropriate behaviour on the part of the other animal, e.g. the adult female behaving sexually at oestrus but eliciting only play from the male. Overt fear is again absent.
3. Conflict between two aroused social tendencies, e.g. attack-escape conflict as in both male and female during their early 'trial' introductions; escape inhibiting sex as in the final introduction, where the female was in oestrus but retained some fear of the male.
4. Fright or thwarting of a strongly aroused escape tendency, e.g. when the principal females were handled for cage transfer or

medical treatment; when the two additional study females, newly arrived and extremely timid, were even approached.

In the first three categories particularly, it is uncertain whether or not the undoubted stimulation of the scrotum or pseudoscrotum has a communication function, informing others of the motivational state of the animal, especially as the evidence of stimulation comes mainly from the animal itself rather than its congeners. The very variability of eliciting situations militates against this interpretation and the author is not convinced that cagemates displayed appropriate responses.

In the final category, however, a glandular secretion was patently visible and/or was clearly shown to be present by the behaviour of other animals. Two examples demonstrate these points. The first concerns the two very wild and timid adult females whose behaviour when approached was characterised by fleeing and protracted freezing; when touched or handled, besides pronounced defensive behaviour, these females would produce a copious secretion from the genital glandular surfaces, at times so plentiful as to flood them and drip to the floor as a clear brownish-yellow fluid. This was not urine.

The second example is provided by two occasions on which the principal adult female was handled in order to swab a head wound with antiseptic: when the animal was returned to the shared cage, the male and young female showed pronounced olfactory interest in her pseudoscrotum and this despite the competition of the strongly-smelling antiseptic, investigated only much later.

The tentative suggestion made here is that the potto may possess an *Angstgeruch* (fear scent) secreted by the external genital glands under these conditions and functioning as a warning signal to others of potential or actual danger. The matter is further considered when discussing a comparable phenomenon in the angwantibo and in the final section of this paper. The term *Angstgeruch* itself is taken from a publication demonstrating the existence of such a substance in the house mouse, *Mus musculus.*[13]

Arctocebus calabarensis

Observations on angwantibos of the subspecies *calabarensis* (see Hill)[4] (Fig. 2) centred on a male and two females and covered a period of 28 months between July 1964 and October 1966. A female infant studied during its first five months provided some further relevant information, as did a third adult female. The adult male was housed initially with one of the principal adult females for seventeen months, then spent five months with the second principal adult female before finally rejoining the first. The male fathered three infants in the course of this whole period.[14]

Figure 2. *Arxtocebus calabarensis calabarensis.* Mother and infant.

1. The external genital glands

The external genitalia and associated glands of this species have been considered in several publications[4–6,11,15] but neither always entirely accurately nor completely, according to the author's own observations. For example, the erected penis is perfectly straight and not curved[5] and a tessellated area adjacent to the vulva of the female appears to have been overlooked. Rather surprisingly, the histological studies say little about the glandular areas and a distinct organ is not mentioned; apocrine glands in the skin of the genitalia are simply described as medium-sized[11] or as larger and more numerous than elsewhere, but not clustered.[15]

The genitalia and their glandular areas in *Arctocebus* are in some important respects quite unlike those of the better-known potto and deserve some attention here. In the male, the major difference concerns the size and position of the naked glandular area of the scrotum: in the angwantibo this is restricted to a small triangle, its apex pointing cephalad, on the posterior slope. Under normal conditions, the surrounding hair of the scrotum so encroaches on the naked triangle that only the closely-packed tessellae are visible, the boundaries of the glandular area being indeterminable.

No mention is found in the literature of a glandular area in the female angwantibo corresponding to that of the male, despite the prominence of this feature in the female potto. However, such an area is in fact present in the angwantibo, though relatively poorly defined. It lies immediately posterior to the posterior labium of the latero-laterally oriented vulva and comprises only a small number of tessellae in very low relief, barely discernible except along the anterior border and midline of the area. The area is most clearly seen when the female is in oestrus, at which time the labia and posterior slope become turgid and the surrounding hair is spread away from the general vulval region.

As in the potto, distinct short-term changes in the conformation of the external genitalia of the male angwantibo have been observed. Under certain circumstances the whole scrotal mass projects appreciably ventrally and caudally as the testes bulge downwards and rearwards. Concomitantly the area of the naked triangle is vastly increased and occupies most of the posterior hemisphere of the scrotal mass; the glandular tessellae become spread out and now lie in the centre of this expanded bare region, surrounded by a border of undifferentiated scrotal skin. This state has been observed (i) immediately after rousing from sleep, when it is probably an effect of sustained warmth, (ii) occasionally when being groomed by the female, (iii) when the male grooms his own genitalia vigorously, (iv) on one occasion soon after the arrival in the nocturnal room of two new adult females, (v) during sexual behaviour on the part of the male, and (vi) during episodes in which passing-over occurs (see below).

2. Passing-over

'Passing-over' is a male behaviour pattern directed towards the female. The male, moving along the upper surface of a horizontal branch on which the female has paused or is moving slowly, literally passes over her: employing three limbs for locomotion and keeping one leg cocked, the male moves forward over the entire length of the female's back in such a manner that his underparts are drawn over the fur of her back. The protuberant scrotal mass brushes or gently rubs against the fur as passing-over occurs. At times the pattern includes a distinct depression of the male's rear body and occasionally he may pause briefly in mid-passage along the female's back.

Passing-over may be performed either singly or several times in quick succession. The movement may be in either direction, starting from the rear end or the head end of the female, though the former is more usual. In some performances of the pattern the whole length of the female's back is not traversed but only some part of it, the male crossing obliquely over her.

It should be emphasised that, simple though the pattern might appear to be, it is a quite distinct and deliberate action and not just the 'accidental' outcome of one animal attempting to pass another on the same branch.

As *Arctocebus* of both sexes possess a urination pattern in which the rear body is depressed to deposit traces of urine on branch surfaces, it sometimes happens that the male, in passing-over, leaves a urine smear on the female's fur. The author is convinced, however, that this is no more than a by-product of passing-over and that the primary purpose of the latter behaviour is the deposition of olfactory substances emanating from the male's specialised scrotal glandular area.

Although not a constant feature of male sexuality, passing-over is most commonly seen when the male is engaged in sexual behaviour; it is thus associated with a number of patterns interpreted as having this motivation, e.g. persistent trailing after the female, repeated sniffing at her genitalia, a special courtship call, and another distinctive display that the author calls the 'chin rest'. When passing-over does occur here it appears during lower intensity sexual behaviour rather than the high intensity activities immediately preceding attempted or actual copulation. In the present study, passing-over tended to be observed when the two principal adult females came into oestrus. However, the adult male also occasionally displayed some sexual interest in the females during their dioestrous intervals and when one of them was pregnant and, at such times, passing-over could again be seen.

That passing-over is not entirely linked with male sexuality is indicated by the male's performance of the pattern in the several hours following his first introduction to the second principal adult female. No sexual behaviour could be recognised on that occasion (a week before the female's oestrus), yet passing-over was recorded twice and the male's genitalia revealed the conspicuous scrotal distension described above.

It would seem quite significant, too, that the male's first three close olfactory investigatory approaches to the female should have involved his sniffing closely at the fur of her *back* in preference to other parts including the genitalia; genital inspection only arose and became frequent later. This suggests the possibility that, on first meeting a strange female, the male may very soon ascertain whether or not she has been marked on the back by the passing-over patterns of another male. Some supporting evidence is provided by the male's behaviour upon eventual reintroduction to his original female: within seconds of being released into the cage he had checked the odour of her back.

When two adult females strange to one another were first introduced, the fur of the back was again the focus of much olfactory investigation, starting from the first two close approaches. The possibility thus arises that females too will check each other for scent-marks on meeting.

In short, the provisional picture that emerges is that males check females, females check each other, but females do not check males, all of which fits in well with the fact that passing-over is a masculine pattern and females have never been seen passing-over the backs of males. Clearly the observations recounted here are very few and, however suggestive, call for verification by planned series of introductions using greater numbers of animals.

3. *Angstgeruch*

A number of episodes witnessed during the course of these studies of captive angwantibos lead the author to suspect the existence of an *Angstgeruch* or fear scent such as has already been suggested might exist in the potto. Moreover, this would appear to be produced by the external genital glands of both sexes (cf. passing-over), so that the glandular area of the female, though ill-defined, would seem to be functional in this context at least.

As in the potto, indications of fear scent production were seen when an animal gave evidence of being more than usually frightened. Sources of such alarm were various but included interactions with conspecifics, unusual events in the nocturnal room and unusual noises in the building. At such times, although no visible secretion could be made out, the behaviour of the cage-mate unequivocally demonstrated the presence of an odoriferous substance on the tessellated glandular structure: the cage-mate would be drawn, often from some distance away, to smell closely that part of the alarmed individual's genitalia.

In both the potto and the angwantibo, the examples given reveal one aspect of these situations that might appear to go against the idea of a fear scent signal, namely, the response of the other animal. In most cases this has been to investigate the scent-producing individual rather than to show flight, freezing, etc. – the responses which might be expected on the alarm signal hypothesis. However, it is not inconceivable that what we are seeing is really a reflection of the captive state: animals long habituated to others in a relatively safe environment might be less responsive to signals of this kind than they would be in the wild.

A clue to what is possibly the true context in which the *Angstgeruch* operates in the two species comes from certain observations of mother-infant interactions in the angwantibo. On a number of occasions the author has been struck by the instant communication of alarm from mother to infant (the infant either on the mother or 'parked'), causing the latter instantly to suspend all activity or immediately attach itself to the mother. Unfortunately, on these occasions it has not been possible completely to exclude other factors such as visual cues, but the already-demonstrated fact that alarmed animals do produce an odoriferous secretion makes it at least possible that the infant was reacting to a specific olfactory warning signal.

Conclusions and discussion

It is not within the scope of this paper to consider mammalian scent-marking in general; marking of one individual by another such as has been reported here is by no means unknown and the reader is referred to a recent review.[16] Nor can the other marking systems (e.g. urinary) present in the Lorisidae be dealt with.

Concerning *Perodicticus* and *Arctocebus*, hypotheses regarding the functions of the external genital glands have usually been based on pure speculation rather than the observation of behaviour, and they range from the general to the specific and from the not improbable to the bizarre.[17]

Thus, in the case of the potto it has been suggested that the glands may be related to social and reproductive activities,[7,9] or may enable the sexes to locate each other at night;[6] a defensive function, the secretion having a repellent effect, has also been advocated.[18] One theory that has been advanced is that the glandular areas function as insect attractants,[19] and another that they serve as 'friction pads' and 'to some extent a clasping organ' in copulation.[5] (In the only reference to the angwantibo, the same author again suggests that the tessellated area of the male's scrotum acts as a 'friction pad' during mating.)[5] Several writers have reported observations leading them to believe that the potto's genital glands are used to mark branches or other objects in the environment.[20–22] Only two of these various hypotheses call for comment at this point.

Concerning a possible anti-predator defensive function, the present author's observations of glandular secretion in severe alarm might, at first sight, appear to be confirmatory; moreover, as has been noted,[18] this would tally with one function of the brachial organ of *Nycticebus*. However, unlike the slow loris, in which in most instances there can be no mistaking the nauseating, repellent nature of the fright secretion's odour, the potto's secretion does not recommend itself for inclusion in that class: in the author's experience the odour either escapes notice or simply lacks the overwhelming quality evident in the loris's discharge.

The author has himself seen both pottos and angwantibos apparently scent-marking branches with their genital glands, but so rarely as to make it unlikely that this could be the glands' function. It is possible to confuse such rubbing with rear body depression during urination and the anus-wiping that sometimes follows defaecation.

In the present paper a distinct behaviour pattern in each of the two species has been described, which results, in the author's view, in the scent-marking of another individual. In *Perodicticus*, the sexes mark one another and females may mark other females with genital-scratching grooming; in *Arctocebus*, only the male marks the female with his passing-over performance. What then is the significance of such olfactory marking in the natural life of these animals?

The evidence of recent field-studies confirms earlier reports that members of both species are solitary, i.e. are usually encountered singly in the wild.[22–25] However, such a designation has implications of a simplicity of social structure which is probably far from real; at the very least one has yet to consider precisely the general relationships between neighbouring individuals of the same or different sex, and the nature of the relationship between male and female at and around the time of mating. Associated with such relationships would be the ranging behaviour of individuals: whether home-ranges are exclusively occupied or overlapping and the significance of overlapping should it occur.

The very existence of scent-marking patterns of the kind described above would seem to indicate a social structure more complex than might at first be supposed and, moreover, that the social structures are different in the two species.

Charles-Dominique indeed has already presented evidence that consort pairing in the potto may be for a matter of months,[23,25] and although females do not form small social groups (cf. *Galago demidovii*),[25,26] that daughters may remain in home-ranges adjacent to those of their mothers. It may be the case in fact that the relationships between such animals are closer and even longer-lasting than these data indicate. It is not difficult to see how mutual scent-marking might be advantageous in such a social structure; not only would individual recognition be facilitated, but the exchanged scent-marks could provide the respective animals with immediate olfactory indicators of their already-established harmonious relationship.

In the absence of more field data concerning the social structure of the angwantibo, it is perhaps unwise to speculate too much on the manner in which it differs from that of the potto. However, evidence to be presented elsewhere does suggest that consort pairing at least is dissimilar and, here, the fact that males alone mark females certainly points to the conclusion that such differences exist.

As passing-over occurs mainly when the female is in oestrus it must be especially important that she bears the male's scent-mark at this time, and a tentative suggestion may be made of one reason why this should be so. Although the females of both species produce oestrous urine and become more restless at oestrus, observations reveal that in *Arctocebus*, additionally, the state is signalled by incredibly piercing, loud whistle calls. The calling behaviour especially might be expected to attract males from surrounding areas to the female, and it is then conceivable that if she carries the newly-applied scent-mark of one particular male it acts in some measure as a stamp or seal of ownership, dissuading others. Once copulation is effected, further mating by other males is mechanically frustrated, by the presence in the vagina of the successful male's copulatory plug.[14]

On the evidence currently available it is with some hesitancy that the *Angstgeruch* signal hypothesis is advanced: it is one thing to show that the genital glandular areas of the two species register alarm, but another to prove a real communication function in nature. However, as indicated earlier, the mother-infant relationship is one context in which such a signal is manifestly appropriate. And it is worth noting that (unlike the rapidly-moving, saltatorial and more social galagos) all the slow-moving lorisid species, in keeping with their stealthy and largely solitary way of life, lack alarm calls in the vocal repertoire.

Summary

The principal function of the external genital glands of *Perodicticus* and *Arctocebus* appears to be the scent-marking of other individuals. In the former species, marking is carried out by both sexes during the performance of 'genital-scratching grooming'; in the latter, only the male marks the female by 'passing-over' her back. The significance of such marking in the natural social life of the two species is discussed.

It is possible, though by no means certain, that an additional function of the glands of both sexes of both species is the secretion of an *Angstgeruch* or fear scent signal substance in appropriate circumstances.

Acknowledgments

This work was carried out while the author was Research Fellow at the Wellcome Institute of Comparative Physiology, Zoological Society of London, Regent's Park, London.

NOTES

1 Manley, G.H. (1967), 'Reproduction in lorisoid primates (Progress report)', *Sci. Report Zool. Soc. Lond.* 1966-67, 21-22.

2 Manley, G.H. (1966), 'Reproduction in lorisoid primates', *Symp. Zool. Soc. Lond.* 15, 493-509.

3 Manley, G.H. (1966), 'Prosimians as laboratory animals', *Symp. Zool. Soc. Lond.* 17, 11-39.

4 Hill, W.C.O. (1953), *Primates:* vol. 1, *Strepsirhini*, Edinburgh.

5 Sanderson, I.T. (1940), 'The mammals of the North Cameroons forest area', *Trans. Zool. Soc. Lond.* 24, 623-747.

6 Hill, W.C.O. (1957), 'Reproductive organs: external genitalia', in Hofer, H., Schultz, A.H. and Starck, D. (eds.), *Primatologia: Handbook of Primatology*, Basel, 630-704.

7 Montagna, W. and Ellis, R.A. (1959), 'The skin of primates: 1. The skin of the potto (*Perodicticus potto*)', *Am. J. Phys. Anthrop.* 17, 137-61.

8 Montagna, W. and Yun, J.S. (1962), 'The skin of primates: 14. Further observations on *Perodicticus potto*', *Am. J. Phys. Anthrop.* 20, 441-9.

9 Ellis, R.A. and Montagna, W. (1963), 'The sweat glands of the Lorisidae', in Buettner-Janusch, J. (ed.), *Evolutionary and Genetic Biology of Primates*, vol. 1, New York, 197-228.

10 Charles-Dominique, P. (1966), 'Glandes préclitoridiennes de *Perodicticus potto*', *Biol. Gabon.* 2, 355-9.

11 Machida, H. and Giacometti, L. (1967), 'The anatomical and histochemical properties of the skin of the external genitalia of the primates', *Folia primat.* 6, 48-69.

12 Manley, G.H., unpublished observations.

13 Müller-Velten, H. (1966), 'Über den Angstgeruch bei der Hausmaus', *Z. vergl. Physiol.* 52, 401-29.

14 Manley, G.H. (1967), 'Gestation periods in the Lorisidae', *Int. Zoo Yb.* 7, 80-1.

15 Montagna, W., Machida, H. and Perkins, E.M. (1966), 'The skin of primates: 33. The skin of the angwantibo (*Arctocebus calabarensis*)', *Am. J. Phys. Anthrop.* 25, 277-90.

16 Ralls, K. (1971), 'Mammalian scent marking', *Science* 171, 443-9.

17 Stümpke, H. (1962), *Bau und Leben der Rhinogradentia*, Stuttgart.

18 Seitz, E. (1969), 'Die Bedeutung geruchlicher Orientierung beim Plumplori *Nycticebus coucang* Boddaert 1785 (Prosimii, Lorisidae)', *Z.f. Tierpsychol.* 26, 73-103.

19 Cowgill, U.M. (1966), '*Perodicticus potto* and some insects', *J. Mammal.* 47, 156-7.

20 Bishop, A. (1964), 'Use of the hand in lower primates', in Buettner-Janusch, J. (ed.), *Evolutionary and Genetic Biology of Primates*, vol. 2, New York, 133-225.

21 Andrew, R.J. (1964), 'The displays of the primates', in Buettner-Janusch, J. (ed.), *Evolutionary and Genetic Biology of Primates*, vol. 2, New York, 227-309.

22 Jewell, P.A. and Oates, J.F. (1969), 'Ecological observations on the lorisoid primates of African lowland forest', *Zool. Afric.* 4, 231-48.

23 Charles-Dominique, P. (1968), 'Réproduction des lorisidés africains', in *Cycles génitaux saisonniers de mammifères sauvages*, 1. *Entretien de Chizé*, Paris, 1-9.

24 Charles-Dominique, P. (1971), 'Eco-éthologie des prosimiens du Gabon', *Biol. Gabon.* 7, 121-228.

25 Charles-Dominique, P. (1971), 'Sociologie chez les lémuriens', *Recherche* 2, 780-1.

26 Charles-Dominique, P. (1972), 'Ecologie et vie sociale de *Galago demidovii* (Fischer 1808; Prosimii)', *Z.f. Tierpsychol. Suppl.* 9, 7-41.

J. HARRINGTON

Olfactory communication in Lemur fulvus

Introduction

Communication by scent seems to be very important to most mammals. Except for aquatic mammals and the anthropoid primates, most mammals are macrosmatic, have specialised cutaneous scent glands, and deposit skin gland secretions, urine or faeces on objects or on other animals, often using specialised body movements. However, little is known about just what is communicated among mammals by scent. There is evidence that mammalian scents function as signals in alarm, territorial defence, breeding, orientation, agonistic behaviour, maternal behaviour, and the identification of individuals, sexes, social groups, and taxa. A great deal of new evidence on these points has appeared in the last five years. But our understanding of the communicative significance of most mammalian scents is still unclear, perhaps mainly because of the experimental difficulties of studying the responses of animals to olfactory signals.

Lemur fulvus, like other prosimians, is a representative mammal in the high development of its olfactory communication. Its sebaceous and apocrine sweat glands are numerous and highly developed, and are concentrated into specialised glandular fields.[1] *L. fulvus* is macrosmatic, and it frequently sniffs other animals and objects in the environment. It has conspicuous scent-marking behaviour. For these reasons *L. fulvus* is a suitable animal for a study of the important and little-understood subject of mammalian olfactory communication.

There is another reason for studying olfactory communication in *L. fulvus*, or in any other prosimian, and that is the particular importance of prosimian behaviour for the understanding of the evolution of behaviour in the primates. Prosimians preserve a number of primitive behavioural as well as anatomical characteristics. Furthermore, in the genera *Lemur* and *Propithecus*, primitive characteristics such as strictly seasonal breeding and well-developed olfactory communication coexist with some characteristics apparently evolved in parallel with the anthropoid primates, persistent bisexual social groups, diurnality, and considerable development of visual and vocal communication. A consideration of the olfactory communication of prosimians in relation to their communication in other sensory modalities and to their general biology, therefore, may throw light on the parallel evolution of behaviour in the anthropoid primates.

The following is a report of a field and laboratory study of olfactory communication in *Lemur fulvus.*

Field observations

L.f. fulvus was observed for 300 hours between February and July 1969 at the Forest of Ankarafantsika in northwestern Madagascar. Most of the observations were made under good conditions (2-50 m. away in deciduous forest) on two groups in which all individuals were known. Both groups had twelve members; their composition is given in Table 1. The breeding season occurred during the course of observations, but it was not marked by an increase in aggression such as occurs in *L. catta.*[2] Most intra-group interactions throughout the period of observations consisted of sitting peacefully in close contact with other animals and grooming each other. The two groups behaved territorially toward each other.

Special note of behaviour patterns which seemed to involve

Table 1 Composition of the *Lemur fulvus fulvus* groups observed

	LE Group	
	Males	*Females*
Adult	Russ (RU)	Yellow Eyes (YE)
	Yellow Eyed Male (YEM)	Pseudo Yellow Eyes (PYE)
	Light Eyes (LE)	Bear (BE)
	Rat (RA)	
Subadult	Young Male (YM)	Russela (RLA)
Juvenile	Patch (PA)	Light Face (LF)
	Mouse (MO)	
	PF Group	
	Males	*Females*
Adult	Light Faced Male (LFM)	Left Slit (LS)
	Ear Gouge (EG)	Foxy (FO)
	Orange Eyes (OE)	Yellow Eyed Female (YEF)
	Burly (BU)	Mary (MA)
Subadult		Pale Face (PF)
Juvenile	Junior (JR)	Black Faced Female (BFF)
		Cleo (CL)

olfactory communication was taken. These consisted mainly of anogenital sniffing and three behaviour patterns which may function as scent-marking: marking with the anogenital region, with the forehead, and with the palms of the hands, although anogenital marking is the only one of the three for which there is direct evidence that an olfactory signal is deposited. Animals sometimes sniff places where others have anogenital marked and may then mark there themselves. Furthermore, both sexes have a prominent patch of naked perianal skin in which apocrine and sebaceous glands are concentrated,[1] and, in addition, males have a copious strong-smelling secretion, easily visible in the field, on the bottom of the scrotum. Urine and faeces are other possible sources of anogenital scent. Places which have been only head-marked or hand-marked are not sniffed by other animals. However, hand and head marking are classified provisionally as scent-marking because they are conspicuous rubbing movements which occur at the same time and in the same situations as anogenital marking, and there are some glands on nearly every part of the body surface.

Nearly all scent-marking occurred in one of four fairly well-defined situations: sexual behaviour, alarm response to a man on the

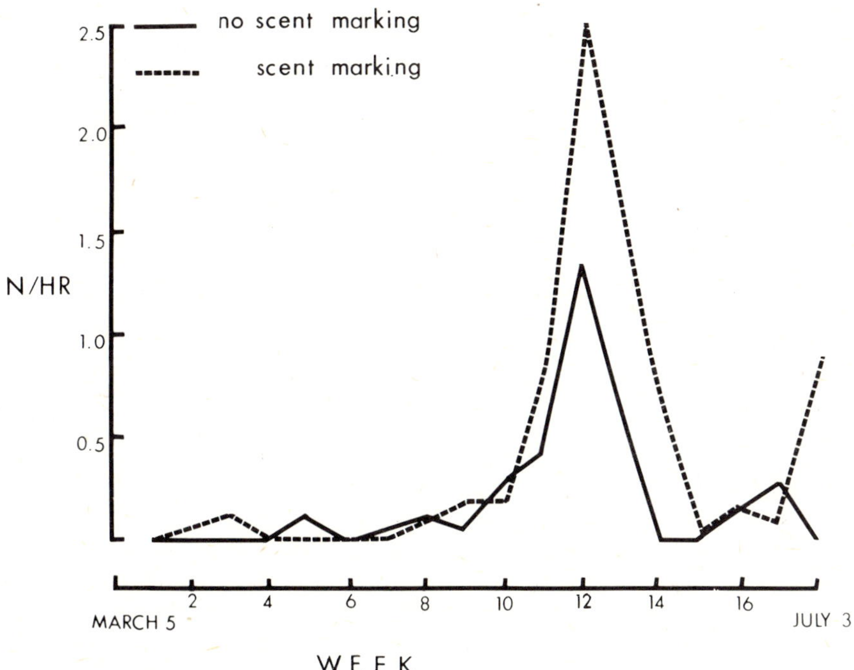

Figure 1. Male sniffs female anogenital scent (LE group).
Broken line: sniffing followed by scent-marking by male.
Solid line: sniffing not followed by scent-marking by male.

Table 2 Scent-marking by different individuals in various situations, for groups LE and PF

	Sex	*Alarm*	*Moving*	*Territorial*	*Play*	*Total*
			Anogenital marking			
Male	59	23	82	2	–	166
Female	18	100	85	6	–	209
Juvenile	–	–	–	–	1	1
Unident.	–	3	18	5	1	27
Total	77	126	185	13	2	403
			Head-marking			
Male	35	6	30	2	–	73
Female	–	1	–	–	–	1
Juvenile	–	–	2	–	–	2
Unident.	–	–	–	–	–	–
Total	35	7	32	2	–	76
			Hand-marking			
Male	10	–	6	–	–	16
Female	–	–	–	–	–	–
Juvenile	–	–	1	–	–	1
Unident.	–	–	–	–	–	–
Total	10	–	7	–	–	17
			Total scent-marking			
Male	104	29	118	4	–	255
Female	18	101	85	6	–	210
Juvenile	–	–	3	–	1	4
Unident.	–	3	18	5	1	27
Total	122	133	224	15	2	496

ground, territorial defence, and undisturbed moving around (Table 2). Most anogenital sniffing was accounted for by one situation – sexual behaviour (Table 3).

As the breeding season approached, males began to approach females and to try to clasp them around the waist, groom them, and sniff their anogenital regions. Usually the female would go away, stopping at several places as she went to anogenital mark or just to sit down. The male would follow and stop to sniff and scent-mark at the same places where the female had anogenital marked or sat. Before mounting, the male sniffed the female's anogenital region intensively and scent-marked the female and nearby objects. Fig. 1 shows that the frequency of sniffing of female anogenital scent by males, and of subsequent scent-marking by males, peaked during week 12 of the study. This was also the week in which other indicators of sexual activity reached a peak.

When the group was alarmed by a man on the ground, they moved

Table 3 Sniffing of anogenital scent in groups LE and PF

1. Sniffs anogenital scent, without scent-marking

	Sniffed				
Sniffers	*Male*	*Female*	*Juvenile*	*Unident.*	*Total*
Male	–	49	4	–	53
Female	11	1	1	–	13
Juvenile	5	2	1	–	8
Unident.	–	–	2	–	2
Total	16	52	8	–	76

2. Sniffs anogenital scent, then scent-marks

	Sniffed				
Sniffers	*Male*	*Female*	*Juvenile*	*Unident.*	*Total*
Male	7	100	–	1	108
Female	5	4	–	–	9
Juvenile	–	1	–	–	1
Unident.	–	–	–	–	–
Total	12	105	–	1	118

around quickly in the trees, swung their tails back and forth, gave loud grunting and shrieking vocalisations, and scent-marked. Sometimes the group mobbed the observer as well, the nearest animal approaching as close as 2 m.

The behaviour of the groups toward each other during territorial encounters was similar to the behaviour of a group toward a man on the ground. The two groups approached each other to within a few metres and gave tail swings, loud grunts and shrieks, and scent-markings.

Finally, many scent-markings were given when the group was not obviously alarmed at anything, but was just moving around in the trees, often just before beginning a long-distance progression.

These observations suggest some things that *might* be communicated by scent in *L. fulvus.* Scents might function for orientation, as alarm signals, as territorial markers, as primer or releaser pheromones in breeding, or for identification of individuals, sexes, social groups, or taxa. However, none of these communicative functions can be established from field observations alone. Field studies of olfactory communication must of necessity rely on observations of sniffing and scent-marking behaviour, which usually give only a suggestion of what is being communicated. Over-interpretation of such observations has resulted in concepts such as that of territorial scent-marking, critically reviewed by Schenkel,[3] for which definitive evidence is lacking. In most cases an experimental approach seems necessary to determine what is communicated by an olfactory signal.

Experiments

In these experiments an attempt has been made to determine whether the scent of *L. fulvus* conveys certain identifying information to other conspecifics; specifically, whether an animal's scent identifies it as an individual, as a male or a female, and as a member of a particular subspecies of *L. fulvus*.

The basic method in these experiments was: (1) to collect scents from animals caged individually, by leaving gauze pads in their cages overnight; (2) to present these pads to other animals caged individually; and (3) to record the responses of these animals to the scented pads. The method of collecting scent did not allow much control over the anatomical source of the scents. Presumably, the pads picked up a mixture of urine, faeces, and various skin-gland secretions. The animals were captives kept at Duke University, North Carolina. Most of them had been born in the United States.

1. *Identification of individuals*

To determine whether scents of two individuals of the same sex and subspecies were discriminated by another individual, the following method was used: a scent receiver was presented with a series of pads, one pad after another, all scented by the same animal, until the scent receiver became habituated to that animal's scent. Then the scent of a second individual was presented. It was anticipated that, if a scent receiver could discriminate between the two scent donors, its response would increase when it was presented with the scent of the second donor after habituation to the scent of the first.

Fig. 2 shows the results of six series of trials in each of which ♂ Oedipus was habituated to the scent of ♀ Sal, then presented with the scent of ♀ Frodo for ten more trials; and six control series in which ♂ Oedipus was habituated to the scent of ♀ Sal, then presented with the scent of ♀ Sal again for ten more trials. The criterion of habituation was five successive trials in each of which the pad was sniffed for 10 sec. or less. The figure shows that there is more sniffing in the ten post-habituation trials when a new scent is presented than when the same scent as before is presented ($p < .05$; 1-tailed Wilcoxon t-test). Table 4 gives the results of all the experiments, representing eleven different combinations of scent receiver and pair of scent donors. In all eleven, there is more sniffing in the post-habituation trials when a new scent is presented than when the same scent as before is presented. In eight of the eleven, the difference is significant at the .05 level. Fig. 3 shows the averaged results of all these experiments. The conclusion is that scents of *L. fulvus* of the same sex and subspecies can be discriminated by other *L. fulvus*.

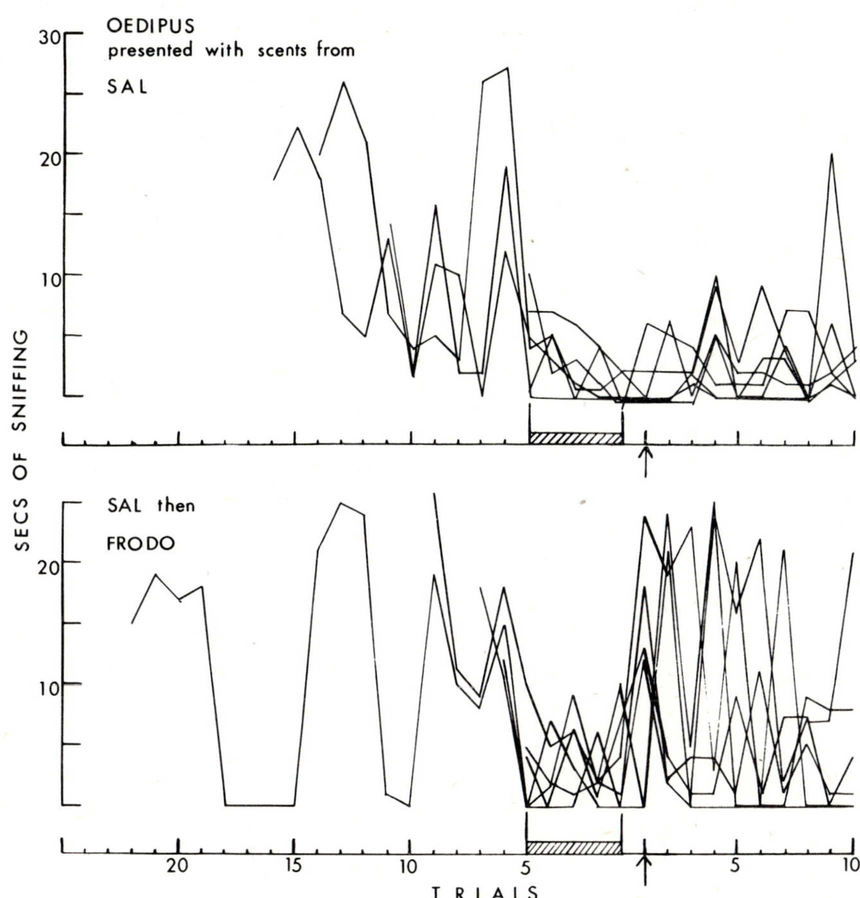

Figure 2. Discrimination of scents of ♀ Sal and ♀ Frodo by ♂ Oedipus. Upper graph shows 6 control series in which ♂ Oedipus was habituated to scent of ♀ Sal, then presented with scent of ♀ Sal for 10 more trials. Lower graph shows 6 experimental series in which ♂ Oedipus was habituated to scent of ♀ Sal, then presented with scent of ♀ Frodo for 10 more trials. Shaded area indicates 5 trials during which the criterion of habituation was reached. Arrow indicates first trial after habituation. Abscissa is numbered from right to left, starting to the left of the arrow, because different series took different numbers of trials to reach habituation.

2. *Identification of sexes*

To determine whether an animal's scent identifies it by sex, two males caged separately were presented with scents of a number of other individuals of both sexes, one scent after another. In addition, one of these males was presented with the scents of the same animals

Table 4 Discrimination of individuals: mean differences in amount of post-habituation sniffing between experimental and control series

Scent receiver	*Pair of scent donors*	*Difference (exptl-cont) (sec)*
Juvenal ♂ r	Thurber ♂ r Kingman ♂ r	*31.7
	Kingman ♂ r Fairchild ♂ r	*53.0
Oedipus ♂ f	Juvenal ♂ r Fairchild ♂ r	46.9
	Hermione ♀ r Calo ♀ r	*52.5
	Sal ♀ f Frodo ♀ f	*46.7
Calo ♀ r	Kingman ♂ r Thurber ♂ r	*69.0
	Kingman ♂ r Fairchild ♂ r	*77.5
	Sal ♀ f Frodo ♀ f	*35.5
Sal ♀ f	Juvenal ♂ r Fairchild ♂ r	22.7
	Hermione ♀ r Calo ♀ r	*82.0
Frodo ♀ f	Hermione ♀ r Calo ♀ r	31.2

N.B. Scent receiver was habituated to scent of donor named first.
* $p < .05$, 1-tailed Wilcoxon test.
r = *L.f. rufus*
f = *L.f. fulvus*

in pairs consisting of a male scent and a female scent of the same subspecies, presented at the same time. Both males were presented with each scent or pair of scents a total of ten times, spread over several days. The experiment was done in March, outside the usual breeding season in the northern hemisphere.

Fig. 4 shows the results for scents presented one at a time. Both males sniffed male scents significantly more than female scents ($p < .05$, Wilcoxon t-test). When the scents were presented in pairs to ♂ Juvenal, the male scent was sniffed more than the female scent in all six pairs (Table 5), and in three of these the difference was significant ($p < .02$, randomisation test). Only ♂ Oedipus showed any appreciable amount of marking in response to the scents; male and female scents did not elicit significantly different amounts of

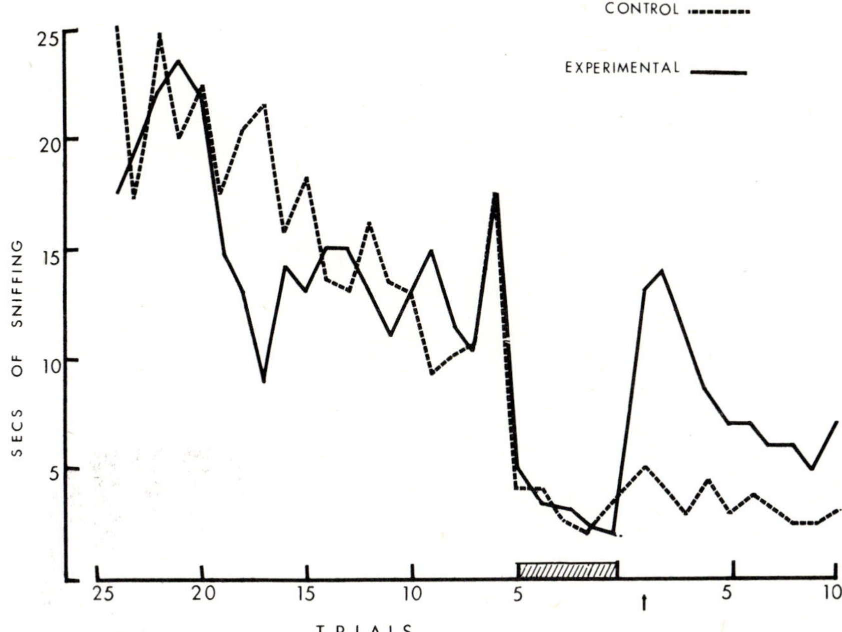

Figure 3. Discrimination of individuals: averaged results of all series. Average of 11 different pairs of graphs like those in Figure 2, representing 11 different combinations of scent receiver and pair of scent donors. Total of 71 experimental series and 71 control series.

marking. The conclusion is that the scent of an adult *L. fulvus* identifies it by sex to conspecifics. This finding was not surprising because, to the human nose, adult male *L. fulvus* have a strong smell, originating from the scrotal secretion, which is very different from the smell of a female.

3. *Identification of subspecies*

Neither of the preceding experiments showed any effect of subspecies on the response to scents. In the experiment on individual identification, scents of pairs of animals of the same sex and subspecies were discriminated by other animals of the same or different subspecies. Furthermore, there were no obvious differences in response to scents of different subspecies. This experiment involved two subspecies, *L.f. fulvus* and *L.f. rufus.*

In the experiment on identification of sexes, male scents were sniffed on the average more than female scents, regardless of subspecies. The marking responses of ♂ Oedipus in this experiment also did not show any effect of subspecies. This experiment involved three subspecies, *L.f. fulvus, L.f. rufus,* and *L.f. albifrons.*

An experiment was done to test further whether there was any

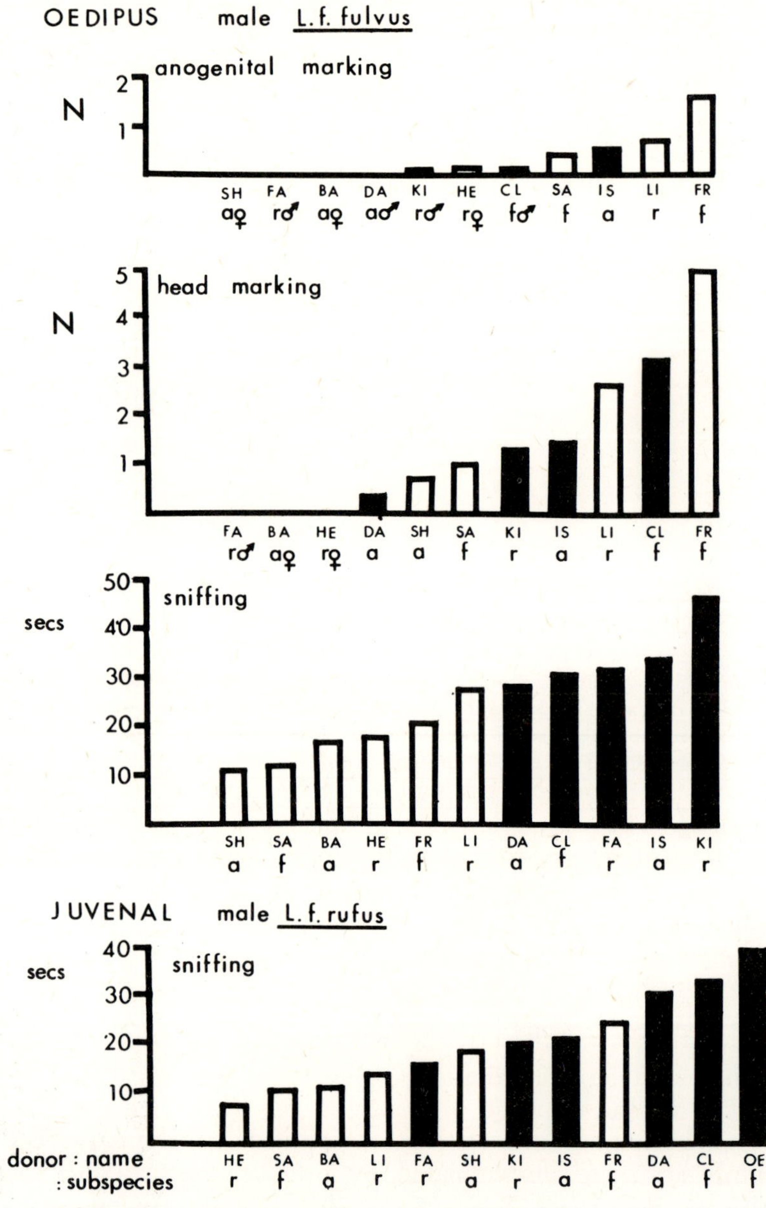

Figure 4. Discrimination of sexes: responses of ♂ Oedipus and ♂ Juvenal to scents presented consecutively. Dark bars = male scents. Abbreviations: HE = ♀ Hermione; SA = ♀ Sal; BA = ♀ Bathsheba; LI = ♀ Lisa; FA = ♂ Fairchild; SH = ♀ Shulamite; KI = ♂ Kingman; IS = ♂ Ishmael; FR = ♀ Frodo; DA = ♂ David; CL = ♂ Clarence; OE = ♂ Oedipus.

f = *L.f. fulvus*
r = *L.f. rufus*
a = *L.f. albifrons*

Table 5 **Mean difference in amount of sniffing of paired male and female scents by ♂ Juvenal**

Scent donors	*Difference (♂-♀) (sec)*
Oedipus ♂ f Sal ♀ f	3.0
Ishmael ♂ a Shulamite ♀ a	*22.0
Fairchild ♂ r Hermione ♀ r	0.2
David ♂ a Bathsheba ♀ a	*22.4
Kingman ♂ r Lisa ♀ r	6.9
Clarence ♂ f Frodo ♀ f	*22.1

* $p < .02$, 2-tailed randomization test for matched pairs.
f = *L.f. fulvus*
r = *L.f. rufus*
a = *L.f. albifrons*

effect of subspecies on the response to the scents. Three males were presented with paired scents of other animals. The two scents in each pair were from members of the same sex. One scent was from an animal of the same subspecies as the scent receiver, and the other scent was from an animal of a different subspecies. Each pair of scents was presented to each male twenty times (except for one pair which was presented to one male only fifteen times, because the animal swallowed a gauze pad on the fifteenth trial). Three subspecies were involved: *L.f. fulvus, L.f. rufus*, and *L.f. albifrons*. The object was to see if the animals responded differently to scents of their own and of other subspecies.

Table 6 shows the results. There were ten different combinations of scent receiver and pair of scent donors. In seven of these, one member of the pair elicited significantly more responses than the other, either sniffing, anogenital marking, or head marking ($p < .05$, signed ranks test). In three of these, the scent receiver responded more to the scent of his own subspecies and in the other four he responded more to the scent of the other subspecies. Thus the receiver may show a significant preference for the scent of one of a pair of animals of the same sex and different subspecies, but the direction of the preference is not determined by subspecies. This experiment, therefore, using three subspecies of *L. fulvus*, does not provide any evidence that the subspecies are identified by their scent.

Table 6 Mean difference in amount of sniffing, anogenital marking, and head-marking given to paired scents of own and other subspecies

		Differences (own/other subspecies)		
Scent receiver	*Scent donors*	*Sniff (sec)*	*Anogenital mark (N)*	*Head-mark (N)*
Juvenal ♂ r	Kingman ♂ r David ♂ a	*-12.4	0.1	-2.5
	Fairchild ♂ r Clarence ♂ f	11.8	0.6	-1.0
	Hermione ♀ r Sal ♀ f	*-7.2	-0.8	-1.5
	Red Child ♀ r Bathsheba ♀ a	4.9	*0.8	6.0
Oedipus ♂ f	Clarence ♂ f Fairchild ♂ r	*-21.5	0.5	3.5
	Sal ♀ f Hermione ♀ r	*-8.8	*-1.1	-1.6
	Frodo ♀ f Shulamite ♀ a	1.3	*1.5	1.4
Ishmael ♂ a	David ♂ a Kingman ♂ r	*28.0		
	Shulamite ♀ a Frodo ♀ f	-2.3		
	Bathsheba ♀ a Red Child ♀ r	-1.9		

* $p < .05$, 2-tailed signed ranks test.
Blank: no response.
f = *L.f. fulvus*
r = *L.f. rufus*
a = *L.f. albifrons*

Discussion

In summary, the scent of *L. fulvus* identifies it to other conspecifics by sex and individual identity. No evidence was found that the scent of *L. fulvus* identifies it by subspecies, among the subspecies *L.f. fulvus, L.f. rufus* and *L.f. albifrons.* Since *L. fulvus*, like *L. catta* and *Propithecus verreauxi*, live in social groups containing several individuals of both sexes, sexually and individually identifying information could be useful to it in a variety of social interactions. The significance of the apparent lack of discrimination among scents of the three subspecies is not clear, because of the uncertain condition of the taxonomy of the *L. fulvus* group, and the lack of knowledge of the geographical distribution of these three subspecies.

The subspecies of *L. fulvus* are distinguished from each other mainly by pelage, and they seem to be very similar in their morphology and behaviour.[4] Scent, then, seems to be another characteristic in which these forms do not differ from each other very much.

The social lemurs have highly developed olfactory communication, but they have considerable development of visual communication as well. They lack the elaborate facial communication of the monkeys and apes, but considerable increase in facial mobility has occurred as compared with the insectivores and the non-social prosimians.[5] There are also several conspicuous and stereotyped body movements, such as tail-swinging, cuffing, head jerking, and scent-marking, which may function as visual signals. The bright colours of these animals also imply some development of visual communication.

The social lemurs, then, are mostly 'olfactory' animals which have gone part of the way toward becoming 'visual' animals. Since the same transition has occurred during the evolution of the monkeys and apes, a consideration of the relationship between olfactory and visual signals in lemurs may provide suggestions as to how this transition might have occurred during the evolution of the anthropoid primates.

Both of the kinds of identifying information that were shown experimentally to be conveyed by the scent of *L. fulvus* are also conveyed quite conspicuously by the visual appearance of the animals. The sexes differ from each other in pelage, very strikingly in some subspecies; and the individual differences in pelage and other characteristics were sufficient for the observer to be able to identify easily nearly all of the individual animals studied in the field. The conveying of the same information by visual and olfactory means may be related to the activity cycles of the social lemurs. They are predominantly diurnal, but are also quite often active at night.

The production of olfactory signals often involves conspicuous body movements, i.e. scent-marking, which may themselves function as visual signals. *L. fulvus* obviously respond to anogenital marking as a visual signal when they go over to an animal which is anogenital marking from up to 5 m. away, and sniff and mark where the other animal has marked. Jolly also remarked on the large visual component in the stink fights and territorial battles of *L. catta.*[2] In these cases, scent-marking functions as both a visual and an olfactory signal. It is possible that scent-marking movements could be further selected for their signal function and evolve as visual signals on their own, without the production of an olfactory signal, as Andrew[5] suggested for presenting in monkeys and apes and Wickler[6,7] suggested for penis displays. If this is so, the scent-marking of lemurs may show how this process may have begun in the ancestors of the anthropoid primates.

A consideration of the general behaviour and ecology of the living prosimians may also suggest some reasons for the general decline in importance of olfaction in the anthropoid primates. The arboreal theory of primate evolution attempts to explain this decline as a result of an arboreal way of life.[8,9] However, prosimians and several other mammalian groups combine arboreality with highly developed olfaction, so arboreality alone does not necessarily result in the regression of olfaction.[10]

The prosimians, however, do have certain behavioural and ecological characteristics by which they differ from at least some of the anthropoids and whose loss in the anthropoids may have been associated with decreased selection for olfactory communication:

1. *Small home-ranges.* The prosimian data are for Lemuriformes, as well as *Galago* spp., which generally have group ranges of 10 hectares or less.[2,11,12] Some monkeys and apes, particularly the more terrestrial forms, have home-ranges one or two orders of magnitude greater. Larger ranges should decrease the usefulness of scent as a means of communication, because there must be a limit to the size of the area a group can effectively mark.
2. *Nocturnality.* Most prosimians are mainly or wholly nocturnal, and even the members of the diurnal genera *Lemur* and *Propithecus* are often active at night. This contrasts with the quite strict diurnality of nearly all monkeys and apes. Scent has obvious advantages as a means of communication for a wholly or partly nocturnal animal.
3. *Seasonal breeding.* The strict seasonal breeding of many prosimians contrasts with the less seasonal breeding of many anthropoid primates. A great deal of olfactory communication in mammals seems to be concerned with the timing of breeding, and therefore a loss of strictly seasonal breeding may be associated with decreased selection for olfactory communication.
4. *Relatively simple agonistic behaviour.* Most of the evidence on this point comes from *L. catta*, the only prosimian for which there are many field observations of agonistic behaviour. Jolly pointed out that *L. catta*, and perhaps the other social Lemuriformes, have a different system for controlling aggression from such primates as baboons and macaques.[2] Agonistic interactions in *L. catta* are almost entirely confined to the short breeding season, are often violent, usually involve only two animals, and involve much olfactory communication. In such primates as chimpanzees, macaques, and baboons, a more complex system is found, in which many agonistic interactions occur in non-sexual situations, elaborate signalling mechanisms have been developed to reduce aggression, more than two animals are often involved in interactions, and in which most signals are visual and acoustic.[13] It seems likely that olfactory

signals, whose coding possibilities in fast social interactions are more limited than those of visual and acoustic signals, are more appropriate for the relatively simple, all-or-none agonistic behaviour of *L. catta* than for the more complex agonistic behaviour of some monkeys and apes. This might also help explain 'why the highly social *Lemur* have not proceeded further with the evolution of facial expression',[5] since a great deal of the facial communication of anthropoid primates is concerned with subtle agonistic interactions.

5. *Oral grooming.* Mutual grooming in prosimians, as in the monkeys and apes, is an important kind of social interaction which seems to function to maintain social bonds. In addition, however, prosimian grooming may be a means of receiving or sending chemical signals, since all lemuriform and lorisiform and some tupaiiform prosimians groom with the tongue and the teeth.[14] The partial replacement of oral by manual grooming in the monkeys and apes may have eliminated an important means of exchanging chemical signals.

Acknowledgments

This paper is based on a PhD dissertation submitted to the Department of Zoology, Duke University. The author wishes to thank the following people for their help during the course of this research: P. Klopfer, J. Buettner-Janusch, J. Bergeron, N. Budnitz, R. Sussman, C. White, and L. Harrington. P. Marler and T. Struhsaker read and commented on the manuscript of this paper. The field work in Madagascar was made possible by the generous co-operation of la Direction des Eaux et Fôrets and le Ministre de l'Agriculture, Gouvernement Malgache; especially MM. Ramanantsoavina, Ramalanjoana, Andriamampianina, Razanajatovo, and Natai. The author also wishes to thank R. Albignac, G. Randrianasolo, and J.-A. Randrianarivelo for valuable advice and help in Madagascar.

Financial support was provided by NIH Grants HD02319 and RR00388, a Sigma Xi grant, a NASA Predoctoral Traineeship, and an NSF Predoctoral Fellowship.

NOTES

1 Montagna, W. (1962), 'The skin of lemurs', *Ann. N.Y. Acad. Sci.* 102, 190-209.

2 Jolly, A. (1966), *Lemur Behavior: A Madagascar Field Study*, Chicago.

3 Schenkel, R. (1966), 'Zum Problem der Territorialität und des Markierens bei Säugern – am Beispiel des Schwarzen Nashorns und des Löwens', *Z.f. Tierpsychol.* 23, 593-626.

4 Nute, P.E. and Buettner-Janusch, J. (1969), 'Genetics of polymorphic transferrins in the genus *Lemur*', *Folia primat.* 10, 181-94.

5 Andrew, R.J. (1964), 'The displays of the primates', in Buettner-Janusch, J. (ed.), *Evolutionary and Genetic Biology of Primates*, vol. 2, New York.

6 Wickler, W. (1966), 'Ursprung und biologische Deutung des Genitalpräsentierens männlicher Primaten', *Z.f. Tierpsychol.* 23, 422-37.

7 Wickler, W. (1967), 'Socio-sexual signals and their intra-specific imitation among primates', in Morris, D. (ed.), *Primate Ethology*, Chicago.

8 Smith, G.E. (1927), *The Evolution of Man*, 2nd ed., London.

9 Jones, F.W. (1926), *Arboreal Man*, London.

10 Cartmill, M. (1970), The Orbits of Arboreal Mammals: a Reassessment of the Arboreal Theory of Primate Evolution, Unpublished PhD Dissertation, University of Chicago.

11 Petter, J.-J. (1962), 'Recherches sur l'écologie et l'éthologie des Lémuriens malgaches', *Mém. Mus. nat. Hist. nat. Paris.*, Sér. A., 27, (1), 1-146.

12 Bearder, S.K. and Doyle, G.A., this volume.

13 Hamburg, D.A. (1971), 'Psychobiological studies of aggressive behaviour', *Nature* 230, 19-23.

14 Simons, E.L. (1962), 'Fossil evidence relating to the early evolution of primate behavior', *Ann. N.Y. Acad. Sci.* 102, 282-94.

A. SCHILLING

A study of marking behaviour in Lemur catta

Introduction

It is generally considered that three forms of marking are practised by the ringtailed lemur[1] (*Lemur catta*), a social prosimian inhabiting the south of Madagascar:

1. Genital marking, performed by both sexes, consisting of rubbing the genital organs on suitable supports and thereby depositing glandular secretions (Fig. 1*a*).
2. Brachial marking, performed only by the male, consisting of rubbing the inner surfaces of the forearms on suitable supports in a rhythmic movement, depositing secretions of the forearm and possibly the brachial gland (Fig. 1*b*).
3. Tail marking, performed by the male, involving rubbing the tail in such a way so as to impregnate it with scent from both arm glands (Fig. 1*c*).

In fact, the latter two forms may be considered analogous, in that the tail may become the object of brachial marking just like a branch. The structures used for marking, and the secretions deposited, are similar in both sexes. In the latter two types of marking behaviour, although there is an apparent difference in sequence and context, the components are generally common to both.

Marking may be considered as a rather stereotyped response to a complex of stimuli which are themselves modulated by ecological and ethological factors:

1. External physiological stimuli: *Lemur catta* possesses good vision and yet at the same time is a macrosmatic animal.
2. Internal physiological and psychological stimuli: phase of

(*a*)

Figure 1. The three different kinds of marking characteristic of *L. catta*. (*a*) genital marking; (*b*) brachial marking; (*c*) tail marking.

oestrous cycle, pregnancy or, in a more general sense, overall excitation level in both sexes.

3. Environmental factors: availability and nature of marking supports, climatic factors, etc.
4. Ethological factors: *Lemur catta* is one of the most social prosimians; marking behaviour could thus be involved in inter-individual and inter-group relationships.

This study was carried out in the gallery forest reserve at Berenty, Madagascar, in which *Lemur catta* is protected. Most of the time was spent following Troops I and II, previously described by Jolly and Sussman.[1,2] However, insufficient effective observations were made to consider this present work more than preliminary.

Morphological structures involved in marking behaviour

1. Genital marking

In males, the bare posterior surface of the scrotum is glandular in nature, a precise description of which has been given by Evans and Goy.[3] In females accessory glands are found on the labiae of the vulva, according to Jolly and Evans and Goy, or on the clitoris, according to Petter.[1,3,4]

(c)

2. *Brachial marking*

Histological studies of epidermal and glandular formations, which are developed or functional only in the males, have recently been carried out (Rumpler, Montagna and Yun, Kneeland).[5,6,7] On the upper part of the internal side of the arm, close to the shoulder, there is a multilobular sebaceous gland which reaches its maximum size (that of a walnut) in July (Petter).[4] On the antero-internal face of the forearm there is a bare cutaneous oval zone, covering a gland of similar size made up of sudoriporic tubules and particular interstitial cells. In females, there is a similar but smaller gland in the same place, in which adipocytes take the place of interstitial cells. In addition, but only in males, the overall zone is internally bordered by an epidermal horn spur. Andriamiandra and Rumpler[8] have shown that the sudoriporic antebrachial structures and sebaceous brachial structures are under testosterone control, a finding which permits the conclusion that they are secondary sexual characteristics.

Marking behaviour

1. *Genital marking* (Fig. 1*a*)

As has been previously noted by Jolly,[1] genital marking is identical in form in the two sexes. The area to be marked is selected, the animal turns about, balances on its forehands, and lifts itself to grab hold of the support above with the hind legs. The tail is curved forward, and the support is marked by the genital glands with an up and down movement, for 2-10 seconds. The trace thus left is strongly

(a) (b)

scented, and produces a dark area several centimetres in length. According to Evans and Goy both brachial and genital secretion deposits remain attractive for several days.[3]

2. Brachial marking of a branch (Fig. 1*b*)

The animal, standing on its hind legs, grasps a suitable branch with one hand, and several centimetres above applies the oval area of the other arm. In a rapid movement, the forearm is pulled back towards the animal. This movement is then usually followed by the other arm, though there may occasionally be repetition, usually only once, on the same side. Up to 15 alternative movements by one male during one bout of marking have been observed; however, the average is about 4 movements. The behaviour has two results: first, the forearm spur cuts into the branch a scar in the form of a comma, the concavity of which indicates the forearm used. This scar is easily visible and about 2 centimetres in size (Figs. 2*a*, 2*b*). Secondly, the scar thus formed contains the scent left by the scraping of the antebrachial gland located near the spur. The secretion of the brachial gland is applied irregularly to the spur, with 'anointing behaviour' described by Jolly.[1] Indeed, by folding the forearms one over the other, the animal brings into contact the brachial gland and the antebrachial structures (the spur itself, or the oval zone). 'Anointing' may or may not be bilateral; it occurs only rarely (0 to 6 times per hour; only once per hour on the average), appearing before

(*c*)

Figure 2. Scars left on marked branches by the forearm spur of the male *L. catta*. (*a*) enlargement of the scar showing its comma form; (*b*) a typical aspect of a marked branch; (*c*) concentration of marked branches in the peripheral 'marking area' (the lengths of scars on the branches are underlined).

brachial marking, seldom before genital marking, or in isolation, without interrupting progression.

3. *Tail marking* (Fig. 1*c*)

This is brachial marking in which the support is the animal's own tail. 'The *Lemur* stands on his hindlegs, tail drawn forward between his legs, and up between his forearms. The spurs enter the fur deeply, and much of the tail must be in contact with the antebrachial gland.'[1]

4. *Associated marking*

About once in every 15 times, the two types of marking are performed successively on the same support, brachial marking almost always preceding genital marking.

Topography and study of marking on external supports

Marking, whether it be genital or brachial, occurs most frequently in the lower zones of a vertical stratification of the forest, on the ground, and on low branches (zones 1 and 2 of Sussman).[2] On the one hand, *Lemur catta* frequently descends to the ground (36% of the time according to Sussman);[2] on the other, it finds numerous fine tender-barked branches, more or less vertical, at low levels, which it can mark more easily than others. It is, therefore, at the lower levels of the gallery forest that most marking behaviour is observed, even though the animal traverses the higher branches, particularly those of the tamarind trees, equally often. It appears that the branches chosen for genital marking can be less vertical and thicker than those used in brachial marking. The few large branches marked were dealt with by genital marking. In fact, the actual nature of the support seems to be less decisive, as a *Lemur catta* has been seen to mark a thin weed with its genitalia, and on another occasion the pad of a cactus.

The type of support most often selected for brachial marking is thin and vertical, or at least upright (compatible with the posture of the animal). The thinness of the branches chosen (2.5 cm. average for 100 measured) may be due less to adaptation for movement of the support than to the nature of such branches: the bark is much more tender when the bush is young. This type of marking has sometimes been observed even on branches with a diameter of 0.5 cm. Fig. 3 gives the results of precise measurements taken on 100 branches marked in different zones of the study area.

It can be seen that marking is more frequent on the thinner branches (approx. 2 cm. diameter) but more extensive on branches of a large diameter (approx. 4 cm. diameter). The number of marks is extremely variable: from a few to several hundreds (3 times 450 marks, once 550). Although these marks become less easily visible upon drying, there remains the stain that forms, either all around or merely on one particular side of the branches, which is visible at several metres and which has an average length of 77 cm. (Fig. 2*b*). This is particularly striking in certain 'marking areas' (zones which are ecologically favourable to marking behaviour because of numerous vertical supports), where marking is frequent and usually occurs above the ground. In this case, the lemur begins marking from just above the ground (50 cm. on the average, never less than 10 cm.), then continues while climbing, or begins after having leaped down to a low position on the bush concerned. The average number of cuts made with the spur, in the course of 75 observed brachial markings, was 4. In the case of stains still visible, it may be inferred that certain branches must have been marked more than a hundred times. There were two important 'marking areas' frequently visited by troop I (home-range = 6 hectares: number of members = 15), which included 4 males, 2 of which were peripheral, 5 females, 2 juveniles, and 4 infants.

The first area (Fig. 2*c*), in the graduated zone which borders the territory of the troop to the north, was a rectangle of about 400 square m., limited by paths and, in the west, by a denser zone of vegetation. It was planted with a variety of fruit trees, and many bushes, rarely exceeding a height of 3 metres. Practically all of these

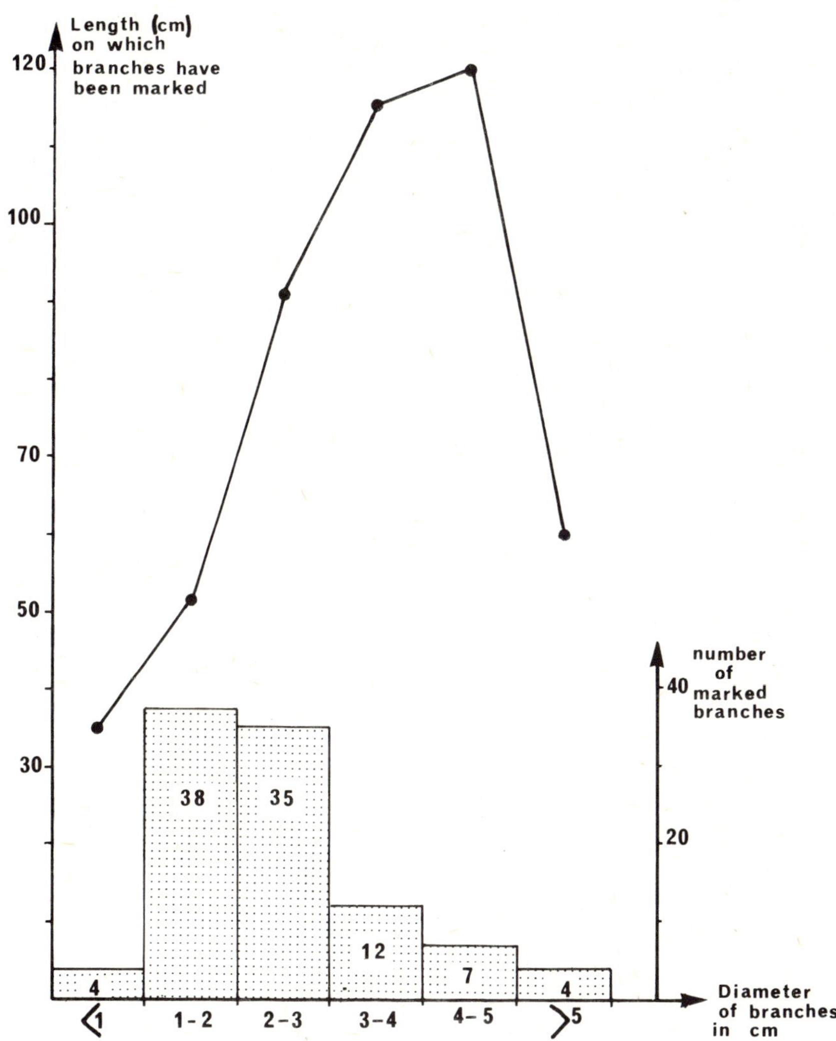

Figure 3. Histogram illustrating the relationship between the length of scars (graph) and the typical diameter of the branches marked (histogram).

trees were marked. The total length of the main stains represents an extent of 33 metres; and as the average density of these is estimated to be around 150 per metre of branch, more than 5000 of them could have been counted on this meagre surface.

The second 'marking area' (see Fig. 4) was located at the centre of the troop territory bordered by the dense gallery forest and a zone planted with cactus. It was an area of sparse undergrowth beneath the cover of large trees, especially tamarinds, where marking also occurred on the ground on low and isolated shrub trees. This zone, which did not need to be defended, was visited daily by the troop, which made incursions into the neighbouring cactus field to feed,

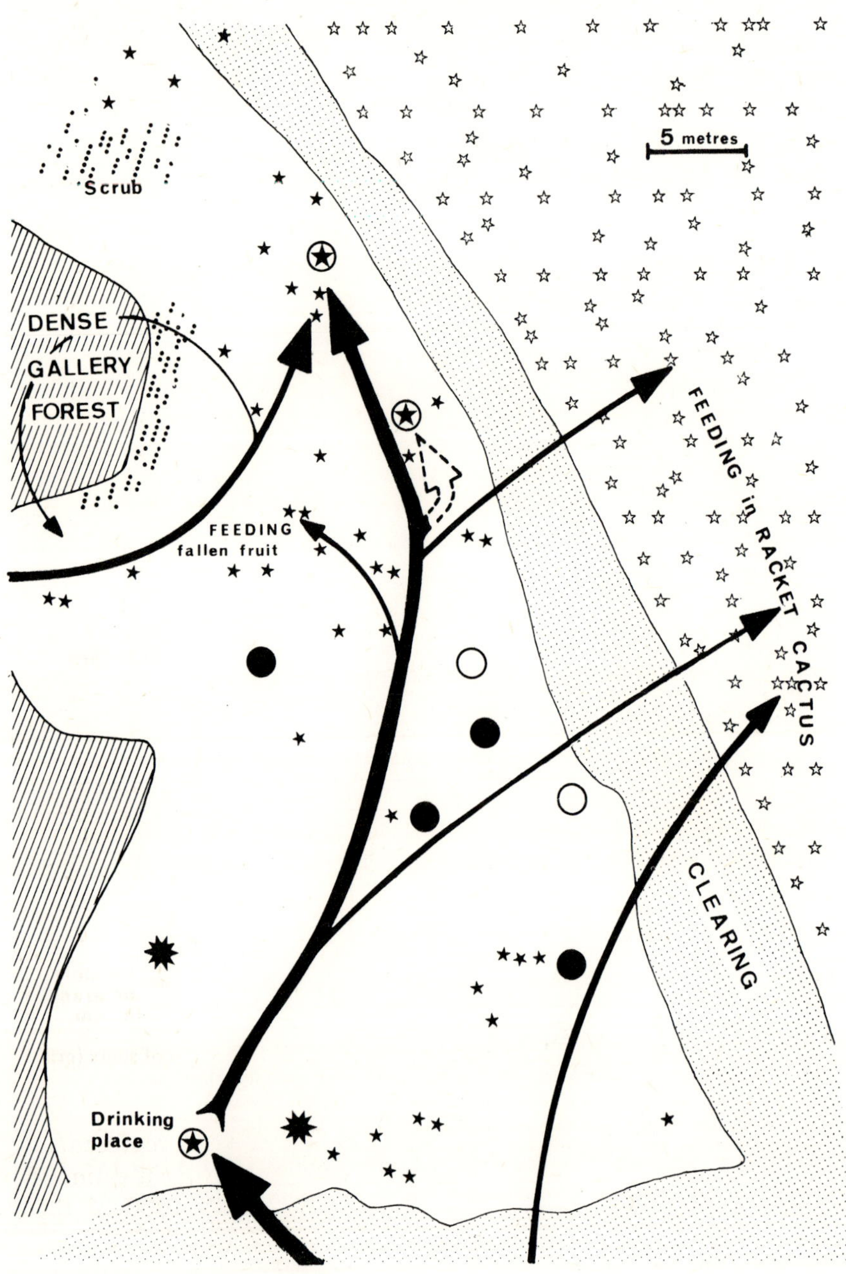

Figure 4. Map of the central 'marking area' of troop I in relation to its daily progression.
small black star = location of marked branches
open circle = tamarind trees
black circle = other large trees
star in circle = used for resting tamarind trees
large asterisk = dead trunk

and utilised certain tamarinds as resting trees for the morning siesta. Very often, at the end of the morning, the lemurs descended from the large trees, or went directly from the treeless boundary area, to play or eat fruit, fallen from the tamarinds, as well as to mark the bushes in this zone. It was not possible to observe a particular orientation of the marks or even a concentration in a precise area, except perhaps in the proximity of a tree in which the trunk contained a cavity filled with water where the troop came regularly to drink, or in an area where the animals found many fallen fruit under one of the resting tamarinds.

Associated behaviour patterns

1. Olfaction

It is evident that marking is associated with olfaction. As in all lemurs, *Lemur catta* possesses well-developed olfactory organs; but olfactory behaviour is far from being systematically associated with that of marking. For example, during a period of two-and-a-half hours of active and continuous marking by a male of troop I, the following was observed: 33 brachial markings preceded by smelling in one case; 60 genital markings preceded by smelling in 22 cases; 2 long smelling episodes without marking. It is probable that the animal is informed of the nature of the mark left by itself or by a conspecific, without stopping its progression or showing any behaviour detectable by the observer. In certain cases, nevertheless, the role of olfaction is evident:

(*a*) Verification of a former brachial mark before marking, either by a male or by a female.
(*b*) Verification of a mark left by a female before 'over-marking' by a male.
(*c*) Verification of marks left by the same animal, sometimes with re-marking.
(*d*) Processing with the nose on branches during a 'marking session' (see below).

2. Other behaviour patterns associated with marking of a support

(*a*) Licking, biting or even scratching of the bark of a branch before brachial marking.
(*b*) One case of faecal ingestion was followed by genital marking.

3. *Behaviour pattern not observed*

Marking of one individual by another, as has been seen in other lemurs, was not observed with *Lemur catta*.

Instances of marking

In a troop of *Lemur catta*, marking occurs either occasionally during progression, when the animal encounters a suitable branch and is stimulated, or more systematically during rest periods. In the latter case, the animal leaves the troop for a 'marking session' lasting two to twenty minutes, with frequent smelling and 'anointing', during which the neighbouring branches are actively marked, before coming back to rest with the troop. Fig. 5 gives an idea of the variability in the frequency of marking by a dominant male (see also Table 1).

Ethological context

1. *Age*

Infants and juveniles of less than one year of age do not mark.

2. *Genital marking by females*

Dominant females, or females carrying young, mark more frequently than other adult females.

3. *Marking and dominance in the males*

It was interesting to compare the frequency of marking by the dominant male of troop I with that of one of the two subordinate males, which were excluded from the troop, but which followed it at a variable distance (from fairly close to 200 metres away). Unfortunately, over several days of observations, only 10 hours of active marking were observed. For this reason the results summarised in the following table (Table 1) are only tentative.

4. *Significance of tail-marking*

As has been observed by Jolly,[1] tail-marking contains a definite aggressive element. All observations of the present study, like those of Sussman,[2] confirm the fact that this type of marking occurs only in an agonistic situation. This may be a simple confrontation between two animals, a fight between two troops, or a conflict situation with other species.

Marking generally takes place between two males, but it can also occur between a male and a female. Marking of the tail is a behaviour

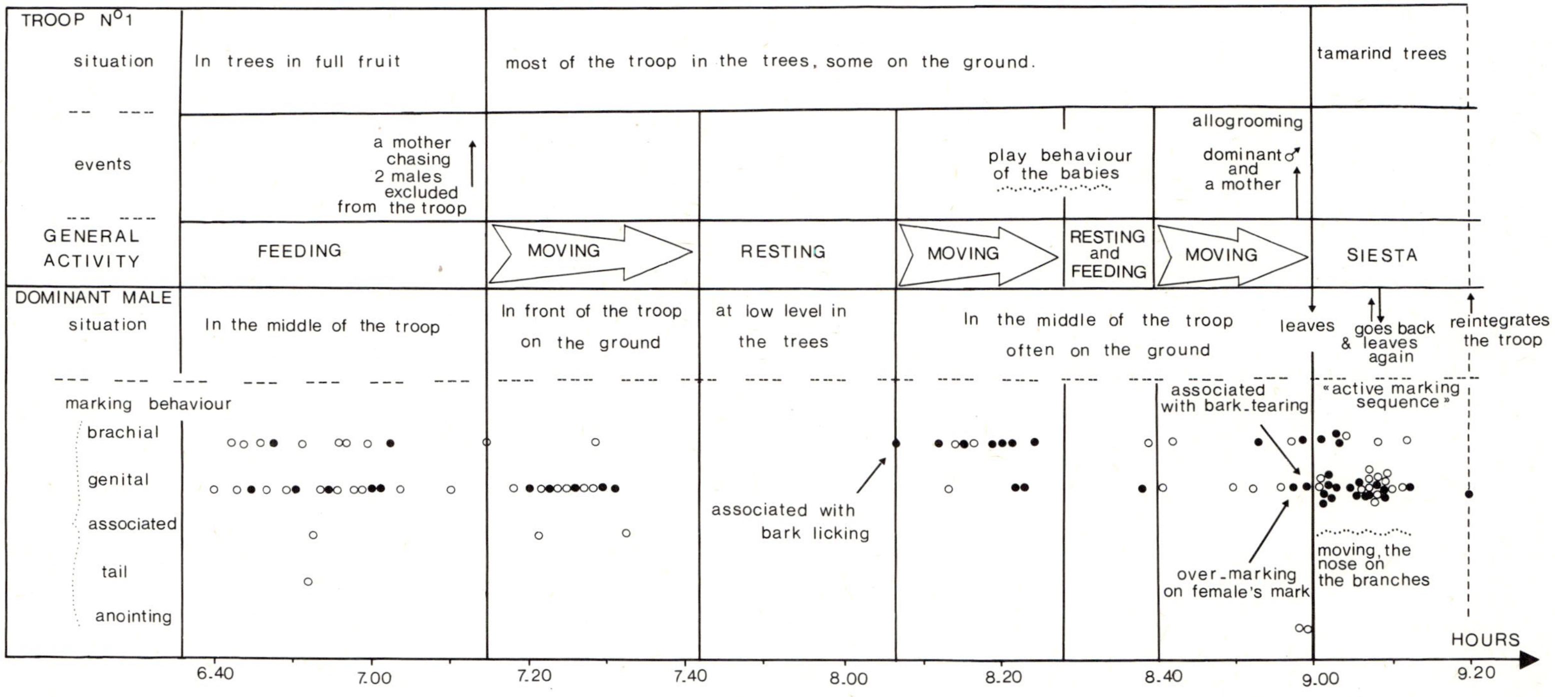

Figure 5. Chart of the marking behaviour of a dominant male, observed over 2½ hours, in relation to the general activity of the rest of the troop.
○ = occurrence of marking behaviour
● = occurrence of marking behaviour after preliminary sniffing

Table 1 Comparison of the frequency of marking between a dominant male and a subordinate male of the same troop

		Marking behaviour				*Anointing behaviour*
		Genital	*Brachial*	*Associated*	*Tail*	
Dominant Male	Number of markings observed in 4½ hours	74	39	8	1	5
	Average no. of markings per hour	16.5	8.7	1.7	0.2	1.1
Subordinate Male	Number of markings observed in 5 hours	43	50	10	2	13
	Average no. of markings per hour	8.6	10.0	1.1	0.3	1.7

pattern which should be included in the range of threats which a *Lemur catta* male, in all probability, uses for assuring its dominance over a conspecific. Ralls[9] has recently provided evidence illustrating the role that marking plays in dominance relationships for numerous mammals. In *Lemur catta*, other forms of behaviour accompanying this threat marking are a sort of grunting, a special high frequency whistling occurring only in this situation, and a characteristic waving of the tail, which is curved in front of the animal and towards the adversary. If the basis of the behaviour is the individual olfactory message which would be made up of the secretions of the brachial and antebrachial glands, Jolly is correct in thinking that 'tail-waving' represents an effective means of dispersing these odours.[1] The waving occurs more rarely than marking of the tail, which is usually sufficient to cause the adversary to flee. It may be a complex agonistic behaviour pattern, olfactory in nature, in which the first sequence involves the second, but is generally sufficient without it.

Marking as an olfactory message

1. Genital secretions of the females

There is no definite proof that the female odour informs the *Lemur catta* male of the stage of the oestrous cycle. However, this hypothesis may be considered, since a high frequency of marking by females in oestrus was observed. Conversely, marking of a support by a male is very often provoked by that of a female. This is exhibited either by over-marking of the female's trace, or by marking close by, either with the genitalia or with the arm.

2. *Brachial secretions*

Since development of the glands of the forelimbs used in marking is a secondary sexual character, it is logical to suppose that brachial marking plays a role in individual recognition, perhaps with respect to the individual dominance hierarchy. The olfactory message could be transmitted directly from one individual to another in the case of tail-marking and tail-waving. It could also be transmitted by means of the marks left on the bark of trees. Naturally, the brachial mark cannot be the only basis for inter-individual relationships, since in the *Lemur catta* troop females are sometimes dominant. Besides, even a dominant male has been observed to retreat before a female (usually a mother) in spite of vigorous marking of the forearm and then of the tail.

3. *Territorial marking*

The role of marking in territorial defence has been demonstrated in numerous mammal species,[10,11] and in particular in certain prosimians.[12] Thiessen and Lindzey have recently demonstrated the hormonal dependency of such behaviour.[13] In *Lemur catta*, all observers underline the abundance of marking in territorial defence. The behaviour, however, does not prove that thc rcsult (thc odour mark) has a determining role in this defence. What must nevertheless be pointed out is the extraordinary abundance of marks left at the 'marking area' located at the periphery of the territory of troop I, and very often the site of territorial disputes with troop II. A territorial encounter between two troops is associated with much marking behaviour, both sexes participating and using all forms of marking (see Fig. 6). It appears to be primarily a means of impressing the opposite troop with the vigour and persistence of the behavioural act itself.

4. *Marking and orientation*

The role of marking in orientation seems to be doubtful, and learning of very precise routes is probably of visual origin. A young male *Lemur catta*, separated from its troop, did not sniff more than normally, but frequently stopped in its progression and uttered a sort of distress mewing, listened attentively to the group's answer, and then guided itself by the sound. Nevertheless, this did not prevent it from marking, mainly with the genitalia, after having carefully sniffed some point on the route where the other members of the troop had previously marked.

Figure 6. Photograph of a 'stink fight' between two troops on the periphery of their territories (courtesy of R.W. Sussman).

Marking without a communicative role

In this case, the marking stimuli are less specific, the behaviour reflecting a general state of excitation of the animal. It is in this light that an attempt is made to explain the 'stink-fight'.[1] At a less intensive level, any state of excitation could be translated into a marking response, being either of endocrine or nervous origin. Marking may be due to a conflict situation, to fear, or it may signify 'friendship' (as Petter suggested it did among tame animals in the Tananarive zoo).[4] Marking behaviour is primarily linked to olfaction, but it must not be forgotten that close connections exist between olfactory structures and those responsible for affective and emotional responses (Papez circuit, hippocampus, amygdaloid body). Marking may be a displacement activity for *Lemur*, as is suggested by some observations, for instance, after fleeing from an observer or an opponent in an agonistic situation, upon hearing a loud noise, falling from a tree after a badly calculated leap, or upon breaking of a branch, etc.

Conclusion

If, in certain cases, marking may seem to be explained as a response to specific external sensory stimuli where the motivation involved and the social role of the message are clear, in other cases the behavioural sequences may be one of the external manifestations of a complex psychophysiological situation which often appears to be the result of a conflict.

There is no proof that certain instances of marking, which appear to be provoked by specific stimuli, would not appear except as the result of a state of conflict or simple excitation which exists within the animal. Lastly, it is perhaps of interest to speculate on the scars left on the branches by *Lemur catta* as being visual signals, as no other primate is known to exhibit such a signal.

Acknowledgments

The author wishes to thank Henri de Heaulme for allowing this study to be carried out at his protected reserve in Berenty, South Madagascar.

NOTES

1 Jolly, A. (1966), *Lemur behavior*, Chicago.

2 Sussman, R.W. (1971), Unpublished doctorate thesis, Duke University. See also this volume.

3 Evans, C.S. and Goy, R.W. (1968), 'Social behaviour and reproductive cycles in captive ring-tailed lemurs (*Lemur catta*)', *J. Zool. Lond.* 156, 181-97.

4 Petter, J.-J. (1962), 'Recherches sur l'écologie et l'éthologie des lémuriens malgaches', *Mém. Mus. nat. Hist. nat. Paris*, sér. A., 27, 1-146.

5 Rumpler, Y. and Andriamiandra, A. (1968), 'Sur l'étude de glandes brachiales et antebrachiales du *Lemur catta*', *C.R.S. Soc. Biol.* 162, 1430; Rumpler, Y. and Andriamiandra, A. (1969), 'Les glandes brachiales et antebrachiales du *Lemur catta*', *Ann. Univ. Madag. Med. et Biol.* 11, 57-66.

6 Montagna, W. and Yun, J.S. (1962), 'The skin of primates: 10. The skin of the ring-tailed lemur (*Lemur catta*)', *Am. J. Phys. Anthrop.* 20, 95-118.

7 Kneeland, J.E. (1966), 'Fine structure of the sweat glands of the antebrachial organ of *Lemur catta*', *Z. Zellforsch. Mikr. Anat.* 73, 521-53.

8 Andriamiandra, A. and Rumpler, Y. (1968), 'Rôle de la testosterone sur le determinisme des glandes brachiales et antebrachiales chez le *Lemur catta*', *C.R.S. Soc. Biol.* 162, 1651.

9 Ralls, K. (1971), 'Mammalian scent marking', *Science* 171, 443-9.

10 Schultze-Westrum, T. (1965), 'Innerartliche Verständigung durch Düfte beim Gleitbeutler *Petaurus breviceps papuanus* Thomas (Marsupialia, Phalangeridae)', *Z. vergl. Physiol.* 50, 151.

11 Mykytowycz, R. (1965), 'Further observations on the territorial function and histology of the submandibular cutaneous (chin) glands in the rabbit *Oryctolagus cuniculus (L.)*', *Anim. Behav.* 13, 400-12; Mykytowycz, R. (1966), 'Observations on odiferous and other glands in the Australian wild rabbit,

Oryctolagus cuniculus (L.), and the hare *Lepus europaeus P.*: 1. The anal gland', *C.S.I.R.O. Wildl. Res.* 11, 11-29; Mykytowycz, R. (1968), 'Territorial marking by rabbits', *Sci. Amer.* 218, 116-26.

12 Ilse, D.S. (1955), 'Olfactory marking of territory in the young male Loris', *Brit. J. Anim. Behav.* 3, 118-20.

13 Thiessen, D.D. and Lindzey, G. (1970*a*), 'Territorial marking in the female Mongolian Gerbil: short-term reactions to hormones', *Hormones and Behav.* 1, 157-60; Thiessen, D.D. and Lindzey, G. (1970*b*), 'The effects of olfactory deprivation and hormones on territorial marking in the male Mongolian Gerbil', *Hormones and Behav.* 1, 315-25.

PART I SECTION D
Physiology of Behaviour in Captivity

A. PETTER-ROUSSEAUX

Photoperiod, sexual activity and body weight variations of Microcebus murinus *(Miller 1777)*

Introduction

Madagascan lemurs breed seasonally.[1] It has often been pointed out that transfer to Europe reverses the breeding season: members of the genus *Lemur* give birth in Madagascar in October, and in zoological gardens of the northern hemisphere in April. The Cheirogaleinae, which are small nocturnal species and include the lesser mouse lemur, *Microcebus murinus*, start mating in September in Madagascar and in April in Paris, under natural local light conditions.

Breeding condition is thus associated with the photoperiod: temperature, humidity and nutrition can be maintained constant in captivity, and these factors consequently have less influence on the observed variations in breeding activity.

It was observed a long time ago that certain cheirogalines, notably those of the genus *Cheirogaleus*, can accumulate fat reserves (under the skin, in the tail and around the viscera) and become torpid during the non-breeding season (viz. the dry season, which is the winter in Madagascar). There is evidence that these animals exhibit an annual cycle affecting the entire body metabolism, and not just reproduction.

In captivity, *Microcebus murinus* exhibit a comparable annual cycle, with fat-accumulation and semi-torpidity alternating with the breeding period. Over the course of several years, we have carried out various experiments to investigate the effects of photoperiodic variations on the physiology of *Microcebus*.

Housing conditions

The animals are housed individually or in groups of 2-3 in cages measuring 60 x 60 cm., each provided with small wooden nest boxes. The animals are fed with fruits, milk, honey, spice-cake, mealworms and fresh meat. Temperature is maintained constant at about 25°C.

Mating condition of females is recognisable from oestrous developments involving swelling of the sexual skin and opening of the vulva, which is otherwise closed.

In the following experiments, the animals were housed in 2 separate rooms:

A. room exposed to the natural daylight conditions of the Paris region.

B. room artificially lit with 'daylight' fluorescent tubes (40 W). In this room the amount of light received was 50-150 lux. The transformation from light to dark and *vice versa* occurred abruptly, without a 'twilight' phase.

Annual variations in body weight

Captive mouse lemurs exhibit regular annual variations in body-weight,[2] which is maximal during the non-breeding season, and minimal during the breeding season. Weight-loss begins 3-4 months before the onset of oestrus. Weight-loss and oestrus occur during the period of increasing day-length, and weight-gain occurs during the period of decreasing day-length; the changeover is abrupt and may be complete within 2 months.

The same phenomenon occurs in animals subjected to the natural daylight conditions of Paris or to an artificial lighting schedule following the Madagascar rhythm; but, as a result of the six-month difference in phase, when animals in one room are in breeding conditions, those in the other are not (Fig. 1).

The magnitude of annual variation in day-length is about 8 hours in natural daylight, but only 2 hours in artificial lighting (Madagascar rhythm). This difference in day-length fluctuation seems to have no effect on the reproductive processes, however.

The increase in weight affects all animals. It is of interest to note that a castrated male exhibited the same annual variation in body-weight, during the first year at least: weight-loss occurred in spring, with a minimum in July, followed by a sudden increase in September/October (Fig. 2).

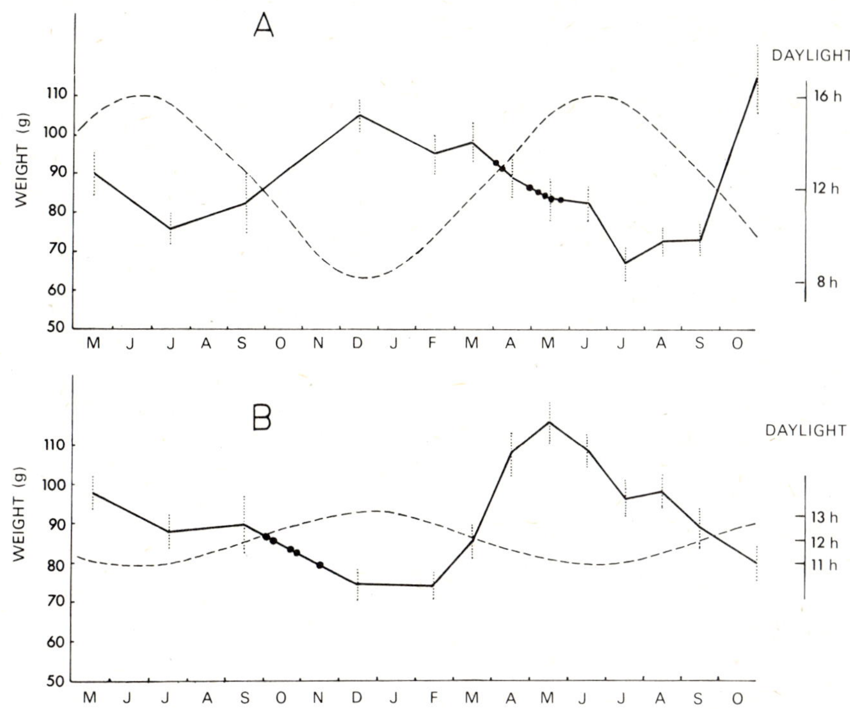

Figure 1. Annual variation in body-weight and oestrus in captive *Microcebus murinus.*
Solid line = mean body-weight of 15 animals.
Dotted line = annual variation in day-length.
Black circles represent the dates of onset of oestrus.
A – natural daylight in Paris
B – artificial Madagascan light regime

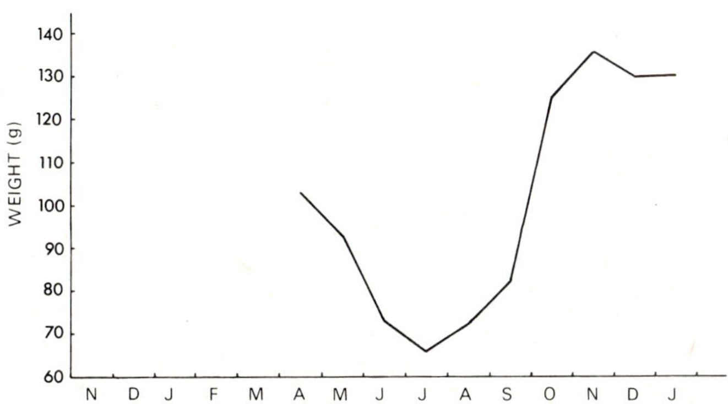

Figure 2. Body-weight variation in a castrated male, from 6 months after operation onwards (natural daylight conditions in Paris).

Adaptation to alteration in the light regime

Transfer from the Madagascan light regime to that of Paris

This transfer generally induces a rapid alteration in the breeding season. The timing of the alteration depends upon the timing of the change.

1. Change at the end of December

Fig. 3 shows the result of moving 7 females from the Madagascan light regime to that of Paris on 21 December. These animals were still in breeding condition in Madagascar, where maximum day-length conditions prevailed. They were suddenly subjected to the very short days of winter in Paris, followed by very rapidly increasing day-lengths. During January and February, their weights increased considerably as a result of the short days; thus, they entered into the non-breeding season earlier than animals under a Madagascan light regime. It was only after March/April that increasing day-length induced first weight-loss and then oestrus, which occurred at the end of April in the most precocious females; that is, only about a fortnight later than in those females under the Paris light regime. All females in this group came into oestrus between the end of April and the end of May.

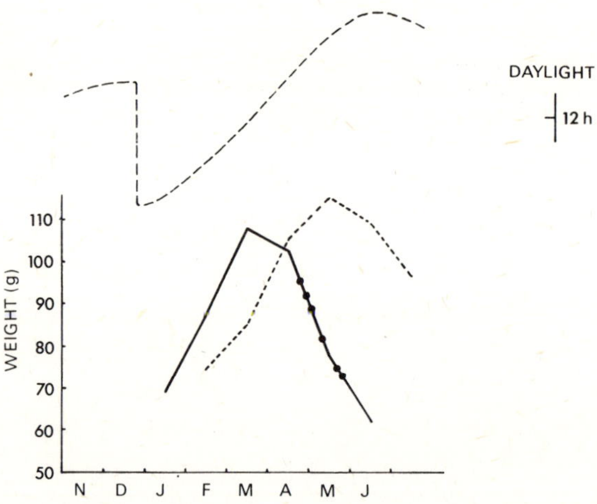

Figure 3. Mean weight and oestrus of 7 mouse lemurs transferred on 21 December, from the Madagascan light regime to natural daylight conditions in Paris (solid line). Black circles represent oestrus onset.
Lower dotted line = body-weight variation of mouse lemurs living under the Madagascan light regime at the same time of the year.
Upper dotted line = variation in day-length during the experiment. (In this and in subsequent figures, the scale for day-length variation is the same as in Fig. 1.)

2. *Change at the beginning of February*

Fig. 4 shows the transfer of 6 females on 7 February from the artificial Madagascan light regime to natural daylight conditions in Paris; none of these females had experienced pregnancy during the previous months. In this case, the effect of short days reinforced the fat-accumulation which would have occurred naturally at this time. The first oestrus occurred at the end of April, as with the preceding group; one female had still not come into oestrus by the end of July.

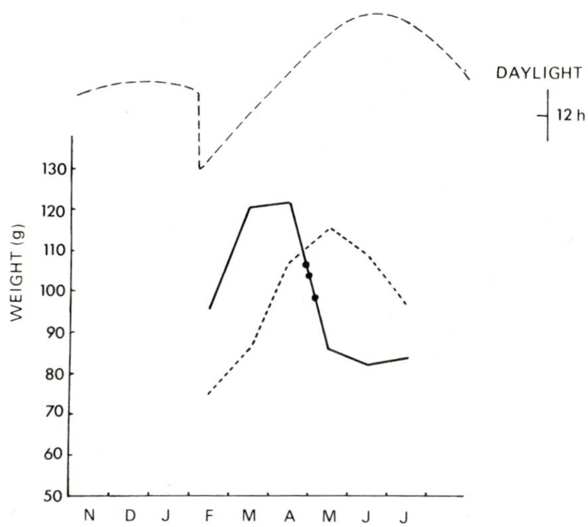

Figure 4. Mean weight and oestrus of 6 mouse lemurs transferred on 7 February from the artificial Madagascan light regime to natural daylight in Paris (solid line). Black circles represent oestrus onset.
Lower dotted line = body weight-variation of mouse lemurs living under the Madagascan light regime at the same time of the year.
Upper dotted line = variation in day-length during the experiment.

3. *Change at the end of April*

Fig. 5 shows the transfer of 6 females on 24 April from the artificial Madagascan light regime to natural daylight conditions in Paris. None of these females had experienced pregnancy during the preceding months. They had already been in the non-breeding season for 2 months. The abrupt change to long and increasing days hastened the weight-loss and induced oestrus, starting in June. The last female came into oestrus on 1 August, that is during the period of decreasing day-lengths, and one female did not come into oestrus at all. Thereafter, the effect of decreasing day-lengths accelerated fat-accumulation, and the breeding season was reduced to 3 months. Thus, in these three cases of transfer (from December to April) almost all of the animals came into oestrus a few months later.

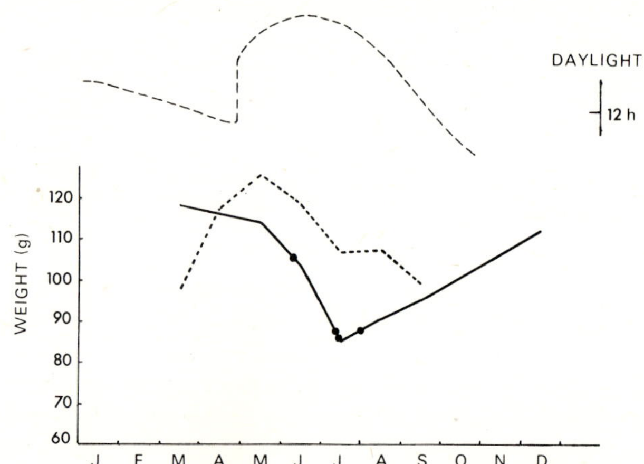

Figure 5. Mean weight and oestrus of 6 mouse lemurs transferred on 24 April from the artificial Madagascan light regime to natural daylight in Paris (solid line).
Black circles represent oestrus onset.
Lower dotted line = body-weight variation of mouse lemurs living under the Madagascan light regime at the same time of the year.
Upper dotted line = variation in day-length during the experiment.

4. *Change at the end of May*

If the transfer is too late, the increasing day-length sequence is too short to induce reproductive activity. Figure 6 shows the effects of actual importation to Paris of 4 females from Madagascar on 27 May. The history of these animals is unknown. Their weight diminished at once under the influence of long day-lengths, but they did not come into oestrus before the onset of the non-breeding season.

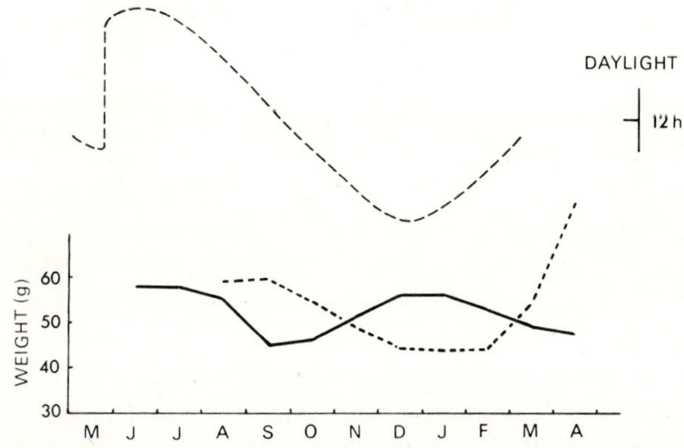

Figure 6. Mean weight and oestrus of 4 mouse lemurs transferred on 27 May from Madagascar to Paris (solid line).
Lower dotted line = body-weight variation of mouse lemurs living under the Madagascan light regime at the same time of the year.
Upper dotted line = variation in day-length during the experiment.

Transportation from the light regime of Paris to that of Madagascar

When transfer is made in the opposite direction, from the light regime of Paris to that of Madagascar, a different problem arises. The decreasing day-lengths of late summer in Paris are longer than the longest day of the Madagascan summer: at the end of the period of increasing day-lengths under the Madagascan regime, the animal will still be subjected to shorter day-lengths than those prevailing prior to transfer. Fig. 7 shows the transfer of 6 females on 3 August; the animals entered earlier into non-breeding condition, increased their body-weights, and were not stimulated to enter breeding condition before November; 3 of them came into oestrus in January. The remainder exhibited a similar weight-loss, but no reproductive activity.

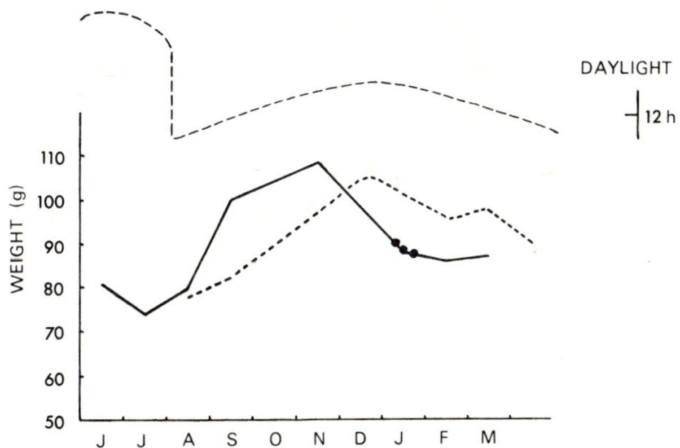

Figure 7. Mean body weight and oestrus of 6 mouse lemurs transferred on 3 August from natural daylight conditions in Paris to the artificial Madagascan light regime (solid line). Black circles represent oestrus onset.
Lower dotted line = body-weight variation of mouse lemurs living in Paris at the same time of the year.
Upper dotted line = variation in day-length during the experiment.

Applications

The transportation to Paris of mouse lemurs trapped in Madagascar during the breeding season results in abrupt, but brief, cessation of activity, and gives rise to the capacity to undergo another breeding period only a few months later. The weight-loss induced by this novel stimulation could have serious effects in cases where the animals do not possess adequate reserves.

It is consequently possible to control the body-weight of the animal by subjecting it to an appropriate lighting regime. Fig. 8 shows an example of such an experiment: a male, imported in December, suffered very severe weight-loss in spring. He was exposed in April to the Madagascan light regime, which resulted in an increase in weight, and in August a reverse transfer subjected him to the natural light regime of Paris. The resultant effect was that of brief breeding season stimulation, followed by renewed increase in weight.

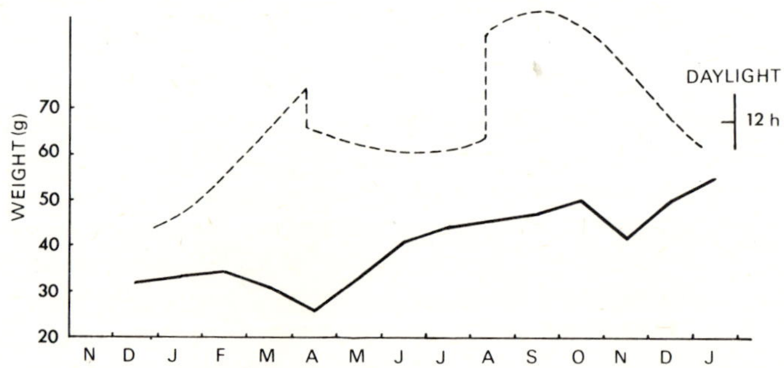

Figure 8. Body-weight of a mouse lemur transferred from the Madagascan light regime to that of Paris on 10 April, followed by the reverse transfer on 10 April (solid line).
Dotted line = variation in day-length during the experiment.

Discussion

These experiments demonstrate that in captive mouse lemurs reproductive activity and body-weight are correlated with variations in day-length.

The body-weights of captive mouse lemurs tend to increase considerably during the non-breeding period. Similar weights are unknown in the wild state, and this considerable increase may possibly be an effect of overfeeding in captivity. Lack of activity could also have this result; but the same phenomenon was observed with mouse lemurs kept in groups of 4 in large cages of 1.50 x 2 m. There is, perhaps, an alteration of nervous control of body-weight related to conditions in captivity.

Seasonal fattening related to the photoperiod is common in hibernating mammals, but it has never been demonstrated in lemurs other than the Cheirogaleinae. However, it is possible that an analogous situation exists in other Madagascan lemurs whose breeding season is similarly related to photoperiod variations. In lorisids, no evidence has yet been produced to demonstrate any effect of photoperiod variation on breeding condition or general metabolism.

Our observations suggest that the induction of a breeding season by an appropriate light regime is not dependent on the magnitude of the variation in day-length. It depends on the fact that days are increasing or decreasing in length; but, in a sudden transfer, increasing days do not have a stimulating effect if they are very much shorter than the days prior to the transfer. If the transfer occurs before the end of the active period, mouse lemurs experience an early non-breeding condition and accumulate fat; some 2 to 3 months later, they are ready to respond to new stimulation. If the transfer occurs later, the reaction is more irregular: some animals react very late, some do not respond at all.

In these experiments, adaptation to a change in light regime is rapid and generally free of difficulty. However, in one case of actual importation (on 1 February) a complete lack of response was observed: of 13 females, only one came into oestrus, and then only after 5 months. The history of these animals was unknown; possibly pregnancy or lactation during the preceding months could have delayed the effect of the stimulation.

NOTES

1 Petter-Rousseaux, A. (1968), 'Cycles génitaux saisonniers des lémuriens malgaches', in Canivenc, R. (ed.), *Cycles génitaux saisonniers des Mammifères sauvages*, Paris.

2 Petter-Rousseaux, A. (1972), 'Application d'un système semestriel de variation de la photopériode chez *Microcebus murinus* (Miller, 1777)', *Ann. Biol. Anim. Bioch. Bioph.* 12, 3.

M. PERRET

Variation of endocrine glands in the lesser mouse lemur, Microcebus murinus

Introduction

Studies of the reproductive and sexual cycles of mammals which show marked seasonal activity permit one to distinguish certain endocrine factors and to obtain a better understanding of their complexity. Little research of this type has been carried out with the Primates, owing to a lack of knowledge of their natural cycles and of the modifications produced by conditions in captivity. The Madagascan lemurs exhibit a seasonal pattern of activity which is maintained in captivity and is determined by the photoperiod.[1] We have therefore tried to determine the endocrine cycles of the lesser mouse lemur, *Microcebus murinus* (Cheirogaleinae), kept in captivity at the Laboratory of General Ecology in Brunoy.

Materials and methods

In captivity, *M. murinus* show a precise annual cycle of activity: during the period of short day-length (from September to February at our Paris latitude) sexual quiescence is complete in both sexes; a reduction in general activity occurs, and this is accompanied by a lowering of the internal body temperature and a gain in body-weight. In the springtime, general activity is stimulated, body temperature rises, animals lose weight and the sexual functions begin. In Paris, under natural daylight conditions, this period of activity lasts from February to August. All the animals used in this study were kept in captivity in Paris, except for six which were sacrificed in Madagascar. The conditions in captivity are constant with respect to temperature and food, but differ with respect both to available space and to the seasonal light-regime (which is either that of Paris or that of Madagascar). However, all dates given in this paper refer to the cycle exhibited by *Microcebus murinus* under the Paris light regime.

The animals were anaesthetised with ether for organ-collection. The endocrine glands (pituitary, thyroid, adrenals and gonads) were collected from different animals throughout the year, together with various other organs (kidneys, lungs, heart, liver, etc.). Fixation was carried out with Elftman's formol sublimate (1957). The glands were embedded in paraffin and cut at 3 to 5 μ. The annual development of the various endocrine glands was quantified by using simple histological procedures (standard staining techniques, measurements of nuclear and cellular diameters, measurements of the different components of each gland). In addition, the technique of immuno-fluorescence was employed to study the pituitary gland. Sections of the pituitary gland were treated with antibodies anti-STH, anti-TSH and anti-LH bovine, anti-HCG and anti-β (1-24) corticotropin. Five categories of cells were identified in the pituitary of *Microcebus murinus*: Cell-types STH, TSH, LTH and LH correspond to the classic descriptions; the corticotropic type, however, contrasts with the majority of observations in other mammals: it is basophilic and PAS-positive, and has been identified as equivalent to the category recognised in almost all other species as secreting gonadotropic hormone.[2] In addition to cytological studies and estimation of cell populations, the variations of chromophil reactivity of the cytoplasm and the intensity of fluorescence permitted estimation of the activity of cell types in the pituitary gland.

Pituitary cycles and target glands

Although the somatotropic cells in the pituitary gland (specifically stained green with luxol blue, using Shanklin's technique, 1959) remain numerous and active all the year, the other categories exhibit annual variations. These modifications have been studied in association with observations made on target glands.

1. *Thyroid function*

The pituitary thyrotropic cells of *M. murinus* are typically cyanophil cells; they are finely granulated and specifically stained by alcian blue, used as prescribed by Herlant (pH 1.5, after permanganate oxydation). They remain few in number throughout the year, but size and chromophily vary markedly: in winter the cells are small and strongly stained, while in spring they become enlarged and the secretory granules are more dispersed. Maximum size is attained in May-June. A slow involution begins in July or August, depending on the individual (decrease in cell size and increase in granulation).

Correlated with these pituitary variations are changes in the thyroid: there is a winter resting stage characterised by flattening of the epithelium and retraction of the colloid; followed by stimulation during springtime, which is indicated by growth of the epithelium and the appearance of absorption vacuoles in the colloid. As for the thyrotropic cells, activity is at its maximum in May-June and then decreases slowly, beginning in July (see Fig. 1).

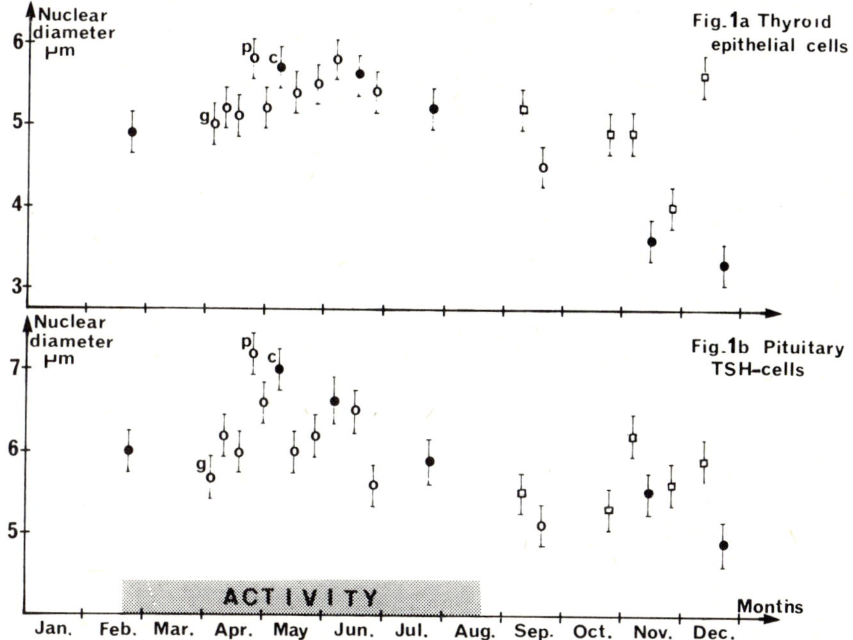

Figure 1. Annual variations of thyroid function.
● = male; o = female; □ = animal with chronic disease; *p* = pro-oestrus; c = castrated; g = pregnant; standard errors are indicated by vertical lines.

Correlation between thyroid activity and TSH-cell activity is highly significant (p=0.01). Thyroid activation occurs in early spring and is maximum in May-June. The resting condition begins in late July and is almost complete in September.

Castration, pregnancy (to a lesser degree), the ovulatory period, or a state of chronic illness are all accompanied by modifications of thyroid activity. It is probable that the steroid hormones (adrenal and gonadal) intervene in the regulation of the thyrotropic hormone, as is the case in most mammals. Observations concerning the thyroids of various hibernating animal species indicate that there is hyperactivity of the thyroid during the period of increased activity in spring. This active state continues in summer throughout the entire reproductive season of these animals (slow-worm, hamster, hedgehog).

2. *Adrenocortical function*

The ACTH-cells have been identified by immunofluorescence with antibody anti-β (1-24) corticotropin. They are stained bright purple by the PAS-Orange G staining technique. Important modifications are evident in the cytoplasm and especially in their number. Observations reveal that there is a relationship between the physiological state of the adrenals and the histological state of the cell-type PAS positive ++. Development of the adrenals is accompanied by a state of pronounced activity of these cells.

During the winter, the ACTH-cells are rare and densely granulated, indicating weak secretory activity; in the adrenal, the cortex is reduced (zona fasciculata unorganised, cells small and heavily stained). Corticotropic function is stimulated during the month of February and is shown by multiplication and degranulation of ACTH-cells at the pituitary level and an increase in volume of the adrenal cortex, the organisation of cords in the zona fasciculata and the appearance of spongiocytes. The zona glomerula does not vary in size during the year and remains about 55 μ in thickness.

During the period of long day-length, there is pronounced corticotropic activity (ACTH-cells are numerous and active; the cortex is well developed, with spongiocytes present). Following this, during the month of July, the adrenal cortex decreases in size to reach a minimum in November. The spongiocytes disappear relatively early, and the size of the nucleus of fascicular cells is reduced from August onwards (see Fig. 2).

Certain animals have been observed to show hyperplasy and hypertrophy of the ACTH-cells, some of which exhibit a state of marked vacuolar degranulation. These animals were a castrated male (contrasting with normal males during the same period); females in oestrus (contrasting with females in di-oestrus), particularly females near the time of ovulation compared to females far from ovulation; and animals which died following a period of more or less extended exhaustion, compared to healthy animals during the same period. Accompanying such hyperactivity of the ACTH-cells (see Fig. 3), there is evident growth of the adrenal cortex.

Study of the adrenals during the oestrous cycle in females indicates that there is an accompanying cycle of corticotropic function. In fact, the adrenal cortex expands very rapidly at the beginning of oestrus and develops to its maximum size by the seventh day of vaginal opening, that is 5 days after ovulation. During this period, the pituitary is almost entirely filled by ACTH-cells showing signs of intense secretory activity (see Fig. 3*c*). The relationships between the adrenal cortex and the physiology of oestrus are certainly very pronounced in *M. murinus*, though they are perhaps accentuated in captivity. In addition, a female treated for 10 days after pro-oestrus

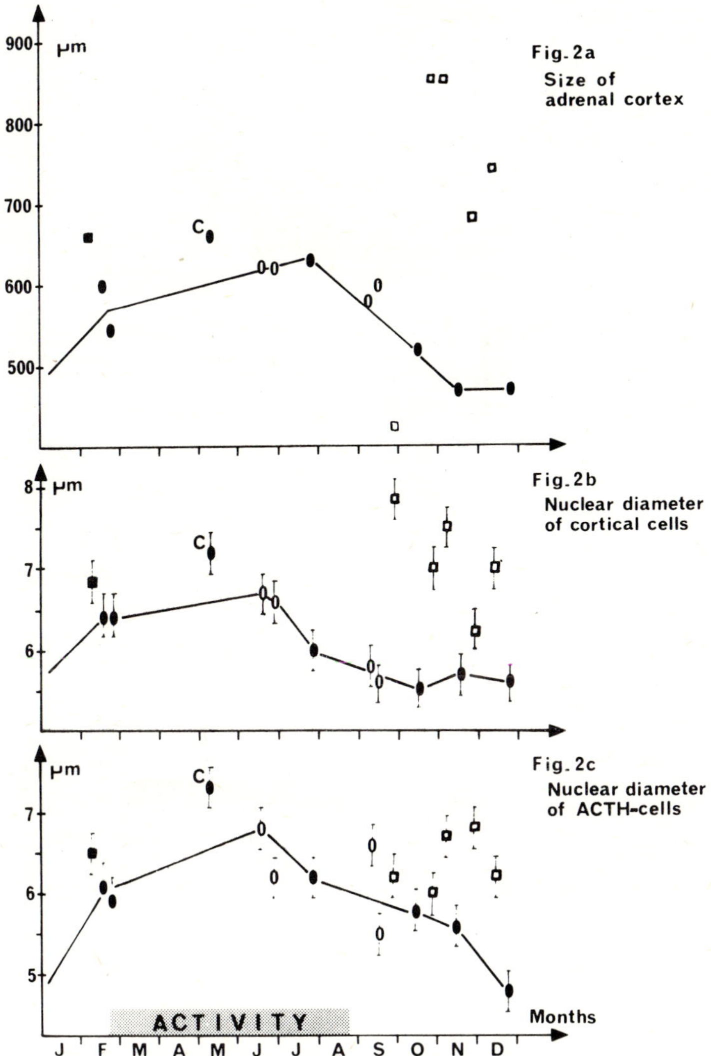

Figure 2. Annual variations of adrenocortical function.
● = male; o = female; □ = animal with chronic disease; C = castrated; standard errors are indicated by vertical lines.

Animals with chronic disease and castrated males show a clear increase of all morphological parameters of adrenocortical function.

with 1 mg./day of hydrocortisone ovulated 5 days late. The ACTH-cells of this female's pituitary were numerous, but inactive, and the adrenal was completely involuted. These observations, in association with the occurrence of adrenal hyperactivity after castration, suggest that corticosteroids play an active role in the sexual physiology of *Microcebus*.

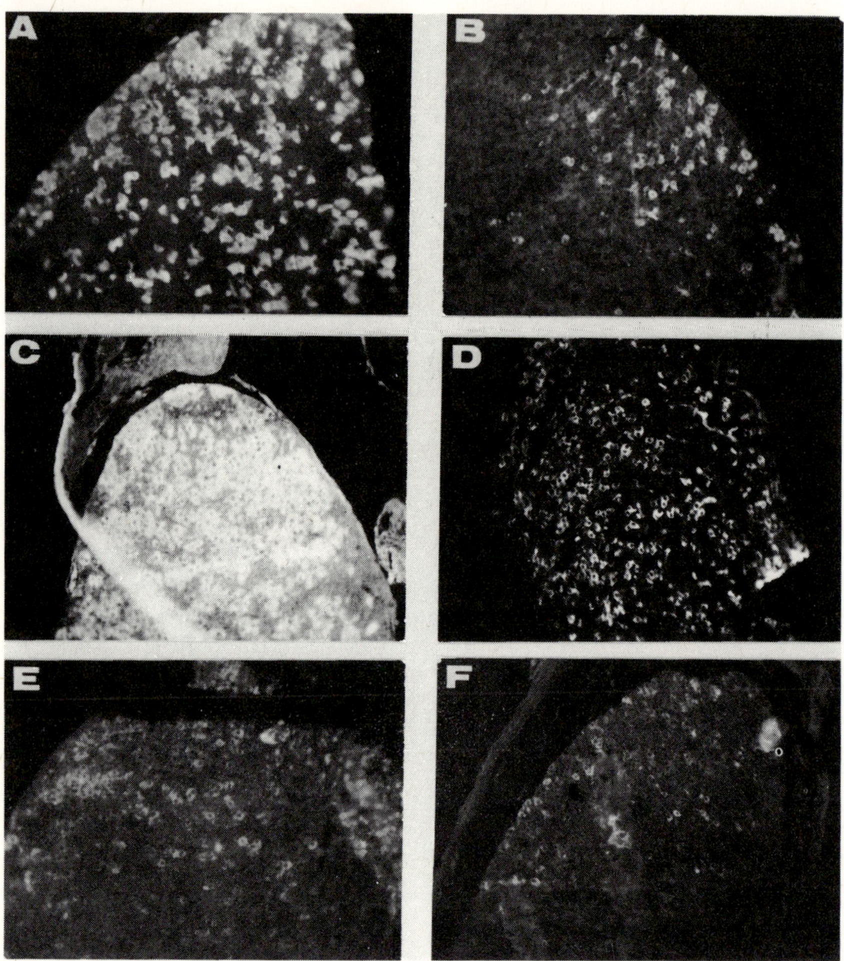

Figure 3. ACTH-cells: immunofluorescence with antibody anti-β (1-24) corticotropin. (*A*) Castrated male during May. (*B*) Normal male during May. (*C*) Female, 3 days after vaginal opening. (*D*) Female in di-oestrus. (*E*) Animal with chronic disease in the winter period. (*F*) Animal in good health in winter period. Hyperactivity is seen with animals *A*, *C* and *E*.

3. *Sexual cycles*

(*a*) *The male.* Annual variations in testis size are very marked in male *M. murinus.* During the winter the testes are just perceptible under the scrotal skin, although they cannot be measured, and during the breeding season they may reach a size of 2 cm. in length and almost as much in width.

In winter, the testes are completely involuted; activation (as for most northern species showing a seasonal sexual cycle) begins in February.[3] During the spring all the germinal elements develop in the seminiferous tubules, excepting the spermatozoa which appear later. Activity is at its maximum in May-June, and the epididymis canal is distended with spermatozoa. Spermatogenesis begins to decrease in activity in July, and regression is complete in September (see Fig. 4).

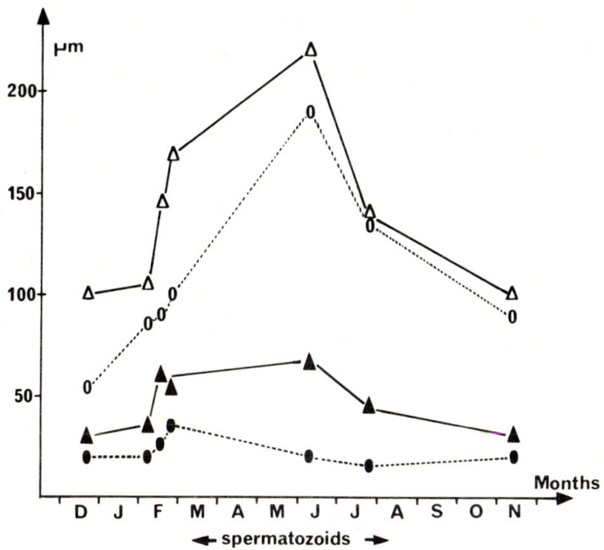

Figure 4. Changes in morphological parameters of testicular and epididymal functions during the year.
△ = diameter of seminiferous tubules;
o = diameter of epididymal tubules;
▲ = height of germinal epithelium;
● = height of epididymal epithelium.

The Leydig cells follow the same cycle of development, activity being maximal in spring. A slight disparity exists between maximum testicular activity and maximum interstitial activity. This separation is also apparent in certain other species showing a seasonal sexual cycle.[4] The curve of development of the gonadotropic LH-cells is, as expected, parallel to that of the Leydig cells (see Fig. 5). Another category of pituitary cells, the prolactin cells, also show annual variations. Easily distinguished by their red colour following treatment with Herlant's tetrachrome, these cells appear at the beginning of the period of activity and attain a maximal activity at the end of the breeding season. This hyperactivity (see Fig. 5) at the end of July is perhaps one of the factors causing reduction of spermatogenesis during this period.

We have not observed hyperactivity of the LH-cells in the

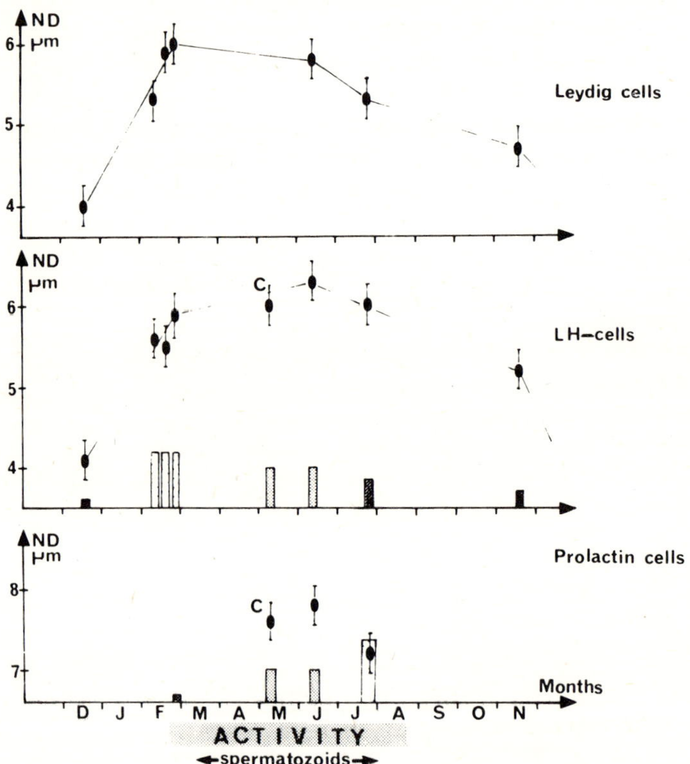

Figure 5. Annual variations of endocrine sexual function in the male. C = castrated; ND = nuclear diameter; standard errors are indicated by vertical lines. The surface of each rectangle is proportional to estimated cell population density; density of cytoplasmic granulation is indicated by the degree of darkness of each rectangle.

Maximum activity is seen in late February for Leydig cells, in May-June of LH-cells and in July for prolactin cells.

castrated male. This male was castrated 2 years ago, and it is possible that the adrenal cortex produces androgens to compensate for castration, and thus plays a normal feed-back role. Castration is accompanied by corticotropic hyperactivity,[5] possibly involving a suppression of the inhibition of ACTH synthesis due to the absence of testosterone, which has been shown to have an inhibitive effect on ACTH-cells and/or corticosteroid metabolism.[6]

(*b*) *The female.* The structure of the ovary and the accessory sex organs varies according to the season and the phase of oestrous cycle. The resting period begins in September and is characterised by degeneration of secondary follicles. Atresia is well advanced by November, and the ovary is invaded by dense conjunctive tissue.

During the oestrous cycle, development of the ovary follows the normal mammalian pattern: at pro-oestrus, each ovary develops one follicle and accessory sex organs enlarge. Follicular development in captive females does not seem to be very pronounced, many follicles showing signs of atresia. The vagina, normally closed, opens for about 10 days at the time of oestrus. Vaginal smears and histological studies have shown that ovulation occurs between the 2nd and 3rd day after vaginal opening. After ovulation, the corpus luteum develops in each ovary and no necrosis is observed. The absence of necrosis in captive females seems to be correlated with the appearance of the corpus luteum. The secretions of the corpus luteum may compensate for the possible lack of gonadotropins in these females.

Interpretation of the activity of the pituitary LH-cells during the oestrous cycle is difficult (see Fig. 6); there does not seem to be a well-defined cycle, but the cells do show variations both in their number and in chromophil reactivity. As in most mammals, the developing follicle would stimulate the formation of LH, and there would be a degranulation of these cells at the time of ovulation, followed by progressive granulation after the formation of the corpus luteum.[7]

The pronounced growth and abundance of ACTH-cells at the time

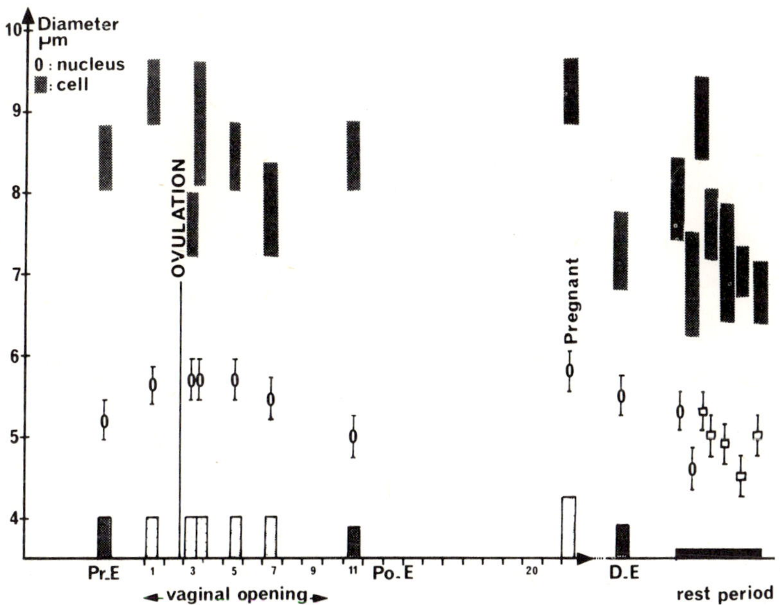

Figure 6. Variations in LH-cells in the female. (□ = animal with chronic disease; Pr.E = pro-oestrus; Po.E = post-oestrus; D.E = di-oestrus; standard errors are indicated by vertical lines.) The rectangles on the abscissa indicate cell population density and density of cytoplasmic granulation, as in Fig. 5. LH-cells show degranulation during ovulation, after which they enter gradually the phase of accumulation of their products.

of ovulation (see Fig. 3*C*) suggest that the cortico-gonadic relationships are particularly important in captive females. It is possible that, for these females, corticosteroid hyperactivity due to stressful conditions of captivity could explain the low rate of reproduction in captive animals. For instance, it is known that an excess of corticosteroids results in ovarian atresia. In addition, a female treated with hydrocortisone shows signs of follicular atresia, in spite of the presence of the corpus luteum.

As for the prolactin cells, they develop at the same time that the corpus luteum appears. The fact that these cells enlarge or regress according to the presence or absence of the corpus luteum could indicate a luteotropic action of prolactin in *M. murinus*, as is the case for mice, rats or hamsters.

Animals with chronic disease

It seemed interesting to include in this paper observations made on animals which died during the winter season (and accordingly exhibited no sexual activity), following a period of extended exhaustion. This is characterised by progressive loss in body-weight, followed by loss of appetite.

Several features are always evident in these animals:

1. *Hyperactivity of corticotropic function*

In the pituitary of these animals, ACTH-cells are numerous (see Fig. 3*E*) with a large nucleus (see Fig. 2*c*) and exhibit signs of great secretory activity (degranulation with cytoplasmic vacuolisation, which may lead to complete disappearance of granulations). Such activity is unusual at this time of the year.

The adrenals are significantly hypertrophied (see Fig. 2*a*) and the cortex accordingly exhibits a characteristic form. Spongiocytes with a large nucleus (see Fig. 2*b*) are visible, but organisation of cords in the zona fasciculata is defective, due to the presence of large PAS-positive regions. On the interior aspect of the Cortex, disorganisation is complete, and the cells remain small and easily stained.

2. *Increase in thyroid function*

In sick animals, greater activity of TSH-cells is observed in the pituitary (see Fig. 1). The thyroid is heterogenous and shows an intermediate state of activity. The follicles are more or less developed, and the colloid (though adhering to the epithelium) has few or no vacuoles. It seems justifiable to attribute this greater thyroid activity to the hypercorticosteroid state which exists in these animals.

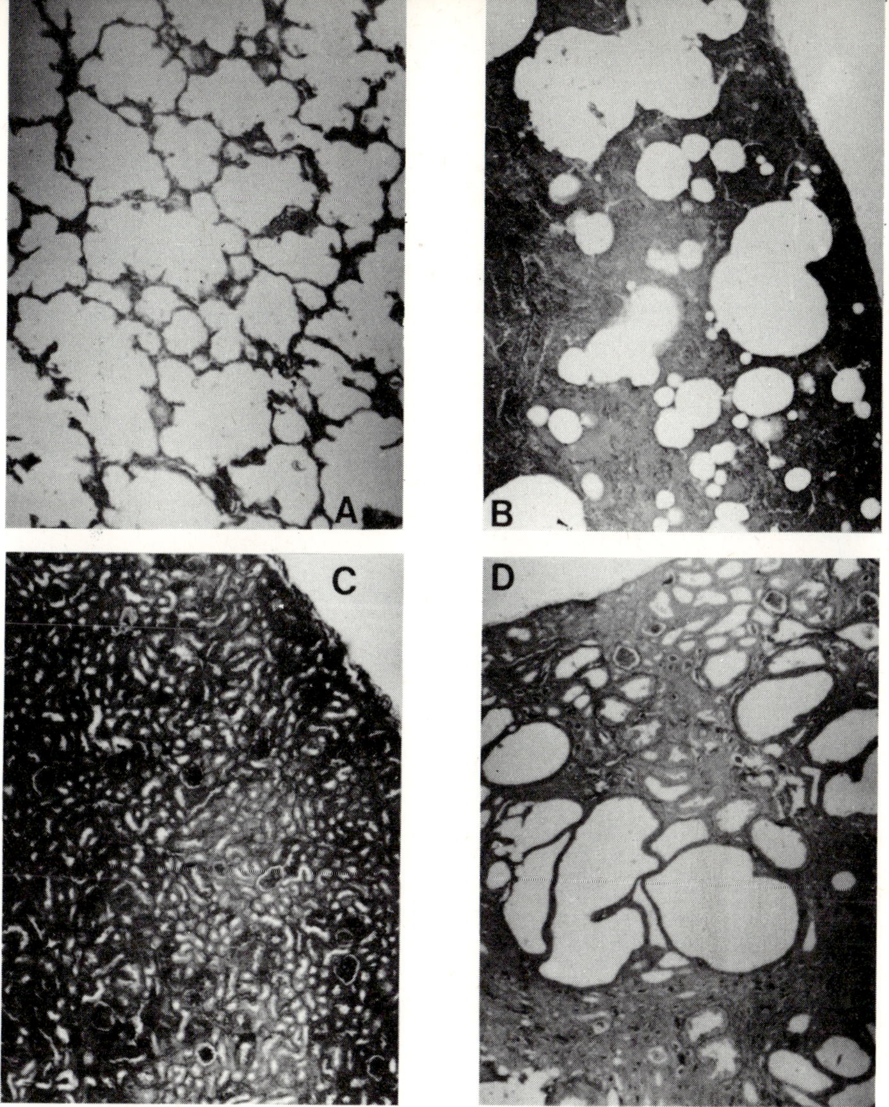

Figure 7. (*A*) Lung of normal animal. (*B*) Lung of animal with chronic disease. (*C*) Kidney of normal animal. (*D*) Kidney of animal with chronic disease.

3. Pulmonary lesions

At the end of a period of extended exhaustion in these animals, signs of more or less accentuated asphyxia have been observed. At necropsy, the lungs show blackish spots or large swellings. Histologically, they are characterised by enlargement of respiratory epithelium, which leads to disappearance of alveolar spaces in the majority of cases (see Fig. 7*B*). There is also exudate-formation in the alveoli, which often leads to haemorrhagic lung oedema. These phenomena (congestion; atelectasis; transudate and exudate formation in alveoli) are undoubtedly the result of a primary state of illness.

4. Kidney lesions

Animals which died after illness, and, to a lesser extent, animals which were in good health but had been in captivity for a long time, had smaller kidneys which exhibited scattered, clear blotches. In such animals, renal structure has more or less disintegrated, particularly at the cortical level. The cortical part of the kidney is invaded by large cavities, which may or may not be filled with PAS-positive mucus, corresponding to dilatated tubules (see Fig. 7*D*).

Other lesions have been observed irregularly in the gastro-intestinal tract and in the genital organs.

These fatal lesions would correspond to a greater secretion of corticosteroids. Studies made by Selye have demonstrated that long-term hyperproduction of adreno-cortical hormones induces a disturbance in physiological states and brings about lesions, particularly at the renal level.[8]

It is probable that such an increase of adrenal activity would occur as a response to stressful conditions of captivity. In any case, the fact that apparently healthy animals in captivity have more developed adrenals than animals captured in the wild indicates that *Microcebus murinus* are submitted to stress under the present conditions of captivity. Indeed, population density in the breeding colony is clearly far greater than that which is observed in the field (20 animals in a room of 6 square m.). Christian has demonstrated that there is an endocrine response to population increase which involves increased activity of the adrenals and a decrease in reproduction.[9] Thus, in captivity, stress (rather than nutritional factors) would be a major factor responsible for the low rate of reproduction and the death of certain animals.

NOTES

1 Petter-Rousseaux, A. (1970), 'Observations sur l'influence de la photopériode sur l'activité sexuelle chez *Microcebus murinus* en captivité', *Ann. Biol. Anim. Bioch. Biophys.* 10, 203-8.

2 Perret, M., Dubois, M.P. and Petter-Rousseaux, A. (1971), 'Les cellules corticotropes dans l'hypophyse d'un lémurien malgache: *Microcebus murinus* (Miller, 1777). Identification par immunofluorescence', *C.R. Acad. Sci.* 272, 2804-7.

3 Canivenc, R. (1968), 'Cycles génitaux des mammifères sauvages', *Entretiens de Chizé* 1, Paris.

4 Girod, C. and Cure, M. (1965), 'Etudes des corrélations hypophyso-testiculaires au cours du cycle annuel chez le hérisson (*Erinaceus europaeus* L.)', *C.R. Acad. Sci.* 261, 257-60.

5 Chester Jones, I. (1957), 'Adrenal gonad relationships', in *The Adrenal Cortex*, London.

6 Kitay, J.L. (1963), 'Pituitary adrenal function in the rat after gonadectomy and gonadal hormones replacement', *Endocrinology* 73, 253-60.

7 Herlant, M. and Ectors, F. (1969), 'Les cellules gonadotropes de l'hypophyse chez le porc', *Z. Zellforsch. Mikrosk. Anat.* 101, 212-31.

8 Selye, H. (1950), *Stress: A treatise based on the concepts of the General-Adaptation-Syndrome and the diseases of adaptation*, Montreal.

9 Christian, J.J. (1955), 'Effect of population size on the adrenal glands and reproductive organs of male mice in populations of fixed size', *Amer. J. Physiol.* 182, 291-300.

D. von HOLST

Social stress in the tree-shrew: its causes and physiological and ethological consequences

Introduction

Darwin based his theory of natural selection on two facts which were already well known in his time:

1. The number of individuals of a species population remains more or less constant over successive generations.
2. Any species is capable of producing many more progeny than are necessary to maintain a given population size.

As possible causes underlying this constancy in numbers of individuals, various factors have been proposed, such as: climate, food shortage, predators, diseases. Without doubt, all these factors can reduce population size; however, they cannot be regarded as the underlying causes of population regulation, because the number of individuals will remain constant even in their absence, at least in some species.[1]

Christian,[2] in 1950, propounded the hypothesis of self-regulation of mammalian populations: increasing density of individuals, according to this concept, leads to an increase in stress, to which animals adapt with hormonal changes. These adaptation reactions can be explained on the basis of the stress concept developed by Selye.[3] They are primarily characterised by increased activation of the hypophyseal-adrenal system and a corresponding reduction of hypophyseal-gonadal function. As a consequence of these endocrine changes, the fertility and vitality of the animals are sharply affected, thus counteracting the increase in population density. Excessive social stress, with subsequent exhaustion of the adrenal system, is even thought to be responsible for the death of animals, as in the spectacular population crashes in rodents (see Christian's summary).[4] Endocrine stress reactions in various social situations are well documented for a great number of mammalian species. However, until now there has been little clarification of the particular stimuli emitted from conspecifics which lead to these adaptation reactions, i.e. whether they are brought about by fighting, encounter frequency (density *per se*), or represent the consequences of dominance relationships.[4,5,6] This lack of clarity results largely from the difficulty of determining the level of activation of the adrenal system and, with it, the stressing effect of a given situation.

The following is a report on laboratory experiments with tree-shrews (*Tupaia belangeri*), which could help to clarify this question. Tree-shrews are small, diurnal mammals, which are distributed throughout South-East Asia. In nature they live singly or in pairs, apparently occupying territories which they defend fiercely against intruders of their own kind.[7,8] They are particularly suited for an investigation into the causes of social stress, because the activation of their sympathetic system can be observed directly.

Results

1. *Tail-ruffling (SST)*

Normally, the hair on the tail of tree-shrews lies flat. With any disturbance, however (such as an unexpected noise, approach of a strange person, fighting with a conspecific), the *Musculi arrectores pilorum* are activated by sympathetic fibres and cause the hair to stand almost vertically, thus giving the tail a bushy appearance (Fig. 1). This ruffling of the tail (SST = *Schwanzsträuben*) enables one to determine solely through visual observation how often and how long a tree-shrew is aroused in a certain situation.

The observer only distinguished between 'ruffled' (angle of hair spreading $> 45°$) and 'not ruffled' (angle $< 45°$) hair, although the *Musculi arrectores pilorum* can react differentially according to the degree of their activation by sympathetic fibres. It was not necessary, however, to distinguish tail ruffling further, since there was normally a clear-cut difference between 'no SST' (Fig. 1*a*) and 'SST' (Fig. 1*b*). This leads to the conclusion that during those social interactions which result in SST, the sympathetic nervous system is always strongly (or even maximally) activated. Without doubt, this results in a simultaneous output of catecholamines by the adrenal medulla, which will maintain or even increase the constriction of the pilomotor muscles. The lasting effect of catecholamines could possibly explain why, according to the disturbing situation, SST always persists for some time after cessation of the releasing stimulus. For example, a fight of few seconds' duration results in many minutes of SST.

Figure 1. The tail of *T. belangeri.* (*a*) Normal appearance. (*b*) Appearance during stress, when sympathetic fibres give the tail a distinctly bushy appearance. This ruffling of the tail is referred to by the author as *Schwanzsträuben* (tail ruffling), abbreviated as SST.

2. *SST-value*

The sum of episodes with SST recorded from an animal has been expressed as percentage of the 12-hour observation day (Fig. 2). This SST-value (% SST) can be taken as an indicator for the total duration of sympathetic arousal of the animal during the day. Observations of animals during their 12-hour activity period have led to the following results:

1. The SST-value may vary in any tree-shrew from 0-100%, depending on the environmental conditions of the animal.
2. The SST-value remains quite constant for months in a constant environment. Any change in the surroundings of an animal results in a corresponding change of the SST-level.
3. The SST-values are always correlated with specific ethological and physiological phenomena, which must be based on corresponding hormonal changes (summarised in Fig. 3).[9] This correlation is the same whether SST is triggered by members of the same species or is attributable to other causes.

3. Direct correlation between SST-values and physiological changes

Female fertility: A female tree-shrew which lives with a male in a large cage (floor area more than 5 square metres) normally shows very little (if any) SST per day. She gives birth every 45 days to 1-4 young (normally two). The young are fed by the mother directly after birth (Fig. 4), thereby increasing their body weight by up to 50%.[10] The mother then leaves the young and returns only once every 48 hours for feeding. The male (or other conspecifics) will

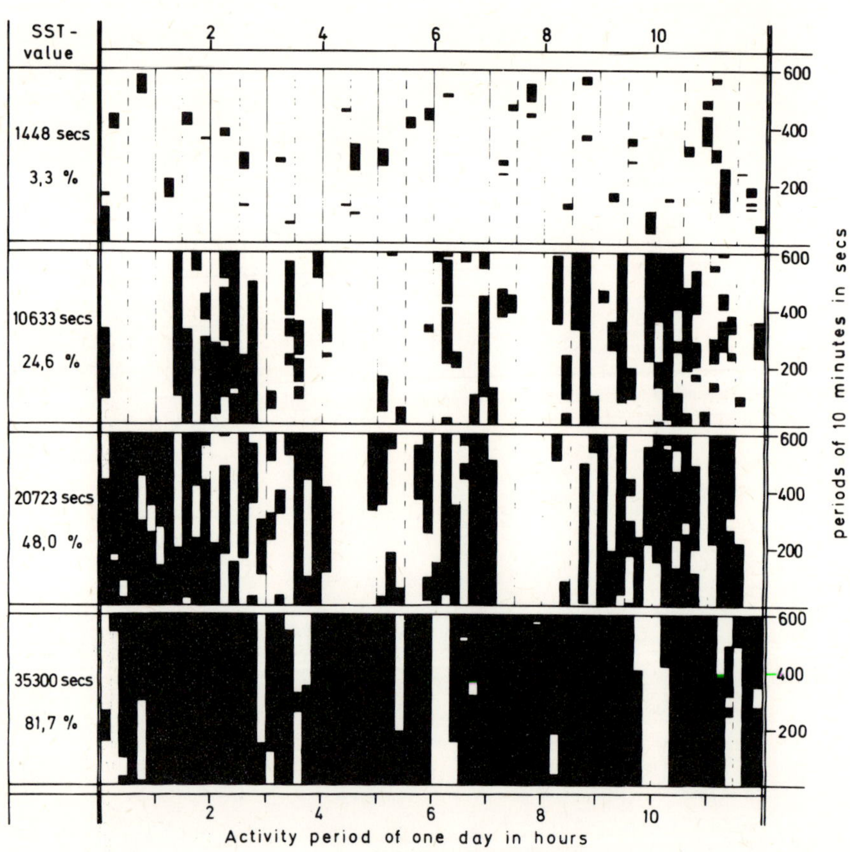

Figure 2. Distribution of times with (black) or without (white) tail-ruffling (SST) in 4 tree-shrews during their periods of activity. Each 12-hour observation day is divided into successive periods of 10 minutes, which are recorded side by side. (To assist interpretation, a vertical line has been drawn for every half-hour interval.) The ordinate indicates the time, in seconds, within each 10-minute period. The episodes of SST are recorded at the appropriate places within the individual periods of 10 minutes. As can be seen, times of SST are distributed more-or-less randomly over the day. The sum of episodes of SST gives the SST value in secs. (or in %) for the 12-hour observation period.

never enter the nest. If for any reason the SST-level of the mother rises above 20% per day, she continues to give birth regularly every 45 days; after the birth, however, any tree-shrew (often the mother) which enters the nest will remove the young and eat them ('cannibalism'; Figs. 5 and 6). This effect is caused by lack of sternal gland secretion with which the female normally marks the young at

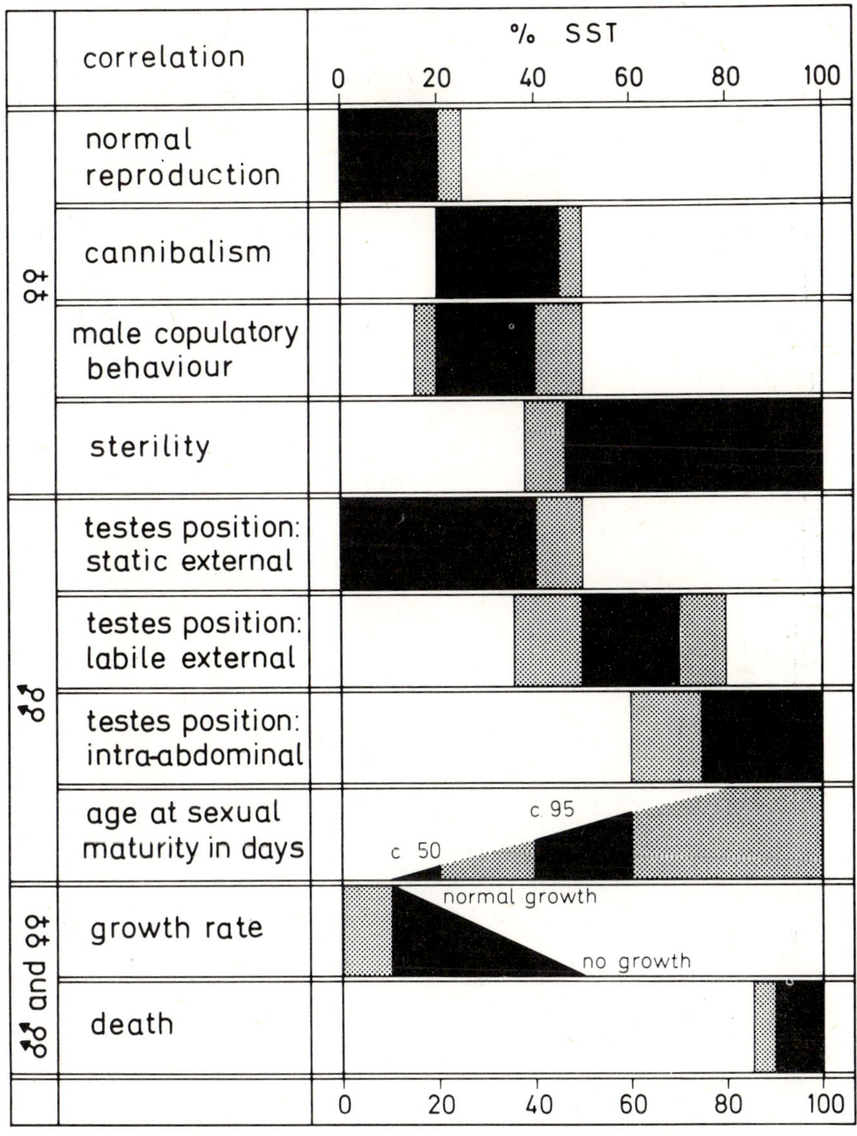

Figure 3. Tabular summary of some physiological and ethological phenomena in tree-shrews, showing their correlation with observed (black) or probable (grey) SST values.

Figure 4. Female *T. belangeri* during the birth of 4 infants.

Figure 5. Tree-shrew mother exhibiting cannibalism. The infant had been well fed just after birth, only 4 hours earlier.

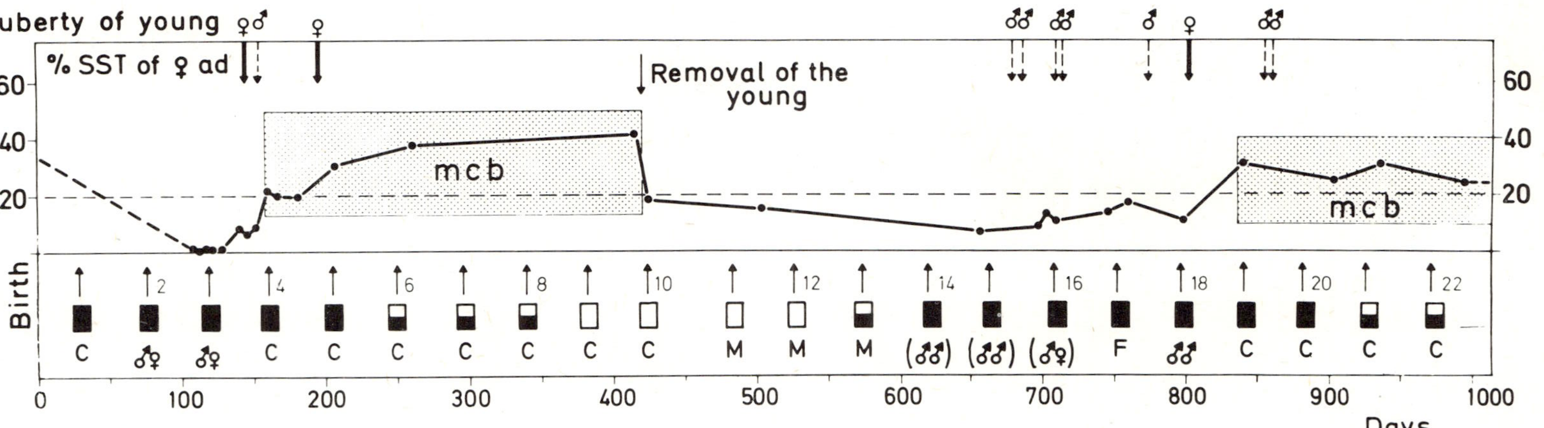

Figure 6. Correlation between the SST-value of a female, her reproduction and the occurrence of male copulatory behaviour (mcb), exhibited by the female. Each arrow on the lower scale indicates a birth. The nourishment of the infants after birth is indicated by the rectangles (black = normal, half-black = reduced, white = no milk given). The sexes of the young which reached sexual maturity are shown. The other litters did not survive because of cannibalism (C) or starvation, which was caused either by reduced maternal milk-production (M) or by disturbance of the maternal 48-hour feeding rhythm (F). The litters 14-16 (in brackets) would have died, as did litter no. 17, if the mother had not been forced to suckle by the experimenter (see text for details). It can be seen that an increase in the maternal SST-value is only brought about when female offspring attain puberty.

birth. At the same SST-level the female shows male copulatory behaviour, which is not seen below this SST-value (Fig. 6). When SST exceeds 50%, it results in sterility, arising from degeneration of ripening follicles in the ovaries.

Male fertility: With males, an increase in SST-values is equally correlated with a decrease in fertility (Figs. 7-9). This can be recognised externally by the state of development of the scrotum and the location of the testes:

(*a*) *Static external position of the testes* (Fig. 7*a*). When SST is below 50%, the testes lie outside the abdomen in a dark-pigmented scrotum, and cannot be squeezed into the body through the inguinal canal. Active spermatogenesis exists (Fig. 8*a*); correspondingly, the epididymes are filled with sperm (Fig. 9*a*).

(*b*) *Labile external position of the testes.* Above 50% SST, the scrotum loses its original compactness and the testes can easily be squeezed into the abdomen. As a consequence of any disturbance (fighting with conspecifics, capture by the experimenter), the testes are withdrawn instantly into the abdomen. This is impossible below 50% SST. When the disturbance ends, the testes return into the scrotum within a few seconds, possibly due to relaxation of the abdominal wall. In this physiological state, males are fertile; their spermatogenesis, however, is markedly lower than in animals with static external position of the testes.

(*c*) *Static intra-abdominal position of the testes* (Fig. 7*b*). If males live in a situation with SST exceeding 70%, the testes are withdrawn into the abdomen within 8-10 days, and remain there. At the same time, the testes and epididymes lose up to 80% of their initial weight. Spermatogenesis ceases completely (Figs. 8*b*, 9*b*). The scrotum is highly reduced, and after about two weeks it is essentially absent (Fig. 7*b*).

Also, as SST rises, the growth rate in young animals is progressively lowered and puberty delayed. All these changes are reversible (see also Fig. 6).

In a situation leading to more than 90% sympathetic arousal per day, within a few days the animal develops muscular twitching, convulsions, coma and dies.

4. *After-effects of situations leading to higher SST-values*

In the instances mentioned so far, there is a direct correlation between SST-values and changes in physiology and behaviour.

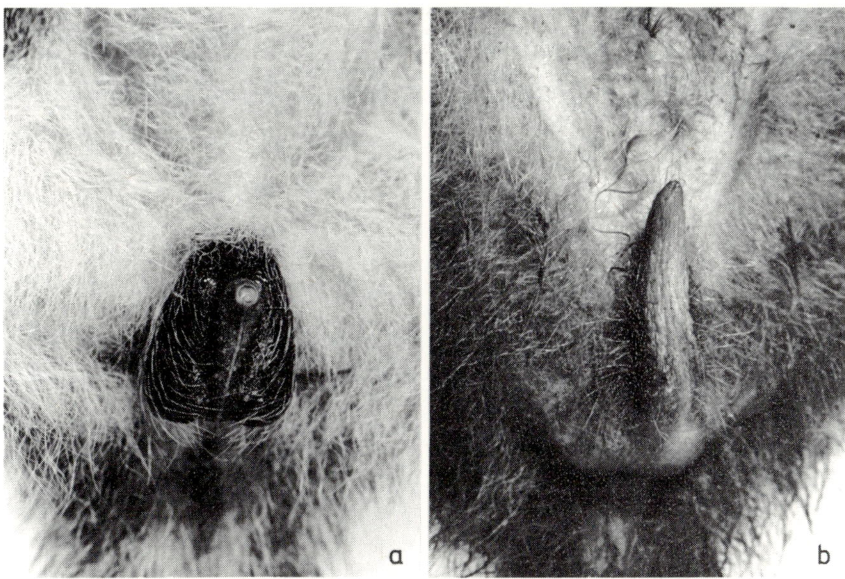

Figure 7. Position of the testes and form of the scrotum in a male tree-shrew. (*a*) Normal scrotum and static external position of the testes. (*b*) A male after 4 weeks of stress exposure leading to more than 70% SST. The scrotum is greatly reduced and the testes are in a static intra-abdominal position.

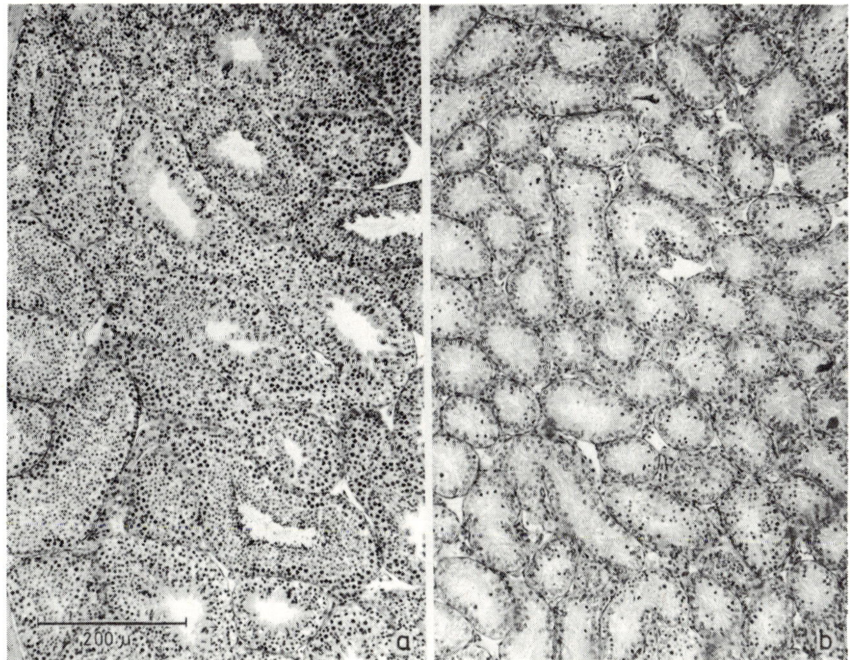

Figure 8. Testes of a control animal (*a*) and a male that had lived for 12 days with SST-values exceeding 70% (*b*). Sections 4 μ thick; staining with haematoxylin/eosin.

However, some disturbances of behavioural patterns and physiological processes can be observed in tree-shrews in the absence of any correlation with the social situation, or the SST-values present at that moment. Rather, they tend to be the result of relatively harmless (low SST-values), but long-lasting stressful situations. This continuous stress can result in after-effects lasting for months, or even lead to persistent physiological and morphological changes.

The following briefly covers some of these consequences, since they show how important the knowledge of an animal's history can be for the determination of the 'normal' behaviour or physiological state of a species.[9]

(*a*) *Body weight of adult males.* Males growing under undisturbed conditions achieve a body weight of more than 260 gms. as adults (at an age of about 6 months). If offspring are exposed to stress stimuli during their development, their growth rate is depressed. As adults, these animals are smaller (and lighter) than controls; even under optimal conditions they weigh only 150-250 gms. Their body weights depend on the duration and intensity of stress during their growth phases. At the same time, these animals differ distinctively in their physiological state from tree-shrews that have grown up undisturbed; their adrenal weight, relative to body weight, is

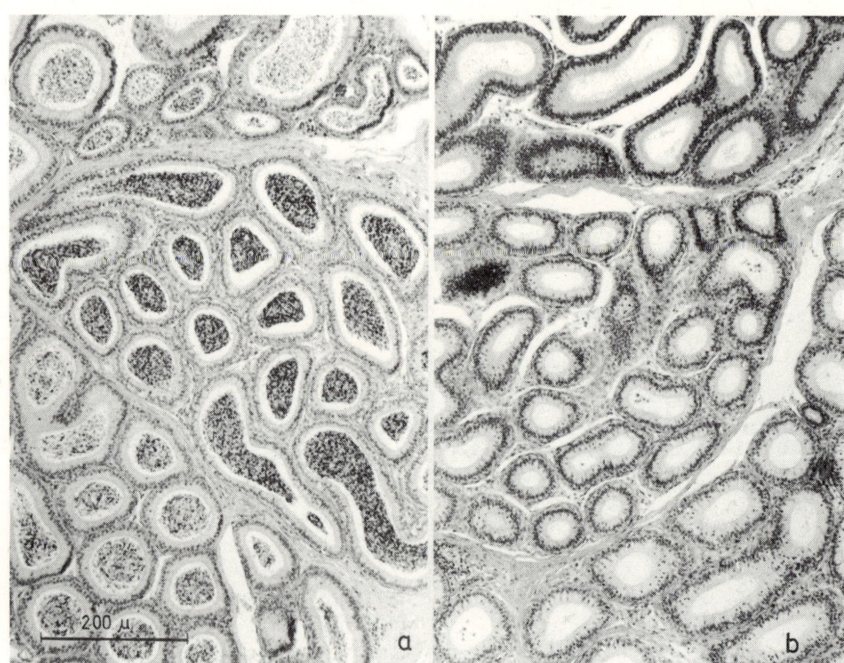

Figure 9. Epididymes of the males whose testes are illustrated in Fig. 8. (See text for details).

much higher than in controls. This indicates a higher basal activity of the adrenal system. Correspondingly, the concentrations of glucocorticoids and glucose in the blood are higher than in controls. Moreover, there are differences in liver glycogen, adrenal ascorbic acid concentration, and androgenic activity, i.e. stress during development results in animals that are morphologically and physiologically different from controls with undisturbed growth.[11]

(*b*) *Mammogenesis.* As mentioned earlier, a female displaying 20-50% SST can give birth to young every 45 days, these being eaten within a few hours after birth. While the first 2-3 litters are fed normally at birth (before they are eaten), subsequent litters are suckled only moderately, and after 4-6 litters, not at all. As far as we know, this is due to a failure of mammogenesis in the mother. When SST of the female falls below 20%, subsequent litters are not eaten, but they die of starvation nevertheless due to a lack of maternal milk production. Not before 2-4 litters have been produced (3-6 months) are the offspring suckled normally after birth (see also Fig. 6). The same effect on milk production is observed when a female has previously lived for many months with more than 50% SST, which induces sterility.

(*c*) *Suckling rhythm of females.* Even when the offspring of a female which had lived with high SST-levels for some time are fed normally again, i.e. after several months (see above), the subsequent litters usually die of starvation, because the normal 48-hour feeding rhythm of the mother is disturbed. Sometimes the mother suckles her young once to several times daily, sometimes normally – every 48 hours; in most cases, however, she does so after 3-4 days. The latter always results in the death of the young due to starvation during the first week, unless they are given an additional meal by the experimenter in the meantime. This 'omitting' of feeding is not due to inadequate milk production and associated tension of the mammary gland, because normal feeding of the young can be initiated regularly if the mother is locked in the nesting box with its young for some time (this was done, for example, with the litters 14-16 in Fig. 6). In experiments with mild stress of 6-9 months duration, the normal 48-hour feeding rhythm of the mother did not appear until 10-13 months after the end of the stressing period (*e.g.* litter no. 18 in Fig. 6).

These persistent after-effects of slight stress could certainly provide an explanation for the bad breeding results of several authors, and for the contradictory data on the 'normal' feeding rhythm in tree-shrews.

Conclusions

There is as yet no plausible physiological explanation of the after-effects mentioned above. The correlation between SST-value and growth, body weight, fertility and health of an animal, however, can be interpreted through a corresponding correlation between SST-value and the hormonal adaptation reactions of an organism to a stressor, as delineated by Selye in the 'general adaptation syndrome'.[3] This assumption is supported by data on adrenal function, i.e. adrenal weight and histology and corticoid levels in the blood. The externally measurable SST can, therefore, be taken as an indicator of the stress of an animal in a certain situation. This makes it possible to recognise through direct observation those social interactions, or stimuli arising from conspecifics, which cause SST and, with them, the detrimental ethological and physiological consequences.

Social causes of stress

Social stress is brought about almost exclusively by the following:

1. Density-effect: its extent depends solely on the number of mature animals of the same sex in the cage (without aggression).
2. Dominance-effect: its extent is based on dominance relationships that require at least one fight between the animals.

1. Density-effect

As mentioned earlier, a female tree-shrew can give birth to young every 45 days. After weaning, at an age of about 30 days, the young live with their parents in a close family group (without possibility of emigration in the laboratory). At night, family members usually sleep in the same nest. During the day as well, there is a strong tendency to rest with conspecifics; this might result in huddling-groups of up to 9 animals, lying in three layers one upon the other (Fig. 10). When meeting, the animals very often show mouth-licking (up to 15 min./day per animal; Fig. 11). Even after puberty, at about 50 days of age, there is no change in this close cohesion between parents and offspring for several months. Agonistic behaviour is completely absent and there are no dominance relationships of any kind. If, however, the young are placed with strange tree-shrews of the same sex, they will be attacked at once; this is not the case before they reach sexual maturity.

Social contact between parents and immatures never, or only very seldom, leads to arousal: therefore, SST-values of the parents remain

Figure 10. Huddling of adult tree-shrews in a family group.

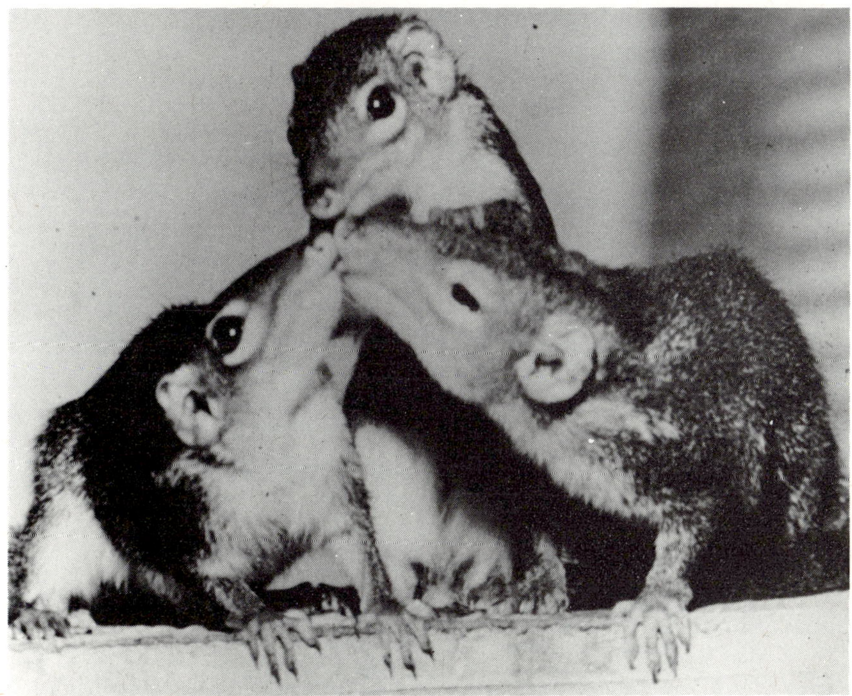

Figure 11. Mouth-licking (greeting) between adult male tree-shrews of a family group.

constant in spite of the increase in number of individuals in the cage. With puberty of the young, however, the SST-values of the parents increase distinctly to a new level: this happens in males only when a young male reaches maturity, and correspondingly in females only at sexual maturity of a young female (Figs. 6, 12).

In females, the addition of only one female to the cage results in a rise of SST above 20%, thereby suppressing further increase of the group because of cannibalism (cage size about 5-7 sq. m.). If a female gives birth only to males, their acquisition of maturity does not influence her. She continues to give birth regularly and raises the young normally until birth, or rather sexual maturity, of a female offspring (e.g. Fig. 6). This has resulted in groups of up to 9 animals. The mere presence of conspecifics, without aggression and dominance relationships, does not, however, lead to SST-values high enough to produce sterility either in males or in females.

2. *Olfactory marking*

The increase in the SST-values in parents does not result from observable visual, acoustic, or tactile stimuli. Instead, the arousing influence is based on some kind of olfactory marking of the young, independent of their immediate presence. Thus, after removal of the young, it takes several days before the SST-values of the parents decrease to the original levels. A corresponding increase in the SST-value also results from marking by strange tree-shrews of the same sex. This becomes apparent when the cage of an animal is inhabited for a short time by a strange conspecific with direct contact between the animals prevented, or when an animal is placed in the empty cage of a strange tree-shrew.

The increase in the SST-levels of the parents at the time of sexual maturity of their young indicates an influence of gonadal hormones. Support for this comes from experiments in which young males, castrated before puberty, do not cause an increase in the SST-value of their father, while castration after puberty results in a decrease in the SST-value of the father to the initial level. When castrated males are injected with testosterone, the SST-value of the father rises (Fig. 12).

The importance of gonadal hormones in the effectiveness of the olfactory marking can also be seen in their influence on the marking behaviour of male tree-shrews (females have yet to be investigated).[12] If a male is introduced into a thoroughly-cleaned experimental cage (PVC material; 70 x 50 x 50 cm. in size) for a fixed period of time (10 minutes in the present experiments), it will mark this cage several times with its sternal gland (Fig. 13). If the experiment is repeated, the number of sternal-markings is more or less the same for any tree-shrew, but there are considerable differences in mean values between the individuals, varying from 3 to

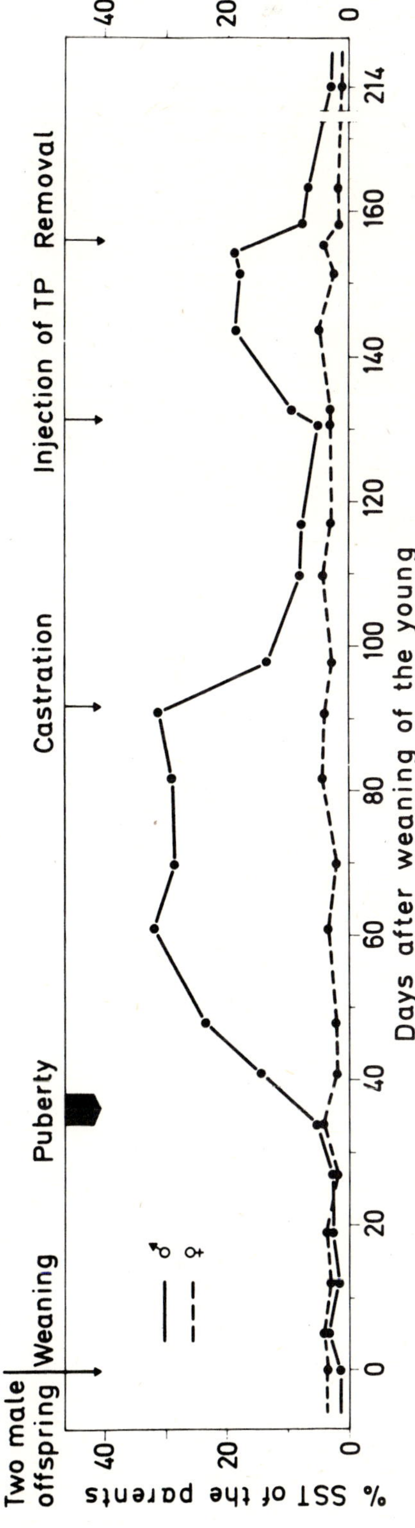

Figure 12. Fluctuations in the SST-levels of the two parents as a function of both normal changes in two male offspring over time (e.g. puberty) and artificially induced changes (e.g. castration, injection of testosterone propionate).

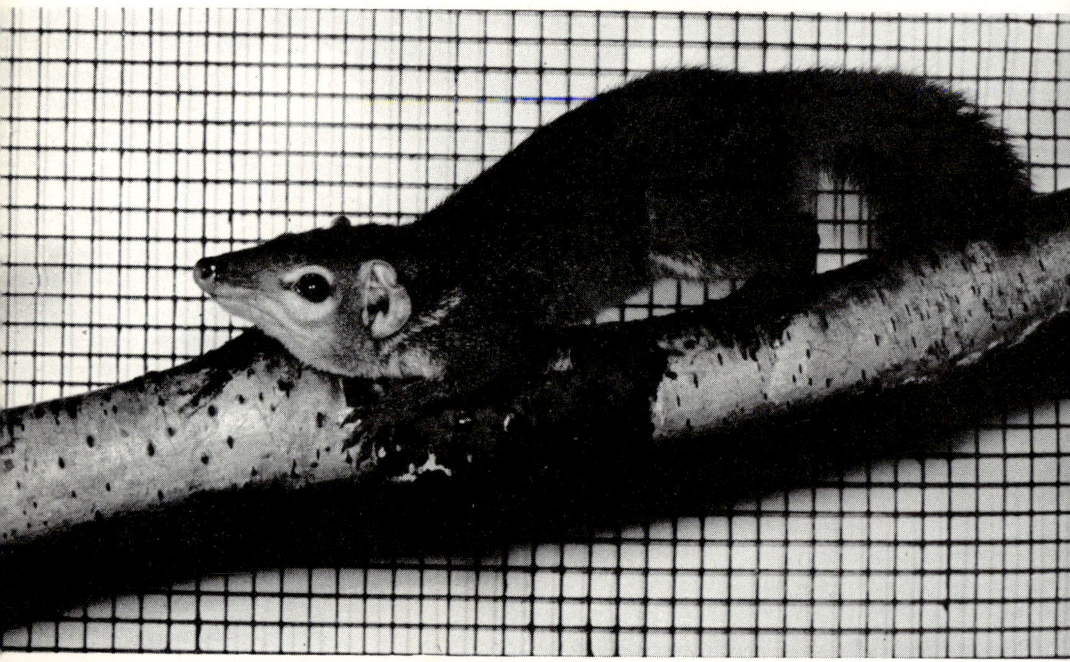

Figure 13. Sternal marking behaviour (chinning) of a male tree-shrew.

26 markings per 10 minute period (see also Fig. 14). These differences are at least to some extent due to corresponding differences in the blood levels of androgens. However, if the experimental cage is contaminated with urine, faeces and secretions of cutaneous glands of a strange mature male, the experimental animal is noticeably aroused, and marks significantly more than in an uncontaminated cage ($p < 0.05$-0.001). The increase in the mean values can be between 50% and 300% in different animals. The markings of immature or castrated males, as well as those of females or members of other species, have no or only very little effect on marking behaviour of males (Fig. 14). However, this does not mean that the secretions or excretions of the latter cannot be recognised or are odourless; males can accurately discriminate between the markings (secretions of sternal glands and urine) of mature, immature and castrated males as well as those of female tree-shrews, and from markings of other species, as has been shown by training experiments.[13]

The substances responsible for the arousing effect of the olfactory markings of mature males (and only of these) are not yet known. They could be the gonadal hormones themselves or their metabolites, excreted with urine or faeces, and/or secretions of certain glands.

The secretion of the male sternal gland, for example, has a highly arousing effect on other males. The production of this secretion is controlled by androgens, as is the marking behaviour which leads to

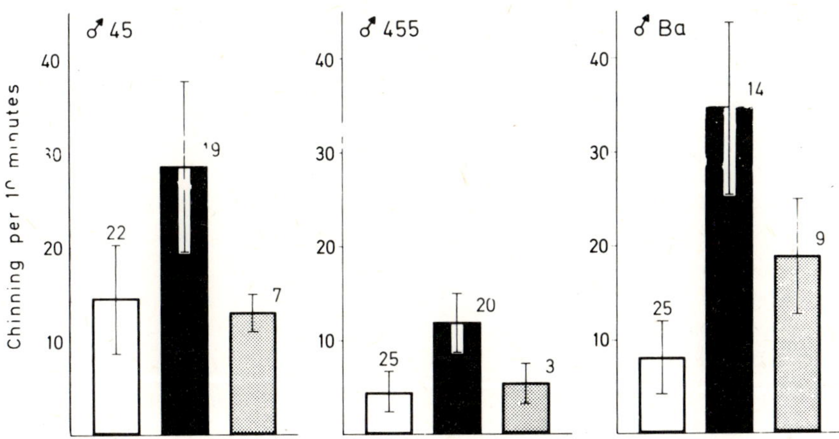

Figure 14. Chinning of adult male tree-shrews and its influence on conspecifics. Each bar indicates the mean value of chinning, with standard deviation and number of experiments, in the following experimental situations: cage uncontaminated (white bar), contaminated by a mature adult male tree-shrew (black bar), or contaminated by a castrated adult male tree-shrew (grey bar).

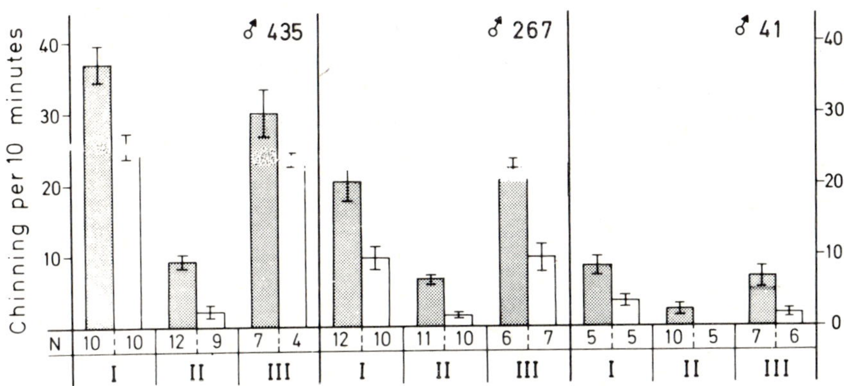

Figure 15. Chinning of male tree-shrews before (I) and after castration, either without (II) or with (III) testosterone-therapy. Experimental cage uncontaminated (white bar) and contaminated by a strange mature male (grey bar). Each bar indicates the mean value with its standard deviation. The number of experiments (N) is shown below each bar.

its distribution (Fig. 15). This does not mean, however, that only the secretion of the male sternal gland is involved. Rather, it is probable that various different (androgen-dependent) odorous substances work together in this arousing effect.

3. Dominance-effect

In the laboratory, as well as in nature, adult tree-shrews of both sexes immediately attack strange conspecifics of the same sex. Even in family groups, sooner or later (after 4-10 months), fighting will break out, usually terminating in less than a few minutes with the subjugation of one animal by the other, especially with males. During fighting, the tail hair of both animals is always maximally erected. Within 10 minutes after fighting has ended, the victor shows no further sign of physiological arousal (no SST) and pays virtually no attention to the submissive animal. The subordinate tree-shrew, however, cowers in a corner, which it leaves only to feed and drink. The hair of its tail is constantly erected. It hardly moves at all, spending more than 90% of the daily activity phase lying motionless in a corner and following the movements of the victor with its head. Approach of the dominant animal very often leads to 'fear squealing'.

During the following days, agonistic encounters between the two animals are very rare (less than a few minutes per day) or completely lacking in larger cages. The submissive animal, nevertheless, remains constantly aroused by the presence of the victor. This permanent arousal leads to a distinct loss in body weight and to death, even without any wounds. In contrast to the density-effect, the dominance-effect depends largely on vision. The olfactory markings or the smell of the victor alone never result in death. If the victor is removed, the submissive animal recovers in a few days, provided that the stressful situation has not persisted too long.

4. Physiological cause of death

The death of subordinate animals, even without wounds, leads to the question of the cause of death. The physiological and ethological changes under milder stress (below 80% SST) can be interpreted, as mentioned earlier, on the basis of Selye's 'general adaptation syndrome' and correspond with results from other species. Death under extreme social stress, however, has not yet been clarified for any species. Christian's postulation of exhaustion of the hypophyseal-adrenal system is based exclusively on adrenal weight and histology or histochemistry (mainly in rodents). It seems doubtful that these data allow a realistic statement as to the function of the adrenals.[14] Also, death either from hypoglycaemic shock or from haemolytic anaemia (snowshoe hare, vole) does not seem likely after more detailed investigation of the data.[15]

Therefore, the cause of death in subordinate tree-shrews was investigated in the following experiment. A cage (floor area 100 x 100 cm.; height 60 cm.) which could be divided by a vertical partition into two compartments of equal size was continuously occupied by a male 'fighter'. A strange male was introduced to the fighter, and it was usually attacked at once and subjugated in less than 3 minutes. The two tree-shrews were then separated from one another by the partition. Every 1 to 2 days, the fight and subsequent submission were repeated by removing the partition.

When the animals were separated by an opaque partition, the subordinate animal recovered as quickly as did the victor (no SST after 10 minutes). Even when a subordinate was defeated several times a day over a week or more, in this situation he lost little weight and never died.

If, however, both animals were separated from each other with a cage-wire partition, so that the subordinate animal was in continuous visual contact with the dominant rival, though immune from actual attack, it showed SST continuously and always died in less than 20 days.

The cause of the permanent arousal (SST), and with it the cause of death, is not the physical parameter of 'fighting', but the continued presence of the victor. The situation, therefore, does not operate as a source of stress through observable social interactions and their physiological results, but through central nervous processes based on experience and learning which take place in the subordinate animal. Correspondingly, there are distinctive differences in the serotonin metabolism in different brain areas between subordinate animals with or without presence of the victor, and controls.[16] Thus, tree-shrews die as a result of a type of stress which we can refer to as 'psycho-social'.

The physiological reactions of tree-shrews under this form of permanent social stress was examined. Some of the subordinates were kept in the stressful situation until death occurred after 2-16 days, but most experimental animals were removed from their cages at various periods after the first defeat, sacrificed and examined for physiological changes.[14,15]

Body weight of all animals decreased markedly from the time of first submission to death. The daily weight-loss was more or less constant for any individual, but there were considerable differences between animals (1.5-6.5% body weight loss per day). The more weight an animal lost per day, the sooner it died ($r = 0.89$; $p < 0.001$). However, death was not caused by exhaustion of available energy reserves of the body, as would occur with hypoglycaemia. In general, the glycogen content of the livers of subordinates was lower ($p < 0.003$; $n = 35$) than the mean control level, but even in dying animals the glycogen content was never exhausted. Also, the blood sugar content of stressed animals was as high as that of controls (90-105 mg./100 ml.).

The adrenal weight of the subordinates increased within 4 days from its initial value (18.0 ± 0.5 mg.; $n = 18$) to a new level (31.7 ± 0.9 mg.; $n = 21$), which was maintained virtually constant until death (Fig. 16). Correspondingly, the concentration of the glucocorticoid hormones in the blood was considerably above the control level. In the course of stress exposure, the haemoglobin content of the blood decreased ($r = 0.75$; $p < 0.001$) due to haemolysis and decrease in erythropoiesis. After 14 days, the mean value was only 50% of the initial value of 19.4 ± 0.2 *g.*/100 ml. blood ($n = 20$). Nevertheless anaemia was not the cause of death.

Non-protein nitrogen content of the blood increased in all tree-shrews during stress-exposure. The increase varied considerably from animal to animal, but all dying tree-shrews had values above 150 mg. per 100 ml. blood. This, together with the symptoms of dying animals (muscular twitching, convulsions, coma, decrease in body temperature) indicates *uraemia* as cause of death. The increase of non-protein nitrogen resulted from a marked decrease in renal blood flow. Thus, the glomeruli in the kidneys of subordinates were distended in comparison with control animals, and contained very few erythrocytes or none at all (Fig. 17). In the Bowman's capsule there was usually accumulated material, indicating increased permeability of the basement membranes. The lumina of the renal

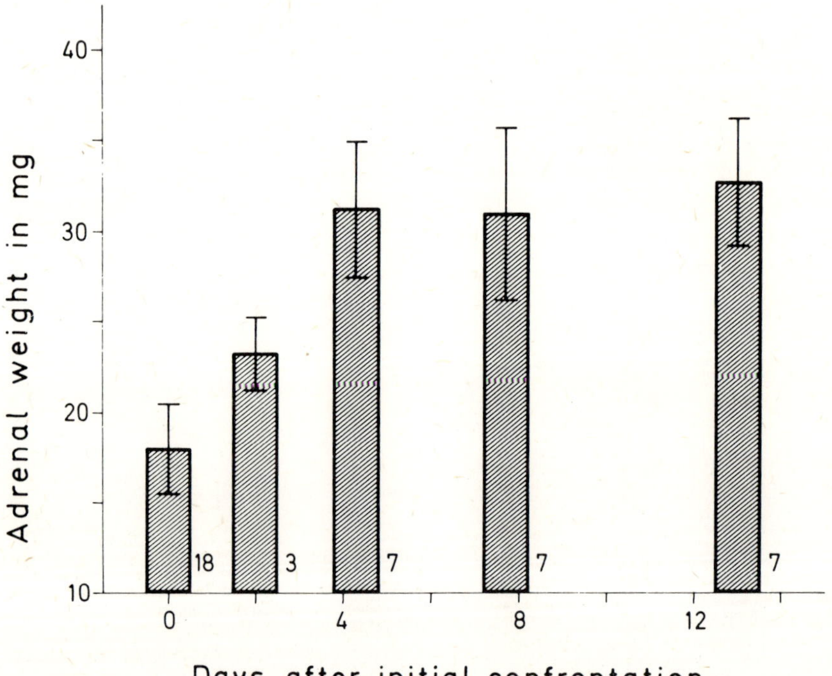

Figure 16. Adrenal weights in male tree-shrews under stress in the form of persistent visual exposure to a dominant conspecific following defeat. Each category (bar) shows the mean value with its standard error and the number of experimental animals. Categories: controls; subordinates after 2, 3-6, 7-10 and 11-16 days of stress, respectively.

tubules were more or less distended in all animals exposed to the stressing situation, while they had a closed appearance in controls (Fig. 17). The extent of decrease of renal blood flow varied from animal to animal, as can be seen histologically; there was a corresponding variation in the increase of non-protein nitrogen in the blood. In extreme cases, the blood urea-N rose nearly as quickly as in controls with surgically impaired renal function. Correspondingly, these subordinates died after 2 days of stress, as do animals without kidneys. This indicates almost complete arrest of renal blood flow due to psychosocial stress.

The exact cause of renal failure is not yet known. Primarily, there is probably constriction of the renal vessels through sympathetic fibres and catecholamines. Their influence alone, however, would not suffice; complementary mechanisms (renin-angiotensin, vasopressin, aldosterone) may be necessary to maintain or even increase the vasoconstriction.

The current concept of adrenal failure under strong social stress does not apply to tree-shrews; here death is brought about by uraemia which results from ischaemia or even anaemia of the kidneys. Whether this also applies to other species is unknown.

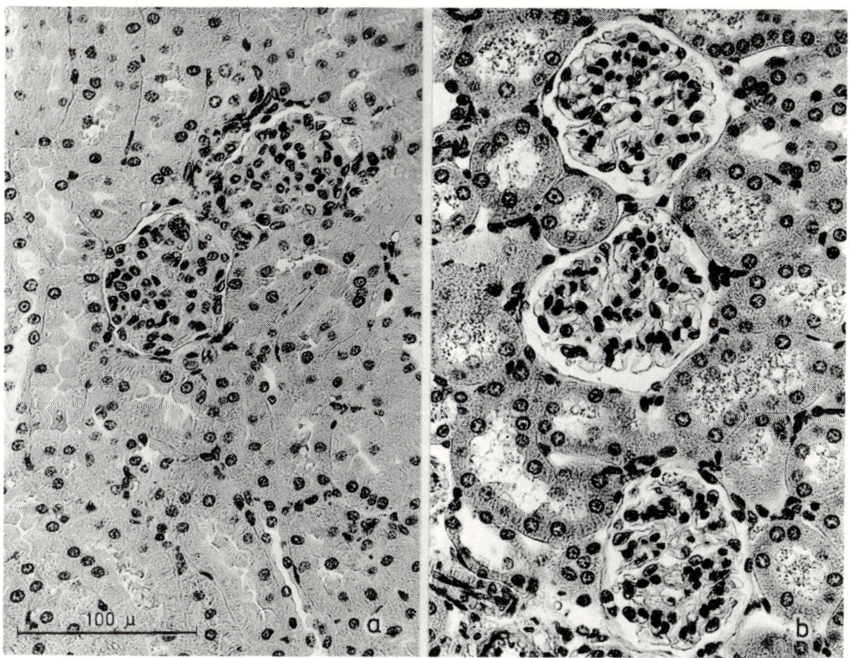

Figure 17. Kidneys of male tree-shrews. Sections 4 μ thick, staining with haematoxylin/eosin. (*a*) Control condition. (*b*) After 2 days of stress exposure: glomerular capillaries are distended, and contain very few erythrocytes or none at all. The lumina of the renal tubules are distended in comparison with controls.

Corresponding investigations of renal function have yet to be carried out in other animal species. However, the possibility seems to be present, since in all species so far studied (rats, rabbits, dogs, cats, monkeys, human beings) it is extremely easy to bring about renal ischaemia through stimulation of renal nerves and certain areas in the central nervous system, or through injection of catecholamines. Furthermore, various indications from the literature suggest that in natural populations too, it is renal function which is disrupted at times of high population densities and/or pronounced social stress (e.g. snowshoe hares, voles, lemmings, mice, prairie dogs, Sika-deer, human beings).[15]

Conclusion

This paper attempts to outline the social situations that lead to stress in tree-shrews, and to give some examples of the physiological consequences of social interactions. Whether and to what extent social stress may be effective in regulating numbers of tree-shrews in their natural habitat, as well as those of other species, cannot yet be discussed without detailed studies in nature. Without doubt, the behavioural situations leading to stress-reactions differ according to the different species and their social structures; the physiological reactions of organisms, however, are basically the same in all mammalian species. Thus, the present findings with tree-shrews show the significance which social interactions, and the resulting social or psychosocial stress, may have in the origin of many ethological and physiological disturbances and diseases (from changes in maternal behaviour to renal disease), possibly in human beings as well as in other mammals.

NOTES

1 Elton, C. (1942), *Voles, Mice and Lemmings*, Oxford.

2 Christian, J.J. (1950), 'The adreno-pituitary system and population cycles in mammals', *J. Mammal.* 31, 247-59.

3 Selye, H. (1950), *The Physiology and Pathology of Exposure to Stress*, Montreal.

4 Christian, J.J. (1963), 'Endocrine adaptive mechanisms and the physiological regulation of population growth', in Mayer, W.V. and van Gelder, R.G. (eds.), *Physiological Mammalogy*, vol. 1, *Mammalian populations*, New York.

5 Barnett, S.A. (1964), 'Social stress: The concept of stress', in Carthy, J.D. and Duddington, C.L. (eds.), *Viewpoints in Biology*, vol. 3, London.

6 Wynne-Edwards, V.C. (1962), *Animal Dispersion in Relation to Social Behaviour*, Edinburgh.

7 Cantor, T. (1846), 'Catalogue of Mammalia inhabiting the Malayan peninsula and islands', *J. Asiat. Soc. Beng.* 15, 171-279.

8 von Holst, D. and Raab, A. (1970), unpublished observations from Thailand (near Bangkok).

9 von Holst, D. (1969), 'Sozialer Stress bei Tupajas (*Tupaia belangeri*). Die

Aktivierung des sympathischen Nervensystems und ihre Beziehung zu hormonal ausgelösten ethologischen und physiologischen Veränderungen', *Z. vergl. Physiol.* 63, 1-58.

10 Martin, R.D. (1968), 'Reproduction and ontogeny in tree-shrews (*Tupaia belangeri*), with reference to their general behaviour and taxonomic relationships', *Z.f. Tierpsychol.* 25, 409-95 and 505-32.

11 Holst, D. von, (1972), 'Die Nebenniere von *Tupaia belangeri*', *J. comp. Physiol.* 78, 274-88.

12 Traum, S. (1972), Das Markierverhalten männlicher *Tupaia belangeri*. Diplomarbeit, Zool. Inst. Univ. München, unpublished.

13 Stralendorff, F.V. (1972), Das Sternaldrüsensekret männlicher *Tupaia belangeri*. Untersuchungen zu seinem olfaktorischen Informationsgehalt und dessen stofflichen Äquivalent. Diplomarbeit, Zool, Inst. Univ. München, unpublished.

14 Holst, D. von (1972), 'Die Funktion der Nebennieren männlicher *Tupaia belangeri*: Nebennierengewicht, Ascorbinsäure und Glucocorticoide bei kurzem und bei andauerndem soziopsychischem Stress', *J. comp. Physiol.* 78, 289-306.

15 Holst, D. von (1972), 'Renal failure as the cause of death in *Tupaia belangeri* exposed to persistent social stress', *J. comp. Physiol.* 78, 236-73.

16 Raab, A. (1971), 'Der Serotoninstoffwechsel in einzelnen Hirnteilen vom Tupaja (*Tupaia belangeri*) bei soziopsychischem Stress', *Z. vergl. Physiol.* 72, 54-66.

Index of authors

General index